INTRODUCTORY GEOMETRIC ALGEBRA
and Geometric Calculus

INTRODUCTORY GEOMETRIC ALGEBRA
and Geometric Calculus

by
Malavika Kumari | Tanvi Mishra

Wordpen Academics
Publishing Science That Matter

ISBN: 978-93-6084-985-6 (Hardback)

Published by : **WORDPEN ACADEMICS**
4736/23, Ansari Road, Darya Ganj,
New Delhi -110 002
Phone: +91-011-35004336
E-mail: info@wordpenacademics.com

Printed at **:** **Replika Press Pvt. Ltd.**

Exclusive Distributor: ***Bio-Green Books, New Delhi***

Preface

"Introductory Geometric Algebra & Geometric Calculus" offers a comprehensive introduction to the powerful mathematical framework of geometric algebra and its extension, geometric calculus. This book serves as an essential resource for students, researchers, and practitioners interested in exploring the elegant and intuitive geometrical underpinnings of algebra, calculus, and mathematical physics.

Traditional algebra and calculus provide powerful tools for solving mathematical problems and describing the physical world. However, these approaches often rely on abstract symbols and operations that can obscure the underlying geometric meanings and relationships. Geometric algebra offers a unified framework that seamlessly integrates algebraic and geometric concepts, providing a more intuitive and geometrically meaningful approach to mathematical modeling and problem-solving.

The primary objective of this book is to provide readers with a gentle yet rigorous introduction to geometric algebra and its applications, covering key topics such as vector spaces, geometric products, multivectors, geometric transformations, and differential operators. Each chapter offers clear explanations, illustrative examples, and practical exercises to help readers develop a solid understanding of geometric algebra and its computational techniques.

Contributions from leading experts in the field explore various applications of geometric algebra and geometric calculus in areas such as computer graphics, robotics, computer vision, physics, engineering, and applied mathematics. Special emphasis is placed on the geometric interpretation of mathematical operations, enabling readers to gain insight into the underlying geometric structures and relationships.

As we delve into the elegant world of geometric algebra and geometric calculus, it becomes apparent that these mathematical frameworks offer powerful tools for understanding and manipulating geometric objects and spaces in a way that is both intuitive and computationally efficient. By embracing geometric algebra, students and researchers can unlock new perspectives, insights, and applications in diverse fields of study.

We extend our sincere gratitude to the contributors whose expertise and dedication have enriched this book. It is our hope that "Introductory Geometric Algebra and Geometric Calculus" will serve as a valuable resource and gateway to further exploration and discovery in the fascinating realm of geometric mathematics.

Malavika Kumari | Tanvi Mishra

Contents

	Preface	v
1.	Introduction	1
2.	Group Theory	13
3.	Ring Theory	33
4.	Galois Theory	69
5.	Topics in Group Theory	79
6.	Module Theory	86
7.	Commutative Algebra	91
8.	Homological Algebra	95
9.	Representation Theory	104
10.	Differential Calculus	122
11.	Tangents and Normals	136
12.	Curvature	153
13.	Asymptotes	184
14.	Change of the Independent Variables	213
15.	Maxima and Minima	225
16.	Singular Point	253
17.	Lagrange's Theorem	301
	Bibliography	321
	Index	118

1

Introduction

ABSTRACT ALGEBRA

In algebra, which is a broad division of mathematics, abstract algebra is the study of algebraic structures. Algebraic structures include groups, rings, fields, modules, vector spaces, lattices, and algebras. The term abstract algebra was coined in the early 20th century to distinguish this area of study from the other parts of algebra.Algebraic structures, with their associated homomor phisms, form mathematical categories.Category theory is a formalism that allows a unified way for expressing properties and constructions that are similar for various structures. Universal algebra is a related subject that studies types of algebraic structures as single objects. For example, the structure of groups is a single object in universal algebra, which is called variety of groups.

History of Abstract Algebra

As in other parts of mathematics, concrete problems and examples have played important roles in the development of abstract algebra.Through the end of the nineteenth century, many – perhaps most – of these problems were in some way related to the theory of algebraic equations. Major themes include:

- Solving of systems of linear equations, which led to linear algebra
- Attempts to find formulas for solutions of general polynomial equations of higher degree that resulted in discovery of groups as abstract manifestations of symmetry
- Arithmetical investigations of quadratic and higher degree forms and diophantine equations, that directly produced the notions of a ring and ideal.

Numerous textbooks in abstract algebra start with axiomatic definitions of various algebraic structures and then proceed to establish their properties. This creates a false impression that in algebra axioms had come first and then served as a motivation and as a basis of further study. The true order of historical development was almost exactly the opposite. For example, the hypercomplex numbers of the nineteenth century had kinematic and physical motivations but challenged comprehension. Most theories that are now recognized as parts of algebra started as collections of disparate facts from various branches of mathematics, acquired a common theme that served as a core around which various results were grouped, and finally became unified on a basis of a common set of concepts. An archetypical example of this progressive synthesis can be seen in the history of group theory.

Early Group Theory

There were several threads in the early development of group theory, in modern language loosely corresponding to number theory, theory of equations, and geometry.Leonhard Euler considered algebraic operations on numbers modulo an integer, modular arithmetic, in his generalization of Fermat's little theorem. These investigations were taken much further by Carl Friedrich Gauss, who considered the structure of multiplicative groups of residues mod n and established many properties of cyclic and more general abelian groups that arise in this way. In his investigations of composition of binary quadratic forms, Gauss explicitly stated the associative law for the composition of forms, but like Euler before him, he seems to have been more interested in concrete results than in general theory. In 1870, Leopold Kronecker gave a definition of an abelian group in the context of ideal class groups of a number field, generalizing Gauss's work; but it appears he did not tie his definition with previous work on groups, particularly permutation groups. In 1882, considering the same question, Heinrich M. Weber realized the connection and gave a similar definition that involved the cancellation property but omitted the existence of the inverse element, which was sufficient in his context (finite groups).

Permutations were studied by Joseph-Louis Lagrange in his 1770 paper Réflexions sur la résolution algébrique des équations (Thoughts on the algebraic solution of equations) devoted to solutions of algebraic equations, in which he introduced Lagrange resolvents. Lagrange's goal was to understand why equations of third and fourth degree admit formulas for solutions, and he identified as key objects permutations of the roots. An important novel step taken by Lagrange in this paper was the abstract view of the roots, i.e. as symbols and not as numbers. However, he did not consider composition of permutations. Serendipitously, the first edition of Edward Waring's Meditationes Algebraicae (Meditations on Algebra) appeared in the same year, with an expanded version published in 1782. Waring proved the fundamental theorem of symmetric polynomials, and specially considered the relation between the roots of a quartic equation and its resolvent cubic. Mémoire sur la résolution des équations (Memoire on the Solving of Equations) of Alexandre Vandermonde (1771) developed the theory of symmetric functions from a slightly different angle, but like Lagrange, with the goal of understanding solvability of algebraic equations.

Kronecker claimed in 1888 that the study of modern algebra began with this first paper of Vandermonde. Cauchy states quite clearly that Vandermonde had priority over Lagrange for this remarkable idea, which eventually led to the study of group theory.

Paolo Ruffini was the first person to develop the theory of permutation groups, and like his predecessors, also in the context of solving algebraic equations. His goal was to establish the impossibility of an algebraic solution to a general algebraic equation of degree greater than four. En route to this goal he introduced the notion of the order of an element of a group, conjugacy, the cycle decomposition of elements of permutation groups and the notions of primitive and imprimitive and proved some important theorems relating these concepts, such as

if G is a subgroup of S_5 whose order is divisible by 5 then G contains an element of order 5. Note, however, that he got by without formalizing the concept of a group, or even of a permutation group. The next step was taken by Évariste Galois in 1832, although his work remained unpublished until 1846, when he considered for the first time what is now called the closure property of a group of permutations, which he expressed as if in such a group one has the substitutions S and T then one has the substitution ST.

The theory of permutation groups received further far-reaching development in the hands of Augustin Cauchy and Camille Jordan, both through introduction of new concepts and, primarily, a great wealth of results about special classes of permutation groups and even some general theorems. Among other things, Jordan defined a notion of isomorphism, still in the context of permutation groups and, incidentally, it was he who put the term group in wide use.

The abstract notion of a group appeared for the first time in Arthur Cayley's papers in 1854. Cayley realized that a group need not be a permutation group (or even finite), and may instead consist of matrices, whose algebraic properties, such as multiplication and inverses, he systematically investigated in succeeding years. Much later Cayley would revisit the question whether abstract groups were more general than permutation groups, and establish that, in fact, any group is isomorphic to a group of permutations.

Modern Algebra

The end of the 19th and the beginning of the 20th century saw a tremendous shift in the methodology of mathematics. Abstract algebra emerged around the start of the 20th century, under the name modern algebra. Its study was part of the drive for more intellectual rigor in mathematics. Initially, the assumptions in classical algebra, on which the whole of mathematics (and major parts of the natural sciences) depend, took the form of axiomatic systems. No longer satisfied with establishing properties of concrete objects, mathematicians started to turn their attention to general theory. Formal definitions of certain algebraic structures began to emerge in the 19th century. For example, results about various groups of permutations came to be seen as instances of general theorems that concern a general notion of an abstract group. Questions of structure and classification of various mathematical objects came to forefront.

These processes were occurring throughout all of mathematics, but became especially pronounced in algebra. Formal definition through primitive operations and axioms were proposed for many basic algebraic structures, such as groups, rings, and fields. Hence such things as group theory and ring theory took their places in pure mathematics. The algebraic investigations of general fields by Ernst Steinitz and of commutative and then general rings by David Hilbert, Emil Artin and Emmy Noether, building up on the work of Ernst Kummer, Leopold Kronecker and Richard Dedekind, who had considered ideals in commutative rings, and of Georg Frobenius and Issai Schur, concerning representation theory of groups, came to define abstract algebra. These developments of the last quarter of the 19th century and the first quarter of 20th century were systematically exposed in Bartel van der Waerden's Moderne algebra, the two-volume monograph published in 1930–1931 that forever changed for the mathematical world the meaning of the word algebra from the theory of equations to the theory of algebraic structures.

Basic Concepts

By abstracting away various amounts of detail, mathematicians have defined various algebraic structures that are used in many areas of mathematics. For instance, almost all systems studied are sets, to which the theorems of set theory apply. Those sets that have a certain binary operation defined on them form magmas, to which the concepts concerning magmas, as well those concerning sets, apply. We can add additional constraints on the algebraic structure, such as associativity (to form semigroups); identity, and inverses (to form groups); and other more complex structures. With additional structure, more theorems could be proved, but the generality is reduced. The "hierarchy" of algebraic objects (in terms of generality) creates a hierarchy of the corresponding theories: for instance, the theorems of group theory may be used when studying rings (algebraic objects that have two binary operations with certain axioms) since a ring is a group over one of its operations. In general there is a balance between the amount of generality and the richness of the theory: more general structures have usually fewer nontrivial theorems and fewer applications.

Examples of algebraic structures with a single binary operation are:

- Magma
- Quasigroup
- Monoid
- Semigroup
- Group

Examples involving several operations include:

- Ring
- Field
- Module
- Vector space
- Algebra over a field
- Associative algebra
- Lie algebra
- Lattice
- Boolean algebra
- Algebraic structure

In mathematics, and more specifically in abstract algebra, an algebraic structure on a set A is a collection of finitary operations on A; the set A with this structure is also called an algebra. Examples of algebraic structures include groups, rings, fields, and lattices. More complex structures can be defined by introducing multiple operations, different underlying sets, or by altering the defining axioms. Examples of more complex algebraic structures include vector spaces, modules, and algebras.

The properties of specific algebraic structures are studied in abstract algebra. The general theory of algebraic structures has been formalized in universal algebra. The language of category theory is used to express and study relationships between different classes of algebraic and non-algebraic objects. This is because it is sometimes possible to find strong connections between some classes of objects, sometimes of different kinds. For example, Galois theory establishes a connection between certain fields and groups: two algebraic structures of different kinds.

INTRODUCTION

Addition and multiplication on numbers are the prototypical example of an operation that combines two elements of a set to produce a third. These operations obey several algebraic laws. For example, $a + (b + c) = (a + b) + c$ and $a(bc) = (ab)c$, both examples of the associative law. Also $a + b = b + a$, and $ab = ba$, the commutative law. Many systems studied by mathematicians have operations that obey some, but not necessarily all, of the laws of ordinary arithmetic. For example, rotations of objects in three-dimensional space can be combined by performing the first rotation and then applying the second rotation to the object in its new orientation. This operation on rotations obeys the associative law, but can fail the commutative law.

Mathematicians give names to sets with one or more operations that obey a particular collection of laws, and study them in the abstract as algebraic structures. When a new problem can be shown to follow the laws of one of these algebraic structures, all the work that has been done on that category in the past can be applied to the new problem. In full generality, algebraic structures may involve an arbitrary number of sets and operations that can combine more than two elements (higher arity), but this article focuses on binary operations on one or two sets. The examples here are by no means a complete list, but they are meant to be a representative list and include the most common structures. Longer lists of algebraic structures may be found in the external links and within Category:Algebraic structures. Structures are listed in approximate order of increasing complexity.

Examples

One set With Operations

Simple structures: no binary operation:

- Set: a degenerate algebraic structure S having no operations.
- Pointed set: S has one or more distinguished elements, often 0, 1, or both.
- Unary system: S and a single unary operation over S.
- Pointed unary system: a unary system with S a pointed set.

Group-like structures: one binary operation. The binary operation can be indicated by any symbol, or with no symbol (juxtaposition) as is done for ordinary multiplication of real numbers.

- Magma or groupoid: S and a single binary operation over S.
- Semigroup: an associative magma.
- Monoid: a semigroup with Identity element.

- Group: a monoid with a unary operation (inverse), giving rise to inverse elements.
- Abelian group: a group whose binary operation is commutative.
- Semilattice: a semigroup whose operation is idempotent and commutative. The binary operation can be called either meet or join.
- Quasigroup: a magma obeying the latin square property. A quasigroup may also be represented using three binary operations.
- Loop: a quasigroup with identity.

Ring-like structures or Ringoids: two binary operations, often called addition and multiplication, with multiplication distributing over addition.

- Semiring: a ringoid such that S is a monoid under each operation. Addition is typically assumed to be commutative and associative, and the monoid product is assumed to distribute over the addition on both sides, and the additive identity satisfies 0/ x = 0 for all x.
- Near-ring: a semiring whose additive monoid is a (not necessarily abelian) group.
- Ring: a semiring whose additive monoid is an abelian group.
- Lie ring: a ringoid whose additive monoid is an abelian group, but whose multiplicative operation satisfies the Jacobi identity rather than associativity.
- Boolean ring:a commutative ring with idempotent multiplication operation.
- Field: a commutative ring which contains a multiplicative inverse for every nonzero element
- Kleene algebras: a semiring with idempotent addition and a unary operation, the Kleene star, satisfying additional properties.
- *-algebra: a ring with an additional unary operation (*) satisfying additional properties.

Lattice structures: two or more binary operations, including operations called meet and join, connected by the absorption law.

- Complete lattice: a lattice in which arbitrary meet and joins exist.
- Bounded lattice: a lattice with a greatest element and least element.
- Complemented lattice: a bounded lattice with a unary operation, complementation, denoted by postfix ¥″. The join of an element with its complement is the greatest element, and the meet of the two elements is the least element.
- Modular lattice: a lattice whose elements satisfy the additional modular identity.
- Distributive lattice: a lattice in which each of meet and join distributes over the other. Distributive lattices are modular, but the converse does not hold.
- Boolean algebra: a complemented distributive lattice. Either of meet or join can be defined in terms of the other and complementation. This can be shown to be equivalent with the ring-like structure of the same name above.
- Heyting algebra: a bounded distributive lattice with an added binary operation, relative pseudo-complement, denoted by infix '!, and governed by the axiomsx/ '!/ x = 1,x (x-y) = xy ,y (x -y) = y ,x- (y-z) = (x-y) (x-z).

Arithmetics: two binary operations, addition and multiplication. S is an infinite set. Arithmetics are pointed unary systems, whose unary operation is injective successor, and with distinguished element 0.

- Robinson arithmetic. Addition and multiplication are recursively defined by means of successor. 0 is the identity element for addition, and annihilates multiplication. Robinson arithmetic is listed here even though it is a variety, because of its closeness to Peano arithmetic.

Peano arithmetic. Robinson arithmetic with an axiom schema of induction. Most ring and field axioms bearing on the properties of addition and multiplication are theorems of Peano arithmetic or of proper extensions thereof.

Two sets with Operations

Module-like structures: composite systems involving two sets and employing at least two binary operations.

- Group with operators: a group G with a set ? and a binary operation $\Omega \times G \rightarrow G$ satisfying certain axioms.
- Module: an abelian group M and a ring R acting as operators on M. The members of R are sometimes called scalars, and the binary operation of scalar multiplication is a function R×M- M, which satisfies several axioms. Counting the ring operations these systems have at least three operations.
- Vector space: a module where the ring R is a division ring or field.
- Graded vector space: a vector space with a direct sum decomposition breaking the space into "grades".

Quadratic space: a vector space V over a field F with a function from V into F satisfying certain properties. Every quadratic space is also an inner product space (see below). Algebra-like structures: composite system defined over two sets, a ring R and an R-module M equipped with an operation called multiplication. This can be viewed as a system with five binary operations: two operations on R, two on M and one involving both R and M.

- Algebra over a ring (also R-algebra): a module over a commutative ring R, which also carries a multiplication operation that is compatible with the module structure. This includes distributivity over addition and linearity with respect to multiplication by elements of R. The theory of an algebra over a field is especially well developed.
- Associative algebra: an algebra over a ring such that the multiplication is associative.
- Nonassociative algebra: a module over a commutative ring, equipped with a ring multiplication operation that is not necessarily associative. Often associativity is replaced with a different identity, such as alternation, the Jacobi identity, or the Jordan identity.
- Coalgebra: a vector space with a "comultiplication" defined dually to that of associative algebras.
- Lie algebra: a special type of nonassociative algebra whose product satisfies the Jacobi identity.•
- Lie coalgebra: a vector space with a "comultiplication" defined dually to that of Lie algebras.
- Graded algebra: a graded vector space with an algebra structure compatible with the grading. The idea is that if the grades of two elements a and b are known, then the grade of ab is known, and so the location of the product ab is determined in the decomposition.

Inner product space: an F vector space V with a sesquilinear binary operation $V \times V \rightarrow F$.

Four or more binary operations:

- Bialgebra: an associative algebra with a compatible coalgebra structure.
- Lie bialgebra: a Lie algebra with a compatible bialgebra structure.
- Hopf algebra: a bialgebra with a connection axiom (antipode).
- Clifford algebra: a graded associative algebra equipped with an exterior product from which may be derived several possible inner products. Exterior algebras and geometric algebras are special cases of this construction.

Hybrid Structures

Algebraic structures can also coexist with added structure of non-algebraic nature, such as partial order or a topology. The added structure must be compatible, in some sense, with the algebraic structure.

- Topological group: a group with a topology compatible with the group operation.
- Lie group: a topological group with a compatible smooth manifold structure.Ordered groups, ordered rings and ordered fields: each type of structure with a compatible partial order.
- Archimedean group: a linearly ordered group for which the Archimedean property holds.
- Topological vector space: a vector space whose M has a compatible topology.
- Normed vector space: a vector space with a compatible norm. If such a space is complete (as a metric space) then it is called a Banach space.
- Hilbert space: an inner product space over the real or complex numbers whose inner product gives rise to a Banach space structure.
- Vertex Operator Algebra
- Von Neumann algebra: a *-algebra of operators on a Hilbert space equipped with the weak operator topology.

Binary Operation

A binary operation is a calculation that combines the arguments x and y to In mathem atics, a binary operation or dyadic operation is a calculation that combines two elements to produce another element. More formally, a binary operation is an operation of arity two. More specifically, a binary operation on a set is a binary operation whose two domains and the codomain are the same set. Examples include the familiar arithmetic operations of addition, subtraction, multiplication. Other examples are readily found in different areas of mathematics, such as vector addition, matrix multiplication and conjugation in groups.

However, a binary operation may also involve several sets. For example, scalar multiplication of vector spaces takes a scalar and a vector to produce a vector, and scalar product takes two vectors to produce a scalar. Binary operations are the keystone of most algebraic structures, that are studied in algebra, and used in all mathematics, such as fields, groups, monoids, rings, algebras, and many more.

Terminology

More precisely, a binary operation on a set S is a map which sends elements of the Cartesian product $S \times S$ to S: Because the result of performing the operation on a pair of elements of S is again an element of S, the operation is called a closed binary operation on S (or sometimes expressed as having the property of closure). If f is not a function, but is instead a partial function, it is called a partial binary operation. For instance, division of real numbers is a partial binary operation, because one can't divide by zero: a/0 is not defined for any real a. However, both in universal algebra and model theory the binary operations considered are defined on all of S × S. Sometimes, especially in computer science, the term is used for any binary function.

Properties and Examples

Typical examples of binary operations are the addition (+) and multiplication (×) of numbers and matrices as well as composition of functions on a single set. For instance,

- On the set of real numbers R, f(a, b) = a + b is a binary operation since the sum of two real numbers is a real number.
- On the set of natural numbers N, f(a, b) = a + b is a binary operation since the sum of two natural numbers is a natural number. This is a different binary operation than the previous one since the sets are different.
- On the set M(2,R) of 2 × 2 matrices with real entries, f(A, B) = A + B is a binary operation since the sum of two such matrices is a 2 × 2 matrix.
- On the set M(2,R) of 2 × 2 matrices with real entries, f(A, B) = AB is a binary operation since the product of two such matrices is a 2 × 2 matrix.
- For a given set C, let S be the set of all functions h : C '! C. Define f : S × S '! S by f(h_1, h_2)(c) = (h_1 '" h_2) (c) = h_1(h_2(c)) for all c " C, the composition of the two functions h_1 and h_2 in S. Then f is a binary operation since the composition of the two functions is again a function on the set C (that is, a member of S).

Many binary operations of interest in both algebra and formal logic are commutative, satisfying f(a, b) = f(b, a) for all elements a and b in S, or associative, satisfying f(f(a, b), c) = f(a, f(b, c)) for all a, b and c in S. Many also have identity elements and inverse elements.

The first three examples above are commutative and all of the above examples are associative.On the set of real numbers R, subtraction, that is, f(a, b) = a " b, is a binary operation which is not commutative since, in general, a " b `" b " a. It is also not associative, since, in general, a " (b " c) `" (a " b) " c; for instance, 1 " (2 " 3) = 2 but (1 " 2) " 3 = "4. On the set of natural numbers N, the binary operation exponentiation, f(a,b) = a^b, is not commutative since, a^b '" b^a (cf. Equation x¸ = yã), and is also not associative since f(f(a, b), c) '" f(a, f(b, c)). For instance, with a = 2, b = 3 and c = 2, f(2^3,2) = f(8,2) = 8^2 = 64, but f(2,3^2) = f(2,9) = 2^9 = 512. By changing the set N to the set of integers Z, this binary operation becomes a partial binary operation since it is now undefined when a = 0 and b is any negative integer. For either set, this operation has a right identity (which is 1) since f(a, 1) = a for all a in the set, which is not an identity (two sided identity) since f(1, b) '" b in general.

Division (/), a partial binary operation on the set of real or rational numbers, is not commutative or associative. Tetration ($\uparrow\uparrow$), as a binary operation on the natural numbers, is not commutative or associative and has no identity element.

Notation

Binary operations are often written using infix notation such as a " b, a + b, a · b or (by juxtaposition with no symbol) ab rather than by functional notation of the form f(a, b). Powers are usually also written without operator, but with the second argument as superscript. Binary operations sometimes use prefix or (probably more often) postfix notation, both of which dispense with parentheses. They are also called, respectively, Polish notation and reverse Polish notation.

Pair and Tuple

A binary operation, ab, depends on the ordered pair (a, b) and so (ab)c (where the parentheses here mean first operate on the ordered pair (a, b) and then operate on the result of that using the ordered pair ((ab), c)) depends in general on the ordered pair ((a, b), c). Thus, for the general, non-associative case, binary operations can be represented with binary trees.

However:

- If the operation is associative, (ab)c = a(bc), then the value of (ab)c depends only on the tuple (a, b, c).
- If the operation is commutative, $(ab)c = a(bc)$, then the value of (ab)c depends only on (a, b, c), where braces indicate multisets.
- If the operation is both associative and commutative then the value of (ab)c depends only on the multiset {a, b, c}.
- If the operation is associative, commutative and idempotent, aa = a, then the value of (ab)c depends only on the set {a, b, c}.

Binary Operations as Ternary Relations

A binary operation f on a set S may be viewed as a ternary relation on S, that is, the set of triples (a, b, f(a,b)) in S × S × S for all a and b in S.

External Binary Operations

An external binary operation is a binary function from K × S to S. This differs from a binary operation on a set in the sense in that K need not be S; its elements come from outside. An example of an external binary operation is scalar multiplication in linear algebra. Here K is a field and S is a vector space over that field. An external binary operation may alternatively be viewed as an action; K is acting on S. The dot product of two vectors maps from S × S to K, where K is a field and S is a vector space over K. It depends on authors whether it is considered as a binary operation.

Universal Algebra

Algebraic structures are defined through different configurations of axioms. Universal algebra abstractly studies such objects. One major dichotomy is between structures that are axiomatized entirely by identities and structures that are not. If all axioms defining a class of algebras are identities, then this class is a variety (not to be confused with algebraic varieties of algebraic geometry). Identities are equations formulated using only the operations the structure allows, and variables that are tacitly universally Universal_quantifier" quantified over the relevant universe.

Identities contain no connectives, existentially quantified variables, or relations of any kind other than the allowed operations. The study of varieties is an important part of universal algebra. An algebraic structure in a variety may be understood as the quotient algebra of term algebra (also called "absolutely free algebra") divided by the equivalence relations generated by a set of identities. So, a collection of functions with given signatures generate a free algebra, the term algebra T. Given a set of equational identities (the axioms), one may consider their symmetric, transitive closure E. The quotient algebra T/E is then the algebraic structure or variety. Thus, for example, groups have a signature containing two operators: the multiplication operator m, taking two arguments, and the inverse operator i, taking one argument, and the identity element e, a constant, which may be considered an operator that takes zero arguments. Given a (countable) set of variables x, y, z, etc. the term algebra is the collection of all possible terms involving m, i, e and the variables; so for example, m(i(x), m(x,m(y,e))) would be an element of the term algebra. One of the axioms defining a group is the identity m(x, i(x)) = e; another is m(x,e) = x. The axioms can be represented as trees. These equations induce equivalence classes on the free algebra; the quotient algebra then has the algebraic structure of a group.

Some structures do not form varieties, because either:

- It is necessary that 0 ‴ 1, 0 being the additive identity element and 1 being a multiplicative identity element, but this is a nonidentity;
- Structures such as fields have some axioms that hold only for nonzero members of S. For an algebraic structure to be a variety, its operations must be defined for all members of S; there can be no partial operations.

Structures whose axioms unavoidably include nonidentities are among the most important ones in mathematics, e.g., fields and division rings. Structures with nonidentities present challenges varieties do not. For example, the direct product of two fields is not a field, because $(1,0)\cdot(0,1)=(0,0)$, but fields do not have zero divisors.

Category Theory

Category theory is another tool for studying algebraic structures (see, for example, Mac Lane 1998). A category is a collection of objects with associated morphisms. Every algebraic structure has its own notion of homomorphism, namely any function compatible with the operation(s) defining the structure. In this way, every algebraic structure gives rise to a category. For example, the category of groups has all groups as objects and all group homomorphisms as morphisms. This concrete category may be seen as a category of sets with added category-theoretic structure. Likewise, the category of topological groups (whose morphisms are the continuous group homomorphisms) is a category of topological spaces with extra structure. A forgetful functor between categories of algebraic structures "forgets" a part of a structure.

There are various concepts in category theory that try to capture the algebraic character of a context, for instance

- algebraic category
- essentially algebraic category
- presentable category
- locally presentable category
- monadic functors and categories

Different Meanings of "Structure"

In a slight abuse of notation, the word "structure" can also refer to just the operations on a structure, instead of the underlying set itself. For example, the sentence, "We have defined a ring structure on the set A," means that we have defined ring operations

on the set A. For another example, the group $(\mathbb{Z},+)$ can be seen as a set $\mathbb{Z}$ that is equipped with an algebraic structure, namely the operation $+$.

Applications

Because of its generality, abstract algebra is used in many fields of mathematics and science. For instance, algebraic topology uses algebraic objects to study topologies. The Poincaré conjecture, proved in 2003, asserts that the fundamental group of a manifold, which encodes information about connectedness, can be used to determine whether a manifold is a sphere or not. Algebraic number theory studies various number rings that generalize the set of integers. Using tools of algebraic number theory, Andrew Wiles proved Fermat's Last Theorem.

In physics, groups are used to represent symmetry operations, and the usage of group theory could simplify differential equations. In gauge theory, the requirement of local symmetry can be used to deduce the equations describing a system. The groups that describe those symmetries are Lie groups, and the study of Lie groups and Lie algebras reveals much about the physical system; for instance, the number of force carriers in a theory is equal to the dimension of the Lie algebra, and these bosons interact with the force they mediate if the Lie algebra is nonabelian.

2

Group Theory

In mathematics, a group is a set equipped with a binary operation that combines any two elements to form a third element in such a way that four conditions called group axioms are satisfied, namely closure, associativity, identity and invertibility. One of the most familiar examples of a group is the set of integers together with the addition operation, but groups are encountered in numerous areas within and outside mathematics, and help focusing on essential structural aspects, by detaching them from the concrete nature of the subject of the study.

Groups share a fundamental kinship with the notion of symmetry. For example, a symmetry group encodes symmetry features of a geometrical object: the group consists of the set of transformations that leave the object unchanged and the operation of combining two such transformations by performing one after the other. Lie groups are the symmetry groups used in the Standard Model of particle physics; Poincaré groups, which are also Lie groups, can express the physical symmetry underlying special relativity; and point groups are used to help understand symmetry phenomena in molecular chemistry.

The concept of a group arose from the study of polynomial equations, starting with Évariste Galois in the 1830s. After contributions from other fields such as number theory and geometry, the group notion was generalized and firmly established around 1870. Modern group theory—an active mathematical discipline—studies groups in their own right.[a[›]] To explore groups, mathematicians have devised various notions to break groups into smaller, better-understandable pieces, such as subgroups, quotient groups and simple groups. In addition to their abstract properties, group theorists also study the different ways in which a group can be expressed

concretely, both from a point of view of representation theory (that is, through the representations of the group) and of computational group theory. A theory has been developed for finite groups, which culminated with the classification of finite simple groups, completed in 2004.[aa[›]] Since the mid-1980s, geometric group theory, which studies finitely generated groups as geometric objects, has become an active area in group theory.

History of Group Theory

The modern concept of an abstract group developed out of several fields of mathematics. The original motivation for group theory was the quest for solutions of polynomial equations of degree higher than 4. The 19th-century French mathematician Évariste Galois, extending prior work of Paolo Ruffini and Joseph-Louis Lagrange, gave a criterion for the solvability of a particular polynomial equation in terms of the symmetry group of its roots (solutions). The elements of such a Galois group correspond to certain permutations of the roots. At first, Galois' ideas were rejected by his contemporaries, and published only posthumously. More general permutation groups were investigated in particular by Augustin Louis Cauchy. Arthur Cayley's On the theory of groups, as depending on the symbolic equation è^{n} = 1 (1854) gives the first abstract definition of a finite group. Geometry was a second field in which groups were used systematically, especially symmetry groups as part of Felix Klein's 1872 Erlangen program. After novel geometries such as hyperbolic and projective geometry had emerged, Klein used group theory to organize them in a more coherent way. Further advancing these ideas, Sophus Lie founded the study of Lie groups in 1884. The third field contributing to group theory was number theory. Certain abelian group structures had been used implicitly in Carl Friedrich Gauss' number-theoretical work Disquisitiones Arithmeticae (1798), and more explicitly by Leopold Kronecker. In 1847, Ernst Kummer made early attempts to prove Fermat's Last Theorem by developing groups describing factorization into prime numbers.

The convergence of these various sources into a uniform theory of groups started with Camille Jordan's Traité des substitutions et des équations algébriques (1870).Walther von Dyck (1882) introduced the idea of specifying a group by means of generators and relations, and was also the first to give an axiomatic definition of an "abstract group", in the terminology of the time. As of the 20th century, groups gained wide recognition by the pioneering work of Ferdinand Georg Frobenius and William Burnside, who worked on representation theory of finite groups, Richard Brauer's modular representation theory and Issai Schur's papers. The theory of Lie groups, and more generally locally compact groups was studied by Hermann Weyl, Élie Cartan and many others. Its algebraic counterpart, the theory of algebraic groups, was first shaped by Claude Chevalley (from the late 1930s) and later by the work of Armand Borel and Jacques Tits.

The University of Chicago's 1960–61 Group Theory Year brought together group theorists such as Daniel Gorenstein, John G. Thompson and Walter Feit, laying the foundation of a collaboration that, with input from numerous other mathematicians, led to the classification of finite simple groups, with the final step taken by Aschbacher and Smith in 2004. This project exceeded previous mathematical endeavours by its sheer size, in both length of proof and number of researchers. Research is ongoing to simplify the proof of this classification. These days, group theory is still a highly active mathematical branch, impacting many other fields.[a[›]]

Definition and Illustration

First example: the integers

One of the most familiar groups is the set of integers $\mathbb{Z}$ which consists of the numbers ..., 4, 3, 2, 1, 0, 1, 2, 3, 4, ..., together with addition.The following properties of integer addition serve as a model for the group axioms given in the definition below.

- For any two integers a andb, the suma + b is also an integer. That is, addition of integers always yields an integer. This property is known as closure under addition.
- For all integers a, b and c, (a + b) + c = a + (b + c). Expressed in words, adding a tob first, and then adding the result to c gives the same final result as adding a to the sum of b and c, a property known as associativity.
- If a is any integer, then 0 + a = a + 0 = a. Zero is called the identity element of addition because adding it to any integer returns the same integer.
- For every integer a, there is an integer b such that a + b = b + a = 0. The integer b is called the inverse element of the integer a and is denoted Ha.

The integers, together with the operation +, form a mathematical object belonging to a broad class sharing similar structural aspects. To appropriately understand these structures as a collective, the following definition is developed.

Definition

The axioms for a group are short and natural... Yet somehow hidden behind these axioms is the monster simple group, a huge and extraordinary mathematical object, which appears to rely on numerous bizarre coincidences to exist. The axioms for groups give no obvious hint that anything like this exists.

Richard Borcherds in Mathematicians: An Outer View of the Inner World A group is a set, G, together with an operation • (called the group law of G) that combines any two elementsa andb to form another element, denoted a • b or ab. To qualify as a group, the set and operation, (G, •), must satisfy four requirements known as the group axioms:

Closure

For all a, b in G, the result of the operation, a • b, is also in G.[b[›]]

Associativity

For all a, b and c in G, (a • b) • c = a • (b • c).

Identity element

There exists an element e in G such that, for every element a in G, the equation e • a = a • e = a holds. Such an element is unique , and thus one speaks of the identity element.

Inverse element

For each a in G, there exists an element b in G, commonly denoted a"1 (or "a, if the operation is denoted "+"), such that a • b = b • a = e, where e is the identity element.

denoted "+"), such that a • b = b • a = e, where e is the identity element. The result of an operation may depend on the order of the operands. In other words, the result of combining element a with element b need not yield the same result as combining element b with element a; the equation

a • b = b • a

may not always be true. This equation always holds in the group of integers under addition, because a + b = b + a for any two integers (commutativity of addition). Groups for which the commutativity equation a • b = b • a always holds are called abelian groups (in honor of Niels Henrik Abel). The symmetry group described in the following section is an example of a group that is not abelian. The identity element of a group G is often written as 1 or 1_G, a notation inherited from the multiplicative identity. If a group is abelian, then one may choose to denote the group operation by + and the identity element by 0; in that case, the group is called an additive group. The identity element can also be written as id.

The set G is called the underlying set of the group (G, •). Often the group's underlying set G is used as a short name for the group (G, •). Along the same lines, shorthand expressions such as "a subset of the group G" or "an element of group G" are used when what is actually meant is "a subset of the underlying set G of the group (G, •)" or "an element of the underlying set G of the group (G, •)". Usually, it is clear from the context whether a symbol like G refers to a group or to an underlying set.

An alternate (but equivalent) definition is to expand the structure of a group to define a group as a set equipped with three operations satisfying the same axioms as above, with the "there exists" part removed in the two last axioms, these operations being the group law, as above, which is a binary operation, the inverse operation, which is a unary operation and maps a to a H 1 , a^{-2} and the identity element, which is viewed as a 0-ary operation.

As this formulation of the definition avoids existential quantifiers, it is generally preferred for computing with groups and for computer-aided proofs. This formulation exhibits groups as a variety of universal algebra. It is also useful for talking of properties of the inverse operation, as needed for defining topological groups and group objects.

Second Example: a Symmetry Group

Two figures in the plane are congruent if one can be changed into the other using a combination of rotations, reflections, and translations. Any figure is congruent to itself. However, some figures are congruent to themselves in more than one way, and these extra congruences are called symmetries. A square has eight symmetries. These are: The elements of the symmetry group of the square (D_4). Vertices are identified by color or number.

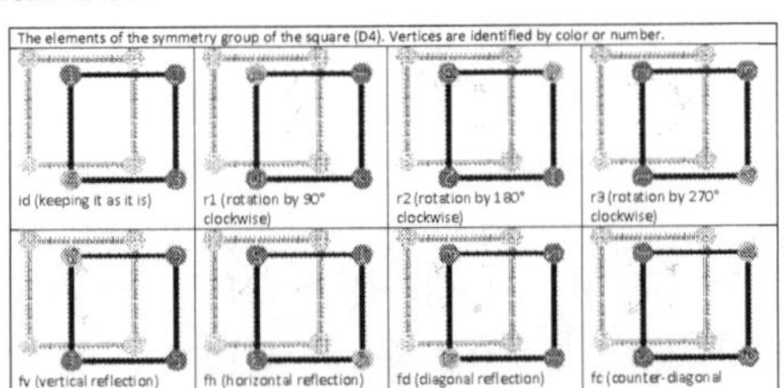

- the identity operation leaving everything unchanged, denoted id;
- rotations of the square around its center by 90° clockwise, 180° clockwise, and 270° clockwise, denoted by r_1, r_2 and r_3, respectively;

- reflections about the vertical and horizontal middle line (f_h and f_v), or through the two diagonals (f_d and f_c).

These symmetries are represented by functions. Each of these functions sends a point in the square to the corresponding point under the symmetry. For example, r_1 sends a point to its rotation 90° clockwise around the square's center, and f_h sends a point to its reflection across the square's vertical middle line. Composing two of these symmetry functions gives another symmetry function. These symmetries determine a group called the dihedral group of degree 4 and denoted D_4. The underlying set of the group is the above set of symmetry functions, and the group operation is function composition. Two symmetries are combined by composing them as functions, that is, applying the first one to the square, and the second one to the result of the first application. The result of performing first a and then b is written symbolically from right to left as

b • a ("apply the symmetry b after performing the symmetry a").The right-to-left notation is the same notation that is used for composition of functions.The group table on the right lists the results of all such compositions possible. For example, rotating by 270° clockwise (r_3) and then reflecting horizontally (f_h) is the same as performing a reflection along the diagonal (f_d). Using the above symbols, highlighted in blue in the group table:

$f_h \bullet r_3 = f_d$.

Group table of D_4

•	**id**	**r_1**	**r_2**	**r_3**	**f_v**	**f_h**	**f_d**	**f_c**
id	id	r_1	r_2	r_3	f_v	f_h	f_d	f_c
r_1	r_1	r_2	r_3	id	f_c	f_d	f_v	f_h
r_2	r_2	r_3	id	r_1	f_h	f_v	f_c	f_d
r_3	r_3	id	r_1	r_2	f_d	f_c	f_h	f_v
f_v	f_v	f_d	f_h	f_c	id	r_2	r_1	r_3
f_h	f_h	f_c	f_v	f_d	r_2	id	r_3	r_1
f_d	f_d	f_h	f_c	f_v	r_3	r_1	id	r_2
f_c	f_c	f_v	f_d	f_h	r_1	r_3	r_2	id

The elements id, r_1, r_2, and r_3 form a subgroup, highlighted in ☐ red (upper left region). A left and right coset of this subgroup is highlighted in ☐ green (in the last row) and ☐ yellow (last column), respectively.

Given this set of symmetries and the described operation, the group axioms can be understood as follows:

1. The closure axiom demands that the composition b • a of any two symmetries a and b is also a symmetry. Another example for the group operation is $r_3 \bullet f_h = f_c$, i.e., rotating 270° clockwise after reflecting horizontally equals reflecting along the counter-diagonal (f_c). Indeed every other combination of two symmetries still gives a symmetry, as can be checked using the group table.
2. The associativity constraint deals with composing more than two symmetries: Starting with three elements a, b and c of D_4, there are two possible ways of using these three symmetries in this order to determine a symmetry of the square. One of these ways is to first compose a andb into a single symmetry, then to compose that symmetry with c. The other way is to first compose b and c, then to compose the resulting symmetry with a. The associativity condition (a • b) • c = a • (b • c)means that these two ways are the same, i.e., a product of many group elements can besimplified in any grouping. For example, (fd o fv) o r2 = fd o (fv o r2) can be checked using the group table at the right

$(f_d \bullet f_v) \bullet r_2$	=	$r_3 \bullet r_2$	=	r_1, which equals
$f_d \bullet (f_v \bullet r_2)$	=	$f_d \bullet f_h$	=	r_1.

not true for all operations. For instance, subtraction of numbers is not associative: (7 " 3) " 2 = 2 is not the same as 7 " (3 " 2) = 6.
" 2 = 2 is not the same as 7 " (3 " 2) = 6.

1. The identity element is the symmetry id leaving everything unchanged: for any symmetry a, performing id after a (or a after id) equals a, in symbolic form,

id • a = a,

a • id = a.

2. An inverse element undoes the transformation of some other element. Every symmetry can be undone: each of the following transformations—identity id, the reflections f_h, f_v, f_d, f_c and the 180° rotation r_2—is its own inverse, because performing it twice brings the square back to its original orientation. The rotations r_3 and r_1 are each other's inverses, because rotating 90° and then rotation 270° (or vice versa) yields a rotation over 360° which leaves the square unchanged. In symbols,

$f_h \bullet f_h = id,$

$r_3 \bullet r_1 = r_1 \bullet r_3 = id.$

In contrast to the group of integers above, where the order of the operation is irrelevant, it does matter in D_4: $f_h \bullet r_1 = f_c$ but $r_1 \bullet f_h = f_d$. In other words, D_4 is not abelian, which makes the group structure more difficult than the integers introduced first.

Elementary consequences of the group axioms

Basic facts about all groups that can be obtained directly from the group axioms are commonly subsumed under elementary group theory. For example, repeated applications of the associativity axiom show that the unambiguity of

a • b • c = (a • b) • c = a • (b • c)

generalizes to more than three factors. Because this implies that parentheses can be inserted anywhere within such a series of terms, parentheses are usually omitted. The axioms may be weakened to assert only the existence of a left identity and left inverses. Both can be shown to be actually two-sided, so the resulting definition is equivalent to the one given above.

Uniqueness of Identity Element and Inverses

Two important consequences of the group axioms are the uniqueness of the identity element and the uniqueness of inverse elements. There can be only one identity element in a group, and each element in a group has exactly one inverse element. Thus, it is customary to speak of the identity, and the inverse of an element. To prove the uniqueness of an inverse element of a, suppose that a has two inverses, denoted b and c, in a group (G, •). Then

b	=	$b \bullet e$		as e is the identity element
	=	$b \bullet (a \bullet c)$		because c is an inverse of a, so $e = a \bullet c$
	=	$(b \bullet a) \bullet c$		by associativity, which allows rearranging the parentheses
	=	$e \bullet c$		since b is an inverse of a, i.e., $b \bullet a = e$
	=	c		for e is the identity element

The term b on the first line above and the c on the last are equal, since they are connected by a chain of equalities. In other words, there is only one inverse element of a. Similarly, to prove that the identity element of a group is unique, assume G is a group with two identity elements e and f. Then e = e • f = f, hence e and f are equal.

Division

In groups, the existence of inverse elements implies that division is possible: given elements a and b of the group G, there is exactly one solution x in G to the equationx • a = b, namely b • $a^{''1}$. In fact, we have (b • $a^{''1}$) • a = b • ($a^{''1}$ • a) = b • e = b.

Uniqueness results by multiplying the two sides of the equation x • a = b by $a^{''1}$. The element b • $a^{''1}$, often denoted b / a, is called the right quotient of b by a, or the result of the right division of b by a. Similarly there is exactly one solution y in G to the equation a • y = b, namely y = $a^{''1}$ • b. This solution is the left quotient of b by a, and is sometimes denoted a \ b. In general b / a anda \ b may be different, but, if the group operation is commutative (that is, if the group is abelian), they are equal. In this case, the group operation is often denoted as an addition, and one talks of subtraction and difference instead of division and quotient. Abstract Algebra A consequence of this is that multiplication by a group element g is a bijection. Specifically, if g is an element of the group G, the function (mathematics) from G to itself that maps h″G to g • h is a bijection. This function is called the left translation by g . Similarly, the right translation by g is the bijection from G to itself, that maps h to h • g. If G is abelian, the left and the right translation by a group element are the same.

Basic Concepts

To understand groups beyond the level of mere symbolic manipulations as above, more structural concepts have to be employed.[c[›]] There is a conceptual principle underlying all of the following notions: to take advantage of the structure offered by groups (which sets, being "structureless", do not have), constructions related to groups have to be compatible with the group operation. This compatibility manifests itself in the following notions in various ways. For example, groups can be related to each other via functions called group homomorphisms. By the mentioned principle, they are required to respect the group structures in a precise sense. The structure of groups can also be understood by breaking them into pieces called subgroups and quotient groups. The principle of "preserving structures"—a recurring topic in mathematics throughout—is an instance of working in a category, in this case the category of groups.

Group Homomorphisms

Group homomorphisms[g[›]] are functions that preserve group structure. A function a: G '! H between two groups (G, •) and (H, ") is called a homomorphism if the equation a(g • k) = a(g) "a(k) holds for all elements g, k in G. In other words, the result is the same when performing the group operation after or before applying the map a. This requirement ensures that $a(1_G) = 1_H$, and also $a(g)^{TM1} = a(g^{TM1})$ for all g in G. Thus a group homomorphism respects all the structure of G provided by the group axioms. Two groups G and H are called isomorphic if there exist group homomorphisms a: G '! H and b: H '! G, such that applying the two functions one after another in each of the two possible orders gives the identity functions of G and H. That is, a(b(h)) = h and b(a(g)) = g for any g in G and h in H. From an abstract point of view, isomorphic groups carry the same information. For example, proving that g • g = 1_G for some element g of G is equivalent to proving that a(g) ™a(g) = 1_H, because applying a to the first equality yields the second, and applying b to the second gives back the first.

Subgroups

Informally, a subgroup is a group H contained within a bigger one, G. Concretely, the identity element of G is contained in H, and whenever h_1 and h_2 are in H, then so are h_1 • h_2 and h_1^{TM1}, so the elements of H, equipped with the group operation on G restricted to H, indeed form a group. In the example above, the identity and the rotations constitute a subgroup R = {id, r_1, r_2, r_3}, highlighted in red in the group table above: any two rotations composed are still a rotation, and a rotation can be undone by (i.e., is inverse to) the complementary rotations 270° for 90°, 180° for 180°, and 90° for 270° (note that rotation in the opposite direction is not defined). The subgroup test is a necessary and sufficient condition for a nonempty subset H of a group G to be a subgroup: it is sufficient to check that $g^{TM1}h$™H for all elements g, h™H. Knowing the subgroups is important in understanding the group as a whole.[d[›]] Given any subset S of a group G, the subgroup generated by S consists of products of elements of S and their inverses. It is the smallest subgroup of G containing S. In the introductory example above, the subgroup generated by r_2 and f_v consists of these two elements, the identity element id and $f_h = f_v$ • r_2. Again, this is a subgroup, because combining any two of these four elements or their inverses (which are, in this particular case, these same elements) yields an element of this subgroup.

Cosets

In many situations it is desirable to consider two group elements the same if they differ by an element of a given subgroup. For example, in D_4 above, once a reflection is performed, the square never gets back to the r_2 configuration by just applying the rotation operations (and no further reflections), i.e., the rotation operations are irrelevant to the question whether a reflection has been performed. Cosets are used to formalize this insight: a subgroup H defines left and right cosets, which can be thought of as translations of H by arbitrary group elements g. In symbolic terms, the left and right cosets of H containing g are gH = {g • h : h™H} and Hg = {h • g : h™H}, respectively.

The left cosets of any subgroup H form a partition of G; that is, the union of all left cosets is equal to G and two left cosets are either equal or have an emptyintersection. The first case $g_1H = g_2H$ happens precisely wheng$_1$"1 • g_2"H, i.e., if the two elements differ by an element of H. Similar considerations apply to the right cosets of H. The left and right cosets of H may or may not be equal. If they are, i.e., for all g in G, gH = Hg, then H is said to be a normal subgroup. In D_4, the introductory symmetry group, the left cosets gR of the subgroup R consisting of the rotations are either equal to R, if g is an element of R itself, or otherwise equal to U = $f_cR = \{f_c, f_v, f_d, f_h\}$ (highlighted in green). The subgroup R is also normal, because f_cR = U = Rf_c and similarly for any element other than f_c. (In fact, in the case of D_4, observe that all such cosets are equal, such that $f_hR = f_vR = f_dR = f_cR$.)

Coset

G is the group (Z/8Z, +), the integers mod 8 under addition. The subgroup H contains only 0 and 4, and is isomorphic to (Z/2Z, +). There are four left cosets of H: H itself, 1 + H, 2 + H, and 3 + H (written using additive notation since this is the additive group). Together they partition the entire group G into equal-size, non-overlapping sets. The index [G : H] is 4. In mathematics, if G is a group, and H is a subgroup of G, and g is an element of G, then gH = is the left coset of H in G ith respect to g, and Hg = is the right coset of H in G with respect to g. Although derived from a subgroup, cosets are not usually themselves subgroups of G, only subsets. For example, it is not necessary for a coset to contain the identity element. A coset may be regarded as a left or a right coset of some subgroup in G. For example, the right coset of H with respect to g, Hg, is also the left coset of the conjugate subgroup g"1Hg with respect to g, g. In other words, a right coset of one subgroup equals a left coset of a conjugate subgroup. It is only when the subgroup (H in the above discussion) is a normal subgroup that they are the same. Hence it is not meaningful to speak of a coset as being left or right unless one first specifies the underlying subgroup. If (and only if) H is a normal subgroup of G, the cosets form a group called the quotient or factor group. If G is an abelian group, the left and right cosets will always be the same. The map gH ¦! (gH)"1 = Hg"1 defines a bijection between the left cosets and the right cosets of H, so the number of left cosets is equal to the number of right cosets. The common value is called the index of H in G. If the group operation is written additively, the notation used changes to g + H or H + g. Cosets are a basic tool in the study of groups; for example they play a central role in Lagrange's theorem.

Examples

C_2Let G = ({"1,1}, ×) be the group formed by {"1,1} under multiplication, which is isomorphic to C_2, and H the trivial subgroup ({1}, ×). Then {"1} = ("1)H = H("1) and {1} = 1H = H1 are the only cosets of H in G. Because its left and right cosets with respect to any element of G coincide, H is a normal subgroup of G.

Integers

Let G be the additive group of the integers, Z = ({..., "2, "1, 0, 1, 2, ...}, +) and H the subgroup (mZ, +) = ({..., "2m, "m, 0, m, 2m, ...}, +) where m is a positive integer. Then the cosets of H in G are the m sets mZ, mZ + 1, ..., mZ + (m " 1), where mZ + a = {..., "2m+a, "m+a, a, m+a, 2m+a, ...}. There are no more than m cosets, because mZ + m = m(Z + 1) = mZ. The coset (mZ + a, +) is the congruence class of a modulo m.

Vectors

Another example of a coset comes from the theory of vector spaces. The elements (vectors) of a vector space form an abelian group under vector addition. It is not hard to show that subspaces of a vector space are subgroups of this group. For a vector space V, a subspace W, and a fixed vector a in V, the sets are called affine subspaces, and are cosets (both left and right, since the group is abelian). In terms of geometric vectors, these affine subspaces are all the "lines" or "planes" parallel to the subspace, which is a line or plane going through the origin.

Definition Using Equivalence Classes

Some authors define the left cosets of H in G to be the equivalence classes under the equivalence relation on G given by x ~ y if and only if $x^{"1}y$ " H. The relation can also be defined by x ~ y if and only if xh=y for some h in H. It can be shown that the relation given is, in fact, an equivalence relation and that the two definitions are equivalent. It follows that any two left cosets of H in G are either identical or disjoint. In other words, every element of G belongs to one and only one left coset and so the left cosets form a partition of G. Corresponding statements are true for right cosets.

Double Cosets

Given two subgroups, H and K of a group G, the double coset of H and K in G are sets of the form HgK = These are the left cosets of K and right cosets of H when H=1 and K=1 respectively.

Notation

Let G be a group with subgroups H and K.

1. denotes the set of left cosets of H in G.
2. denotes the set of right cosets of H in G.
3. denotes the set of double cosets of H and K in G.

General Properties

The identity is in precisely one left or right coset, namely H itself. Thus H is both a left and right coset of itself. A coset representative is a representative in the equivalence class sense. A set of representatives of all the cosets is called a transversal. There are other types of equivalence relations in a group, such as conjugacy, that form different classes which do not have the properties discussed here.

Index of a subgroup

All left cosets and all right cosets have the same order (number of elements, or cardinality in the case of an infinite H), equal to the order of H (because H is itself a coset). Furthermore, the number of left cosets is equal to the number of right cosets and is known as the index of H in G, written as [G : H]. Lagrange's theorem allows us to compute the index in the case where G and H are finite, as per the formula:

This equation also holds in the case where the groups are infinite, although the meaning may be less clear.

Cosets and Normality

If H is not normal in G, then its left cosets are different from its right cosets. That is, there is an a in G such that no element b satisfies aH = Hb. This means that the partition of G into the left cosets of H is a different partition than the partition of G into right cosets of H. (Some cosets may coincide. For example, if a is in the center of G, then aH = Ha.) On the other hand, the subgroup N is normal if and only if gN = Ng for all g in G. In this case, the set of all cosets form a group called the quotient group G / N with the operation " defined by (aN) " (bN) = abN. Since every right coset is a left coset, there is no need to distinguish "left cosets" from "right cosets".

Applications

- Cosets of Q in R are used in the construction of Vitali sets, a type of non-measurable set.
- Cosets are central in the definition of the transfer.
- Cosets are important in computational group theory. For example, Thistlethwaite's algorithm for solving Rubik's Cube relies heavily on cosets.
- Coset leaders are used in decoding received data in linear error-correcting codes

Quotient Groups

In some situations the set of cosets of a subgroup can be endowed with a group law, giving a quotient group or factor group. For this to be possible, the subgroup has to be normal. Given any normal subgroup N, the quotient group is defined by

G / N = {gN, g ∈ G}, "G modulo N". This set inherits a group operation (sometimes called coset multiplication, or coset addition) from the original group G: (gN) • (hN) = (gh)N for all g and h in G. This definition is motivated by the idea (itself an instance of general structural considerations outlined above) that the map G '! G / N that associates to any element g its coset gN be a group homomorphism, or by general abstract considerations called universal properties. The coset eN = N serves as the identity in this group, and the inverse of gN in the quotient group is $(gN)^{"1}$ = $(g^{"1})N$.[e[›]]

Group table of the quotient group D_4 / R

Group table of the quotient group D_4 / R		
•	**R**	**U**
R	*R*	*U*
U	*U*	*R*

The elements of the quotient group D_4 / R are R itself, which represents the identity, and U = f_vR. The group operation on the quotient is shown at the right. For example, U • U = f_vR • f_vR = (f_v • f_v)R = R. Both the subgroup R = {id, r_1, r_2, r_3}, as well as the corresponding quotient are abelian, whereas D_4 is not abelian. Building bigger groups by smaller ones, such as D_4 from its subgroup

R and the quotient D_4 / R is abstracted by a notion called semidirect product. Quotient groups and subgroups together form a way of describing every group by its presentation: any group is the quotient of the free group over the generators of the group, quotiented by the subgroup of relations. The dihedral group D_4, for example, can be generated by two elements r and f (for example, r = r_1, the right rotation and f = f_v the vertical (or any other) reflection), which means that every symmetry of the square is a finite composition of these two symmetries or their inverses. Together with the relations

$r^4 = f^2 = (r \bullet f)^2 = 1$, the group is completely described. A presentation of a group can also be used to construct the Cayley graph, a device used to graphically capture discrete groups.

Sub- and quotient groups are related in the following way: a subset H of G can be seen as an injective map H '! G, i.e., any element of the target has at most one element that maps to it. The counterpart to injective maps aresurjective maps (every element of the target is mapped onto), such as the canonical map G '! G / N.[y[›]] Interpreting subgroup and quotients in light of these homomorphisms emphasizes the structural concept inherent to these definitions alluded to in the introduction. In general, homomorphisms are neither injective nor surjective. Kernel and image of group homomorphisms and the first isomorphism theorem address this phenomenon.

Examples and Applications

Examples and applications of groups abound. A starting point is the group Z of integers with addition as group operation, introduced above. If instead of addition multiplication is considered, one obtains multiplicative groups. These groups are predecessors of important constructions in abstract algebra.

Groups are also applied in many other mathematical areas. Mathematical objects are often examined by associating groups to them and studying the properties of the corresponding groups. For example, Henri Poincaré founded what is now called algebraic topology by introducing the fundamental group. By means of this connection, topological properties such as proximity and continuity translate into properties of groups.[i[›]] For example, elements of the fundamental group are represented by loops. The second image at the right shows some loops in a plane minus a point. The blue loop is considered null-homotopic (and thus irrelevant), because it can be continuously shrunk to a point. The presence of the hole prevents the orange loop from being shrunk to a point.

The fundamental group of the plane with a point deleted turns out to be infinite cyclic, generated by the orange loop (or any other loop winding once around the hole). This way, the fundamental group detects the hole. In more recent applications, the influence has also been reversed to motivate geometric constructions by a group-theoretical background.[j[›]] In a similar vein, geometric group theory employs geometric concepts, for example in the study of hyperbolic groups. Further branches crucially applying groups include algebraic geometry and number theory. In addition to the above theoretical applications, many practical applications of groups exist. Cryptography relies on the combination of the abstract group theory approach together with algorithmical knowledge obtained in computational group theory, in particular when implemented for finite groups. Applications of group theory are not restricted to mathematics; sciences such as physics, chemistry and computer science benefit from the concept.

Numbers

Many number systems, such as the integers and the rationals enjoy a naturally given group structure. In some cases, such as with the rationals, both addition and multiplication operations give rise to group structures. Such number systems are predecessors to more general algebraic structures known as rings and fields. Further abstract algebraic concepts such as modules, vector spaces and algebras also form groups.

Integers

The group of integers $\mathbb{Z}$ under addition, denoted (Z , +) $(\mathbb{Z},+)$, has been described above. The integers, with the operation of multiplication instead of addition, (Z , Å") {\displaystyle \left(\mathbb {Z} ,\cdot \right)} $(\mathbb{Z},\bullet)$ do not form a group. The closure, associativity and identity axioms are satisfied, but inverses do not exist: for example, a = 2 is an integer, but the only solution to the equation a · b = 1 in this case is b = 1/2, which is a rational number, but not an integer. Hence not every element of $\mathbb{Z}$ has a (multiplicative) inverse.

Rationals

The desire for the existence of multiplicative inverses suggests considering fractions Fractions of integers (with b nonzero) are known as rational numbers.[l[›]] The set of all such irreducible fractions is commonly denoted Q {\displaystyle \mathbb {Q} } $\mathbb{Q}$. There is still a minor obstacle for (Q , Å") {\displaystyle \left(\mathbb {Q} ,\cdot \right)} $(\mathbb{Q},\bullet)$, the rationals with multiplication, being a group: because the rational number 0 does not have a multiplicative inverse (i.e., there is no x such that x · 0 = 1), (Q , Å") {\displaystyle \left(\mathbb {Q} ,\cdot \right)} $(\mathbb{Q},\bullet)$ is still not a group.

However, the set of all nonzero rational numbers Q " { 0 } = { q " Q #" q `" 0 } {\displaystyle \mathbb {Q} \setminus \left\{0\right\}=\left\{q\in \mathbb {Q} \mid q\neq 0\right\}} $\mathbb{Q}\setminus\{0\}=\{q\in\mathbb{Q}\mid q\neq 0\}$ does form an abelian group under multiplication, generally denoted Q " {\displaystyle \mathbb {Q} ^{*}} $\mathbb{Q}^*$.[m[›]] Associativity and identity element axioms follow from the properties of integers. The closure requirement still holds true after removing zero, because the product of two nonzero rationals is never zero. Finally, the inverse of a/b is b/a, therefore the axiom of the inverse element is satisfied.

The rational numbers (including 0) also form a group under addition. Intertwining addition and multiplication operations yields more complicated structures called rings and—if division is possible, such as in which occupy a central position in abstract algebra. Group theoretic arguments therefore underlie parts of the theory of those entities.

Modular arithmetic

In modular arithmetic, two integers are added and then the sum is divided by a positive integer called the modulus. The result of modular addition is the remainder of that division. For any modulus, n, the set of integers from 0 to n " 1 forms a group under modular addition: the inverse of any element a isn " a, and 0 is the identity element. This is familiar from the addition of hours on the face of a clock: if the hour hand is on 9 and is advanced 4 hours, it ends up on 1, as shown at the right. This is expressed by saying that 9 + 4 equals 1 "modulo 12" or, in symbols, 9 + 4 a" 1 modulo 12.

The group of integers modulo n is written or Z / n $\mathbb{Z}/n\mathbb{Z}$. For any prime numberp, there is also the multiplicative group of integers modulo p. Its elements are the integers 1 to p ™ 1. The group operation is multiplication modulo p. That is, the usual product is divided by p and the remainder of this division is the result of modular multiplication. For example, if p = 5, there are four group elements 1, 2, 3, 4. In this group, 4 · 4 = 1, because the usual product 16 is equivalent to 1, which divided by 5 yields a remainder of 1. for 5 divides 16 ™ 1 = 15, denoted 16 a™ 1 (mod 5). The primality of p ensures that the product of two integers neither of which is divisible by p is not divisible by peither, hence the indicated set of classes is closed under multiplication.o[›] The identity element is 1, as usual for a multiplicative group, and the associativity follows from the corresponding property of integers. Finally, the inverse element axiom requires that given an integer a not divisible by p, there exists an integer b such that a · b a" 1 (mod p), i.e., p divides the difference a · b " 1.

The inverse b can be found by using Bézout's identity and the fact that the greatest common divisorgcd(a, p) equals 1. In the case p = 5 above, the inverse of 4 is 4, and the inverse of 3 is 2, as 3 · 2 = 6 a" 1 (mod 5). Hence all group axioms are fulfilled. Actually, this example is similar to (Q " { 0 } , Å") {\displaystyle \left(\mathbb above: it consists of exactly those elements in $\mathbb{Z}/p\mathbb{Z}$ that have a multiplicative inverse. These groups are denoted $F_p^\times$. They are crucial to public-key cryptography.p[›]

Cyclic groups

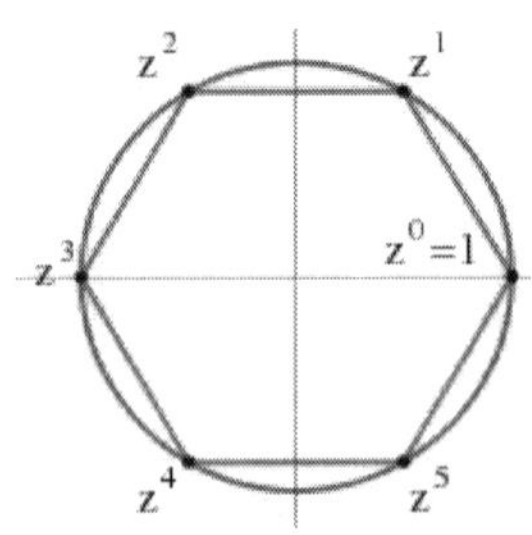

The 6th complex roots of unity form a cyclic group. z is a primitive element, but z^2 is not, because the odd powers of z are not a power of z^2. A cyclic group is a group all of whose elements are powers of a particular element a. In multiplicative notation, the elements of the group are: ..., $a^{"3}$, $a^{"2}$, $a^{"1}$, a^0 = e, a, a^2, a^3, ...,

wherea^2 means a • a, and $a^{"3}$ stands for $a^{"1}$ • $a^{"1}$ • $a^{"1}$ = (a • a • a)$^{"1}$ etc.h[›] Such an element a is called a generator or a primitive element of the group. In additive notation, the requirement for an element to be primitive is that each element of the group can be written as ..., "a"a, "a, 0, a, a+a, ...

In the groups Z/nZ introduced above, the element 1 is primitive, so these groups are cyclic. Indeed, each element is expressible as a sum all of whose terms are 1. Any cyclic group with n elements is isomorphic to this group. A second example for cyclic groups is the group of n-th complex roots of unity, given by complex numbersz satisfying $z^n = 1$. These numbers can be visualized as the vertices on a regular n-gon, as shown in blue at the right for n = 6. The group operation is multiplication of complex numbers. In the picture, multiplying with z corresponds to a counter-clockwise rotation by 60°. Using some field theory, the group $F_p^{\times}$ can be shown to be cyclic: for example, if p = 5, 3 is a generator since $3^1 = 3$, $3^2 = 9$ a" 4, 3^3 a" 2, and 3^4 a" 1.

Some cyclic groups have an infinite number of elements. In these groups, for every non-zero element a, all the powers of a are distinct; despite the name "cyclic group", the powers of the elements do not cycle. An infinite cyclic group is isomorphic to (Z, +), the group of integers under addition introduced above. As these two prototypes are both abelian, so is any cyclic group. The study of finitely generated abelian groups is quite mature, including the fundamental theorem of finitely generated abelian groups; and reflecting this state of affairs, many group-related notions, such as center and commutator, describe the extent to which a given group is not abelian.

Symmetry groups

Symmetry groups are groups consisting of symmetries of given mathematical objects—be they of geometric nature, such as the introductory symmetry group of the square, or of algebraic nature, such as polynomial equations and their solutions. Conceptually, group theory can be thought of as the study of symmetry.Symmetries in mathematics greatly simplify the study of geometrical or analytical objects. A group is said to act on another mathematical object X if every group element performs some operation on X compatibly to the group law. In the rightmost example below, an element of order 7 of the (2,3,7) triangle group acts on the tiling by permuting the highlighted warped triangles (and the other ones, too). By a group action, the group pattern is connected to the structure of the object being acted on.

In chemical fields, such as crystallography, space groups and point groups describe molecular symmetries and crystal symmetries. These symmetries underlie the chemical and physical behavior of these systems, and group theory enables simplification of quantum mechanical analysis of these properties. For example, group theory is used to show that optical transitions between certain quantum levels cannot occur simply because of the symmetry of the states involved. Not only are groups useful to assess the implications of symmetries in molecules, but surprisingly they also predict that molecules sometimes can change symmetry. The Jahn-Teller effect is a distortion of a molecule of high symmetry when it adopts a particular ground state of lower symmetry from a set of possible ground states that are related to each other by the symmetry operations of the molecule.

Likewise, group theory helps predict the changes in physical properties that occur when a material undergoes a phase transition, for example, from a cubic to a tetrahedral crystalline form. An example is ferroelectric materials, where the change from a paraelectric to a ferroelectric state occurs at the Curie temperature and is related to a change from the high-symmetry paraelectric state to the lower symmetry ferroelectric state, accompanied by a so-called soft phonon mode, a vibrational lattice mode that goes to zero frequency at the transition. Such spontaneous symmetry breaking has found further application in elementary particle physics, where its occurrence is related to the appearance of Goldstone bosons.

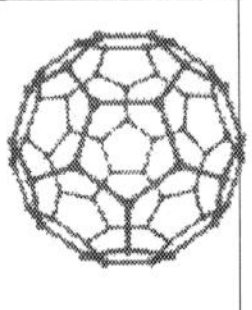		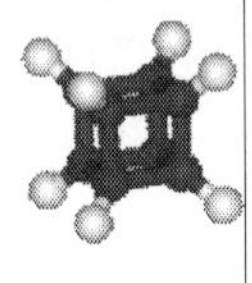	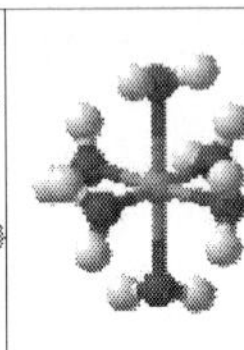	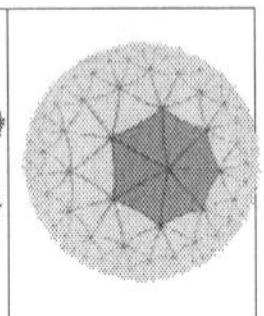
Buckminsterfulleren e displays icosahedral symmetry, though the double bonds reduce this to pyritohedral symmetry.	Ammonia, NH_3. Its symmetry group is of order 6, generated by a 120° rotation and a reflection.	CubaneC_8H_8 features octahedral symmetry.	Hexaaquacopper(II) complex ion, $[Cu(OH_2)_6]^{2+}$. Compared to a perfectly symmetrical shape, the molecule is vertically dilated by about 22% (Jahn-Teller effect).	The (2,3,7) triangle group, a hyperbolic group, acts on this tiling of the hyperbolic plane.

Finite symmetry groups such as the Mathieu groups are used in coding theory, which is in turn applied in error correction of transmitted data, and in CD players. Another application is differential Galois theory, which characterizes functions having antiderivatives of a prescribed form, giving group-theoretic criteria for when solutions of certain differential equations are well-behaved.[u] Geometric properties that remain stable under group actions are investigated in (geometric)invariant theory.

General linear Group and Representation Theory

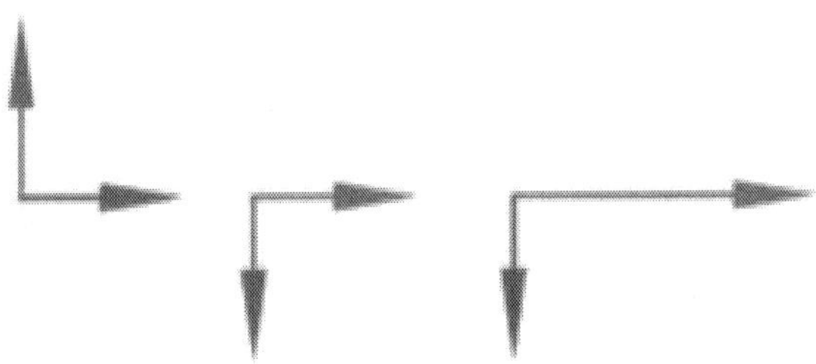

Two vectors (the left illustration) multiplied by matrices (the middle and right illustrations). The middle illustration represents a clockwise rotation by 90°, while the right-most one stretches the x-coordinate by factor 2. Matrix groups consist of matrices together with matrix multiplication. The general linear groupGL(n, R) consists of all invertiblen-by-n matrices with real entries. Its subgroups are referred to as matrix groups or linear groups. The dihedral group example mentioned above can be viewed as a (very small) matrix group. Another important matrix group is the special orthogonal groupSO(n). It describes all possible rotations in n dimensions. Via Euler angles, rotation matrices are used in computer graphics.

Representation theory is both an application of the group concept and important for a deeper understanding of groups. It studies the group by its group actions on other spaces. A broad class of group representations are linear representations, i.e., the group is acting on a vector space, such as the three-dimensional Euclidean spaceR^3. A representation of G on an n-dimensional real vector space is simply a group homomorphism ñ: G '! GL(n, R) from the group to the general linear group. This way, the group operation, which may be abstractly given, translates to the multiplication of matrices making it accessible to explicit computations.Given a group action, this gives further means to study the object being acted on.On the other hand, it also yields information about the group. Group representations are an organizing principle in the theory of finite groups, Lie groups, algebraic groups and topological groups, especially (locally) compact groups.

Galois groups

Galois groups were developed to help solve polynomial equations by capturing their symmetry features. For example, the solutions of the quadratic equation$ax^2 + bx + c = 0$ are given by

$$x = \frac{-b \pm \sqrt{b^2 - 4ac}}{2a}$$

Exchanging "+" and """ in the expression, i.e., permuting the two solutions of the equation can be viewed as a (very simple) group operation. Similar formulae are known for cubic and quartic equations, but do not exist in general for degree 5 and higher. Abstract properties of Galois groups associated with polynomials (in particular their solvability) give a criterion for polynomials that have all their solutions expressible by radicals, i.e., solutions expressible using solely addition, multiplication, and roots similar to the formula above. The problem can be dealt with by shifting to field theory and considering the splitting field of a polynomial. Modern Galois theory generalizes the above type of Galois groups to field extensions and establishes—via the fundamental theorem of Galois theory—a precise relationship between fields and groups, underlining once again the ubiquity of groups in mathematics.

Finite Groups

A group is called finite if it has a finite number of elements. The number of elements is called the order of the group. An important class is the symmetric groups S_N, the groups of permutations of N letters. For example, the symmetric group on 3 letters S_3 is the group consisting of all possible orderings of the three letters ABC, i.e., contains the elements ABC, ACB, BAC, BCA, CAB, CBA, in total 6 (factorial of 3) elements. This class is fundamental insofar as any finite group can be expressed as a subgroup of a symmetric group S_N for a suitable integer N, according to Cayley's theorem. Parallel to the group of symmetries of the square above, S_3 can also be interpreted as the group of symmetries of an equilateral triangle. The order of an element a in a group G is the least positive integer n such that $a^n = e$, where a^n represents

$$\underbrace{a \ldots a}_{n\ factors},$$

i.e., application of the operation • to n copies of a. (If • represents multiplication, then a^n corresponds to the nth power of a.) In infinite groups, such an n may not exist, in which case the order of a is said to be infinity. The order of an element equals the order of the cyclic subgroup generated by this element. More sophisticated counting techniques, for example counting cosets, yield more precise statements about finite groups: Lagrange's Theorem states that for a finite group G the order of any finite subgroup Hdivides the order of G. The Sylow theorems give a partial converse. The dihedral group (discussed above) is a finite group of order 8. The order of r_1 is 4, as is the order of the subgroup R it generates (see above). The order of the reflection elements f_v etc. is 2. Both orders divide 8, as predicted by Lagrange's theorem. The groups $F_p^{\times}$ above have order p " 1.

Classification of Finite Simple Groups

Mathematicians often strive for a complete classification (or list) of a mathematical notion. In the context of finite groups, this aim leads to difficult mathematics. According to Lagrange's theorem, finite groups of order p, a prime number, are necessarily cyclic (abelian) groups Z_p. Groups of order p^2 can also be shown to be abelian, a statement which does not generalize to order p^3, as the non-abelian group D_4 of order $8 = 2^3$ above shows. Computer algebra systems can be used to list small groups, but there is no classification of all finite groups.[q] An intermediate step is the classification of finite simple groups.[r] A nontrivial group is called simple if its only normal subgroups are the trivial group and the group itself.[s] The Jordan–Hölder theorem exhibits finite simple groups as the building blocks for all finite groups.

Listing all finite simple groups was a major achievement in contemporary group theory. 1998 Fields Medal winner Richard Borcherds succeeded in proving the monstrous moonshine conjectures, a surprising and deep relation between the largest finite simple sporadic group—the "monster group"—and certain modular functions, a piece of classical complex analysis, and string theory, a theory supposed to unify the description of many physical phenomena.

Groups with Additional Structure

Many groups are simultaneously groups and examples of other mathematical structures. In the language of category theory, they are group objects in a category, meaning that they are objects (that is, examples of another mathematical structure) which come with transformations (called morphisms) that mimic the group axioms. For example, every group (as defined above) is also a set, so a group is a group object in the category of sets.

Topological Groups

The unit circle in the complex plane under complex multiplication is a Lie group and, therefore, a topological group. It is topological since complex multiplication and division are continuous. It is a manifold and thus a Lie group, because every small piece, such as the red arc in the figure, looks like a part of the real line (shown at the bottom).

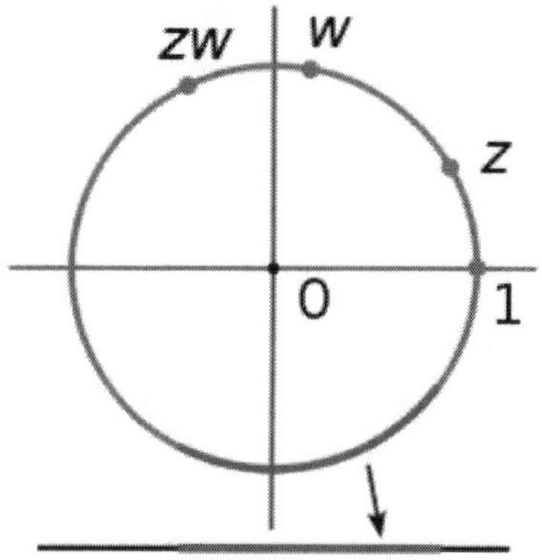

Some topological spaces may be endowed with a group law. In order for the group law and the topology to interweave well, the group operations must be continuous functions, that is, g • h, and $g^{''1}$ must not vary wildly if g and h vary only little. Such groups are called topological groups, and they are the group objects in the category of topological spaces. The most basic examples are the realsR under addition, (R″ {0}, ·), and similarly with any other topological field such as the complex numbers or p-adic numbers. All of these groups are locally compact, so they have Haar measures and can be studied via harmonic analysis. The former offer an abstract formalism of invariant integrals.Invariance means, in the case of real numbers for example:

$$\int_{\mathbb{R}} f(x)dx = \int_{\mathbb{R}} f(x+c)dx$$

for any constant c. Matrix groups over these fields fall under this regime, as do adele rings and adelic algebraic groups, which are basic to number theory. Galois groups of infinite field extensions such as the absolute Galois group can also be equipped with a topology, the so-called Krull topology, which in turn is central to generalize the above sketched connection of fields and groups to infinite field extensions. An advanced generalization of this idea, adapted to the needs of algebraic geometry, is the étale fundamental group.

Lie groups

Lie groups (in honor of Sophus Lie) are groups which also have a manifold structure, i.e., they are spaces looking locally like some Euclidean space of the appropriate dimension. Again, the additional structure, here the manifold structure, has to be compatible, i.e., the maps corresponding to multiplication and the inverse have to be smooth. A standard example is the general linear group introduced above: it is an open subset of the space of all n-by-n matrices, because it is given by the inequality

det (A) ‴ 0,

whereA denotes an n-by-n matrix.

Lie groups are of fundamental importance in modern physics: Noether's theorem links continuous symmetries to conserved quantities.Rotation, as well as translations in space and timeare basic symmetries of the laws of mechanics. They can, for instance, be used to construct simple models—imposing, say, axial symmetry on a situation will typically lead to significant simplification in the equations one needs to solve to provide a physical description.v[›] Another example are the Lorentz transformations, which relate measurements of time and velocity of two observers in motion relative to each other. They can be deduced in a purely group-theoretical way, by expressing the transformations as a rotational symmetry of Minkowski space. The latter serves—in the absence of significant gravitation—as a model of space time in special relativity. The full symmetry group of Minkowski space, i.e., including translations, is known as the Poincaré group. By the above, it plays a pivotal role in special relativity and, by implication, for quantum field theories.Symmetries that vary with location are central to the modern description of physical interactions with the help of gauge theory.

Generalizations

Group-like structures					
	Totality[α]	**Associativity**	**Identity**	**Invertibility**	**Commutativity**
Semigroupoid	Unneeded	Required	Unneeded	Unneeded	Unneeded
Small Category	Unneeded	Required	Required	Unneeded	Unneeded
Groupoid	Unneeded	Required	Required	Required	Unneeded
Magma	Required	Unneeded	Unneeded	Unneeded	Unneeded
Quasigroup	Required	Unneeded	Unneeded	Required	Unneeded
Loop	Required	Unneeded	Required	Required	Unneeded
Semigroup	Required	Required	Unneeded	Unneeded	Unneeded
Inverse Semigroup	Required	Required	Unneeded	Required	Unneeded
Monoid	Required	Required	Required	Unneeded	Unneeded
Group	Required	Required	Required	Required	Unneeded
Abelian group	Required	Required	Required	Required	Required

^αClosure, which is used in many sources, is an equivalent axiom to totality, though defined differently

In abstract algebra, more general structures are defined by relaxing some of the axioms defining a group. For example, if the requirement that every element has an inverse is eliminated, the resulting algebraic structure is called a monoid. The natural numbersN (including 0) under addition form a monoid, as do the nonzero integers under multiplication (Z″ {0}, ·), see above. There is a general method to formally add inverses to elements to any (abelian) monoid, much the same way as (Q″ {0}, ·) is derived from (Z″ {0}, ·), known as the Grothendieck group. Groupoids are similar to groups except that the composition a • b need not be defined for all a andb. They arise in the study of more complicated forms of symmetry, often in topological and analytical structures, such as the fundamental groupoid or stacks. Finally, it is possible to generalize any of these concepts by replacing the binary operation with an arbitrary n-ary one (i.e., an operation taking n arguments). With the proper generalization of the group axioms this gives rise to an n-ary group. The table gives a list of several structures generalizing groups.

3

Ring Theory

In algebra, ring theory is the study of rings—algebraic structures in which addition and multiplication are defined and have similar properties to those operations defined for the integers. Ring theory studies the structure of rings, their representations, or, in different language, modules, special classes of rings (group rings, division rings, universal enveloping algebras), as well as an array of properties that proved to be of interest both within the theory itself and for its applications, such as homological properties and polynomial identities. Commutative rings are much better understood than noncommutative ones. Algebraic geometry and algebraic number theory, which provide many natural examples of commutative rings, have driven much of the development of commutative ring theory, which is now, under the name of

Commutative algebra, a major area of modern mathematics. Because these three fields (algebraic geometry, algebraic number theory and commutative algebra) are so intimately connected it is usually difficult and meaningless to decide which field a particular result belongs to. For example, Hilbert's Nullstellensatz is a theorem which is fundamental for algebraic geometry, and is stated and proved in terms of commutative algebra. Similarly, Fermat's last theorem is stated in terms of elementary arithmetic, which is a part of commutative algebra, but its proof involves deep results of both algebraic number theory and algebraic geometry.

Noncommutative rings are quite different in flavour, since more unusual behavior can arise. While the theory has developed in its own right, a fairly recent trend has sought to parallel the commutative development by building the theory of certain classes of noncommutative rings in a geometric fashion as if they were rings of functions on (non-existent) 'noncommutative spaces'. This trend started in the 1980s with the development of noncommutative geometry and with the discovery of quantum groups. It has led to a better understanding of noncommutative rings, especially noncommutative Noetherian rings. For the definitions of a ring and basic concepts and their properties, see ring (mathematics). The definitions of terms used throughout ring theory may be found in the glossary of ring theory.

COMMUTATIVE RINGS

A ring is called commutative if its multiplication is commutative. Commutative rings resemble familiar number systems, and various definitions for commutative rings are designed to formalize properties of the integers. Commutative rings are also important in algebraic geometry. In commutative ring theory, numbers are often replaced by ideals, and the definition of the prime ideal tries to capture the essence of prime numbers. Integral domains, non-trivial commutative rings where no two non-zero elements multiply to give zero, generalize another property of the integers and serve as the proper realm to study divisibility. Principal ideal domains are integral domains in which every ideal can be generated by a single element, another property shared by the integers. Euclidean domains are integral domains in which the Euclidean algorithm can be carried out. Important examples of commutative rings can be constructed as rings of polynomials and their factor rings. Summary: Euclidean domain =>principal ideal domain =>unique factorization domain =>integral domain =>Commutative ring.

Algebraic geometry

Algebraic geometry is in many ways the mirror image of commutative algebra. This correspondence started with Hilbert's Nullstellensatz that establishes a one-to-one correspondence between the points of an algebraic variety, and the maximal ideals of its coordinate ring. This correspondence has been enlarged and systematized for translating (and proving) most geometrical properties of algebraic varieties into algebraic properties of associated commutative rings.Alexander Grothendieck completed this by introducing schemes, a generalization of algebraic varieties, which may be built from any commutative ring. More precisely, the spectrum of a commutative ring is the space of its prime ideals equipped with Zariski topology, and augmented with a sheaf of rings. These objects are the "affine schemes" (generalization of affine varieties), and a general scheme is then obtained by "gluing together" (by purely algebraic methods) several such affine schemes, in analogy to the way of constructing a manifold by gluing together the charts of an atlas.

Algebraic geometry

Algebraic geometry is a branch of mathematics, classically studying zeros of multivariate polynomials. Modern algebraic geometry is based on the use of abstract algebraic techniques, mainly from commutative algebra, for solving geometrical problems about these sets of zeros. The fundamental objects of study in algebraic geometry are algebraic varieties, which are geometric manifestations of solutions of systems of polynomial equations. Examples of the most studied classes of algebraic varieties are: plane algebraic curves, which include lines, circles, parabolas, ellipses, hyperbolas, cubic curves like elliptic curves, and quartic curves like lemniscates and Cassini ovals. A point of the plane belongs to an algebraic curve if its coordinates satisfy a given polynomial equation. Basic questions involve the study of the points of special interest like the singular points, the inflection points and the points at infinity. More advanced questions involve the topology of the curve and relations between the curves given by different equations.

Algebraic geometry occupies a central place in modern mathematics and has multiple conceptual connections with such diverse fields as complex analysis, topology and number theory. Initially a study of systems of polynomial equations in several variables, the subject of algebraic geometry starts where equation solving leaves off, and it becomes even more important to understand the intrinsic properties of the totality of solutions of a system of equations, than to find a specific solution; this leads into some of the deepest areas in all of mathematics, both conceptually and in terms of technique. In the 20th century, algebraic geometry split into several subareas.

- The mainstream of algebraic geometry is devoted to the study of the complex points of the algebraic varieties and more generally to the points with coordinates in an algebraically closed field.
- Real algebraic geometry is the study of the real points of an algebraic variety.
- Diophantine geometry and, more generally, arithmetic geometry is the study of the points of an algebraic variety with coordinates in fields that are not algebraically closed and occur in algebraic number theory, such as the field of rational numbers, number fields,
  finite fields, function fields, and p-adic fields.
- A large part of singularity theory is devoted to the singularities of algebraic varieties.
- Computational algebraic geometry is an area that has emerged at the intersection of algebraic geometry and computer algebra, with the rise of computers. It consists mainly of algorithm design and software development for the study of roperties of explicitly given algebraic varieties.

Much of the development of the mainstream of algebraic geometry in the 20th century occurred within an abstract algebraic framework, with increasing emphasis being placed on "intrinsic" properties of algebraic varieties not dependent on any particular way of embedding the variety in an ambient coordinate space; this parallels developments in topology, differential and complex geometry. One key achievement of this abstract algebraic geometry is Grothendieck's scheme theory which allows one to use sheaf theory to study algebraic varieties in a way which is very similar to its use in the study of differential and analytic manifolds. This is obtained by extending the notion of point.

In classical algebraic geometry, a point of an affine variety may be identified, through Hilbert's Nullstellensatz, with a maximal ideal of the coordinate ring, while the points of the corresponding affine scheme are all prime ideals of this ring. This means that a point of such a scheme may be either a usual point or a subvariety. This approach also enables a unification of the language and the tools of classical algebraic geometry, mainly concerned with complex points, and of algebraic number theory. Wiles' proof of the longstanding conjecture called Fermat's last theorem is an example of the power of this approach.

Basic notions

Zeros of simultaneous polynomials

In classical algebraic geometry, the main objects of interest are the vanishing sets of collections of polynomials, meaning the set of all points that simultaneously satisfy one or more polynomial equations. For instance, the two-dimensional sphere of radius 1 in three-dimensional Euclidean space R^3 could be defined as the set of all points (x,y,z) with

$$x^2+y^2+z^2-1=0.$$

A "slanted" circle in R^3 can be defined as the set of all points (x,y,z) which satisfy the two polynomial equations

$$x^2+y^2+z^2-1=0,$$

$$x+y+z=0,$$

Affine varieties

First we start with a field k. In classical algebraic geometry, this field was always the complex numbers C, but many of the same results are true if we assume only that k is algebraically closed. We consider the affine space of dimension n over k, denoted $A^n(k)$ (or more simply A^n, when k is clear from the context). When one fixes a coordinate system, one may identify $A^n(k)$ with k^n. The purpose of not working with k^n is to emphasize that one "forgets" the vector space structure that k^n carries.

A function f : A^n '! A^1 is said to be polynomial (or regular) if it can be written as a polynomial, that is, if there is a polynomial p in $k[x_1,\ldots,x_n]$ such that $f(M) = p(t_1,\ldots,t_n)$ for every point M with coordinates $(t_1,\ldots,t_n)$ in A^n. The property of a function to be polynomial (or regular) does not depend on the choice of a coordinate system in A. When a coordinate system is chosen, the regular functions on the affine n-space may be identified with the ring of polynomial functions in n variables over k. Therefore, the set of the regular functions on A^n is a ring, which is denoted $k[A^n]$. We say that a polynomial vanishes at a point if evaluating it at that point gives zero. Let S be a set of polynomials in $k[A^n]$. The vanishing set of S (or vanishing locus or zero set) is the set V(S) of all points in A^n where every polynomial in S vanishes. Symbolically,

$$V(S)=\{(t_1,\ldots,t_n)\mid p(t_1,\ldots,t_n)=0 \text{ for all } p\in S\}.$$

A subset of A^n which is V(S), for some S, is called an algebraic set. The V stands for variety (a specific type of algebraic set to be defined below). Given a subset U of A^n, can one recover the set of polynomials which generate it? If U is any subset of A^n, define I(U) to be the set of all polynomials whose vanishing set contains U. The I stands for ideal: if two polynomials f and g both vanish on U, then f+g vanishes on U, and if h is any polynomial, then hf vanishes on U, so I(U) is always an ideal of the polynomial ring $k[A^n]$.

Two natural questions to ask are:

- Given a subset U of A^n, when is U = V(I(U))?
- Given a set S of polynomials, when is S = I(V(S))?

The answer to the first question is provided by introducing the Zariski topology, a topology on A^n whose closed sets are the algebraic sets, and which directly reflects the algebraic structure of $k[A^n]$. Then U = V(I(U)) if and only if U is an algebraic set or equivalently a Zariski-closed set. The answer to the second question is given by Hilbert's Nullstellensatz. In one of its forms, it says that I(V(S)) is the radical of the ideal generated by S. In more abstract language, there is a Galois connection, giving rise to two closure operators; they can be identified, and naturally play a basic role in the theory; the example is elaborated at Galois connection.

For various reasons we may not always want to work with the entire ideal corresponding to an algebraic set U. Hilbert's basis theorem implies that ideals in $k[A^n]$ are always finitely generated. An algebraic set is called irreducible if it cannot be written as the union of two smaller algebraic sets. Any algebraic set is a finite union of irreducible algebraic sets and this decomposition is unique. Thus its elements are called the irreducible components of the algebraic set. An irreducible algebraic set is also called a variety. It turns out that an algebraic set is a variety if and only if it may be defined as the vanishing set of a prime ideal of the polynomial ring. Some authors do not make a clear distinction between algebraic sets and varieties and use irreducible variety to make the distinction when needed.

Regular functions

Just as continuous functions are the natural maps on topological spaces and smooth functions are the natural maps on differentiable manifolds, there is a natural class of functions on an algebraic set, called regular functions or polynomial functions. A regular function on an algebraic set V contained in A^n is the restriction to V of a regular function on A^n. For an algebraic set defined on the field of the complex numbers, the regular functions are smooth and even analytic. It may seem unnaturally restrictive to require that a regular function always extend to the ambient space, but it is very similar to the situation in a normal topological space, where the Tietze extension theorem guarantees that a continuous function on a closed subset always extends to the ambient topological space.

Just as with the regular functions on affine space, the regular functions on V form a ring, which we denote by k[V]. This ring is called the coordinate ring of V. Since regular functions on V come from regular functions on A^n, there is a relationship between the coordinate rings. Specifically, if a regular function on V is the restriction of two functions f and g in $k[A^n]$, then f ″ g is a polynomial function which is null on V and thus belongs to I(V). Thus k[V] may be identified with $k[A^n]/I(V)$.

Morphism of affine varieties

Using regular functions from an affine variety to A^1, we can define regular maps from one affine variety to another. First we will define a regular map from a variety into affine space: Let V be a variety contained in A^n. Choose m regular functions on V, and call them $f_1, ..., f_m$. We define a regular map f from V to A^m by letting $f = (f_1, ..., f_m)$. In other words, each f_i determines one coordinate of the range of f. If V2 is a variety contained in A^m, we say that f is a regular map from V to V2 if the range of f is contained in V2 . The definition of the regular maps apply also to algebraic sets. The regular maps are also called morphisms, as they make the collection of all affine algebraic sets into a category, where the objects are the affine algebraic sets and the morphisms are the regular maps. The affine varieties is a subcategory of the category of the algebraic sets.

Given a regular map g from V to V2 and a regular function f of k[V2], then f ‴ g ″ k[V]. The map f ′! f ‴ g is a ring homomorphism from k[V2] to k[V]. Conversely, every ring homomorphism from k[V2] to k[V] defines a regular map from V to V2 . This defines an equivalence of categories between the category of algebraic sets and the opposite category of the finitely generated reduced k-algebras. This equivalence is one of the starting points of scheme theory.

Rational function and birational equivalence

In contrast to the preceding sections, this section concerns only varieties and not algebraic sets. On the other hand, the definitions extend naturally to projective varieties (next section), as an affine variety and its projective completion have the same field of functions. If V is an affine variety, its coordinate ring is an integral domain and has thus a field of fractions which is denoted k(V) and called the field of the rational functions on V or, shortly, the function field of V. Its elements are the restrictions to V of the rational functions over the affine space containing V. The domain of a rational function f is not V but the complement of the subvariety (a hypersurface) where the denominator of f vanishes. As with regular maps, one may define a rational map from a variety V to a variety V′. As with the regular maps, the rational maps from V to V′ may be identified to the field homomorphisms from k(V′) to k(V). Two affine varieties are birationally equivalent if there are two rational functions between them which are inverse one to the other in the regions where both are defined. Equivalently, they are birationally equivalent if their function fields are isomorphic. An affine variety is a rational variety if it is birationally equivalent to an affine space. This means that the variety admits a rational parameterization. For example, the circle of equation x 2 + y 2 " 1 = 0 {\displaystyle x^{2}+y^{2}-1=0}

$x^2 + y^2 - 1 = 0$ is a rational curve, as it has the parameterization

$$x = \frac{2t}{1+t^2}$$

$$y = \frac{1-t^2}{1+t^2},$$

which may also be viewed as a rational map from the line to the circle.

The problem of resolution of singularities is to know if every algebraic variety is birationally equivalent to a variety whose projective completion is nonsingular (see also smooth completion). It was solved in the affirmative in characteristic 0 by Heisuke Hironaka in 1964 and is yet unsolved in finite characteristic.

Projective variety

Just as the formulas for the roots of second, third, and fourth degree polynomials suggest extending real numbers to the more algebraically complete setting of the complex numbers, many properties of algebraic varieties suggest extending affine space to a more geometrically complete projective space. Whereas the complex numbers are obtained by adding the number i, a root of the polynomial $x^2 + 1$, projective space is obtained by adding in appropriate points "at infinity", points where parallel lines may meet. To see how this might come about, consider the variety V(y " x^2). If we draw it, we get a parabola. As x goes to positive infinity, the slope of the line from the origin to the point (x, x^2) also goes to positive infinity. As x goes to negative infinity, the slope of the same line goes to negative infinity.

Compare this to the variety V(y " x^3). This is a cubic curve. As x goes to positive infinity, the slope of the line from the origin to the point (x, x^3) goes to positive infinity just as before. But unlike before, as x goes to negative infinity, the slope of the same line goes to positive infinity as well; the exact opposite of the parabola. So the behavior "at infinity" of V(y ″ x^3) is different from the behavior "at infinity" of V(y ″ x^2). The consideration of the projective completion of the two curves, which is their prolongation "at infinity" in the projective plane, allows us to quantify this difference: the point at infinity of the parabola is a regular point, whose tangent is the line at infinity, while the point at infinity of the cubic curve is a cusp. Also, both curves are rational, as they are parameterized by x, and the Riemann-Roch theorem implies that the cubic curve must have a singularity, which must be at infinity, as all its points in the affine space are regular.

Thus many of the properties of algebraic varieties, including birational equivalence and all the topological properties, depend on the behavior "at infinity" and so it is natural to study the varieties in projective space. Furthermore, the introduction of projective techniques made many theorems in algebraic geometry simpler and sharper: For example, Bézout's theorem on the number of intersection points between two varieties can be stated in its sharpest form only in projective space. For these reasons, projective space plays a fundamental role in algebraic geometry. Nowadays, the projective space P^n of dimension n is usually defined as the set of the lines passing through a point, considered as the origin, in the affine space of dimension n + 1, or equivalently to the set of the vector lines in a vector space of dimension n + 1. When a coordinate system has been chosen in the space of dimension n + 1, all the points of a line have the same set of coordinates, up to the multiplication by an element of k. This defines the homogeneous coordinates of a point of P^n as a sequence of n + 1 elements of the base field k, defined up to the multiplication by a nonzero element of k (the same for the whole sequence).

A polynomial in n + 1 variables vanishes at all points of a line passing through the origin if and only if it is homogeneous. In this case, one says that the polynomial vanishes at the corresponding point of P^n. This allows us to define a projective algebraic set in P^n as the set $V(f_1, ..., f_k)$, where a finite set of homogeneous polynomials $\{f_1, ..., f_k\}$ vanishes. Like for affine algebraic sets, there is a bijection between the projective algebraic sets and the reduced homogeneous ideals which define them. The projective varieties are the projective algebraic sets whose defining ideal is prime. In other words, a projective variety is a projective algebraic set, whose homogeneous coordinate ring is an integral domain, the projective coordinates ring being defined as the quotient of the graded ring or the polynomials in n + 1 variables by the homogeneous (reduced) ideal defining the variety. Every projective algebraic set may be uniquely decomposed into a finite union of projective varieties. The only regular functions which may be defined properly on a projective variety are the constant functions. Thus this notion is not used in projective situations. On the other hand, the field of the rational functions or function field is a useful notion, which, similarly to the affine case, is defined as the set of the quotients of two homogeneous elements of the same degree in the homogeneous coordinate ring.

Computational algebraic geometry

One may date the origin of computational algebraic geometry to meeting EUROSAM'79 (International Symposium on Symbolic and Algebraic Manipulation) held at Marseille, France in June 1979. At this meeting,

- Dennis S. Arnon showed that George E. Collins's Cylindrical algebraic decomposition (CAD) allows the computation of the topology of semi-algebraic sets,
- Bruno Buchberger presented the Gröbner bases and his algorithm to compute them,
- Daniel Lazard presented a new algorithm for solving systems of homogeneous polynomial equations with a computational complexity which is essentially polynomial in the expected number of solutions and thus simply exponential in the number of the unknowns. This algorithm is strongly related with Macaulay's multivariate resultant.

Since then, most results in this area are related to one or several of these items either by using or improving one of these algorithms, or by finding algorithms whose complexity is simply exponential in the number of the variables. A body of mathematical theory complementary to symbolic methods called numerical algebraic geometry has been developed over the last several decades. The main computational method is homotopy continuation. This supports, for example, a model of floating point computation for solving problems of algebraic geometry.

1Gröbner basis

A Gröbner basis is a system of generators of a polynomial ideal whose computation allows the deduction of many properties of the affine algebraic variety defined by the ideal.

Given an ideal I defining an algebraic set V:

- V is empty (over an algebraically closed extension of the basis field), if and only if the Gröbner basis for any monomial ordering is reduced to {1}.
- By means of the Hilbert series one may compute the dimension and the degree of V from any Gröbner basis of I for a monomial ordering refining the total degree.
- If the dimension of V is 0, one may compute the points (finite in number) of V from any Gröbner basis of I (see Systems of polynomial equations).
- A Gröbner basis computation allows one to remove from V all irreducible components which are contained in a given hypersurface.
- A Gröbner basis computation allows one to compute the Zariski closure of the image of V by the projection on the k first coordinates, and the subset of the image where the projection is not proper.
- More generally Gröbner basis computations allow one to compute the Zariski closure of the image and the critical points of a rational function of V into another affine variety.

Gröbner basis computations do not allow one to compute directly the primary decomposition of I nor the prime ideals defining the irreducible components of V, but most algorithms for this involve Gröbner basis computation. The algorithms which are not based on Gröbner bases use regular chains but may need Gröbner bases in some exceptional situations. Gröbner bases are deemed to be difficult to compute. In fact they may contain, in the worst case, polynomials whose degree is doubly exponential in the number of variables and a number of polynomials which is also doubly exponential. However, this is only a worst case complexity, and the complexity bound of Lazard's algorithm of 1979 may frequently apply. Faugère F5 algorithm realizes this complexity, as it may be viewed as an improvement of

Lazard's 1979 algorithm. It follows that the best implementations allow one to compute almost routinely with algebraic sets of degree more than 100. This means that, presently, the difficulty of computing a Gröbner basis is strongly related to the intrinsic difficulty of the problem.

Gröbner basis

In mathematics, and more specifically in computer algebra, computational algebraic geometry, and computational commutative algebra, a Gröbner basis is a particular kind of generating set of an ideal in a polynomial ring $K[x_1, .., x_n]$ over a field K. A Gröbner basis allows many important properties of the ideal and the associated algebraic variety to be deduced easily, such as the dimension and the number of zeros when it is finite. Gröbner basis computation is one of the main practical tools for solving systems of polynomial equations and computing the images of algebraic varieties under projections or rational maps. Gröbner basis computation can be seen as a multivariate, non-linear generalization of both Euclid's algorithm for computing polynomial greatest common divisors, and Gaussian elimination for linear systems.

Gröbner bases were introduced in 1965, together with an algorithm to compute them (Buchberger's algorithm), by Bruno Buchberger in his Ph.D. thesis. He named them after his advisor Wolfgang Gröbner. In 2007, Buchberger received the Association for Computing Machinery's Paris Kanellakis Theory and Practice Award for this work. However, the Russian mathematician Nikolai Günther had introduced a similar notion in 1913, published in various Russian mathematical journals. These papers were largely ignored by the mathematical community until their rediscovery in 1987 by Bodo Renschuch et al. An analogous concept for multivariate power series was developed independently by Heisuke Hironaka in 1964, who named them standard bases. This term has been used by some authors to also denote Gröbner bases. The theory of Gröbner bases has been extended by many authors in various directions. It has been generalized to other structures such as polynomials over principal ideal rings or polynomial rings, and also some classes of non-commutative rings and algebras, like Ore algebras.

Monomial ordering

All operations related to Gröbner bases require the choice of a total order on the monomials, with the following properties of compatibility with multiplication. For all monomials M, N, P,

1. M d" N ú' M P d" N P {\displaystyle M\leq N\Longleftrightarrow MP\leq NP}
2. M d" M P {\displaystyle M\leq MP} .

A total order satisfying these condition is sometimes called an admissible ordering.

These conditions imply Noetherianity, which means that every strictly decreasing sequence of monomials is finite.

Although Gröbner basis theory does not depend on a particular choice of an admissible monomial ordering, three monomial orderings are specially important for the applications:

- Lexicographical ordering, commonly called lex or plex (for pure lexical ordering).
- Total degree reverse lexicographical ordering, commonly called degrevlex.

- Elimination ordering, lexdeg.

Gröbner basis theory was initially introduced for the lexicographical ordering. It was soon realised that the Gröbner basis for degrevlex is almost always much easier to compute, and that it is almost always easier to compute a lex Gröbner basis by first computing the degrevlex basis and then using a "change of ordering algorithm". When elimination is needed, degrevlex is not convenient; both lex and lexdeg may be used but, again, many computations are relatively easy with lexdeg and almost impossible with lex. Once a monomial ordering is fixed, the terms of a polynomial (product of a monomial with its nonzero coefficient) are naturally ordered by decreasing monomials (for this order). This makes the representation of a polynomial as an ordered list of pairs coefficient–exponent vector a canonical representation of the polynomials. The first (greatest) term of a polynomial p for this ordering and the corresponding monomial and coefficient are respectively called the leading term, leading monomial and leading coefficient and denoted, in this article, lt(p), lm(p) and lc(p).

Reduction

The concept of reduction, also called multivariate division or normal form computation, is central to Gröbner basis theory. It is a multivariate generalization of the Euclidean division of univariate polynomials. In this section we suppose a fixed monomial ordering, which will not be defined explicitly. Given two polynomials f and g, one says that f is reducible by g if some monomial m in f is a multiple of the leading monomial lm(g) of g. If m happens to be the leading monomial of f then one says that f is lead-reducible by g. If c is the coefficient of m in f and m = q lm(g), the one-step reduction of f by g is the operation that associates to f the polynomial red 1 a (f,g)=f" c lc a (g) q g. The main properties of this operation are that the resulting polynomial does not contain the monomial m and that the monomials greater than m (for the monomial ordering) remain unchanged. This operation is not, in general, uniquely defined; if several monomials in f are multiples of lm(g), then one may choose arbitrarily which one to reduce. In practice, it is better to choose the greatest one for the monomial ordering, because otherwise subsequent reductions could reintroduce the monomial that has just been removed.

Given a finite set G of polynomials, one says that f is reducible or lead-reducible by G if it is reducible or lead-reducible, respectively, by an element g of G. If it is the case, then one defines red 1 a (f, G) = red 1 a (f, g). The (complete) reduction of f by G consists in applying iteratively this operator red 1, which is irreducible by G. It is called a normal form of f by G. In general this form is not uniquely defined (this is not a canonical form); this non-uniqueness is the starting point of Gröbner basis theory.

For Gröbner basis computations, except at the end, it is not necessary to do a complete reduction: a lead-reduction is sufficient, which saves a large amount of computation. The definition of the reduction shows immediately that, if h is a normal form of f by G, then we have f = h + " g " G q g g, are polynomials. In the case of univariate polynomials, if G consists of a single element g, then h is the remainder of the Euclidean division of f by g, q_g is the quotient and the division algorithm is exactly the process of lead-reduction. For this reason, some authors use the term multivariate division instead of reduction.

Formal definition

A Gröbner basis G of an ideal I in a polynomial ring R over a field is a generating set of I characterized by any one of the following properties, stated relatively to some monomial order:

- the ideal generated by the leading terms of polynomials in I equals the ideal generated by the leading terms of G;
- the leading term of any polynomial in I is divisible by the leading term of some polynomial in G;
- the multivariate division of any polynomial in the polynomial ring R by G gives a unique remainder;
- the multivariate division by G of any polynomial in the ideal I gives the remainder 0.

All these properties are equivalent; different authors use different definitions depending on the topic they choose. The last two properties allow calculations in the factor ring R/I with the same facility as modular arithmetic. It is a significant fact of commutative algebra that Gröbner bases always exist, and can be effectively computed for any ideal given by a finite generating subset.

Multivariate division requires a monomial ordering, the basis depends on the monomial ordering chosen, and different orderings can give rise to radically different Gröbner bases. Two of the most commonly used orderings are lexicographic ordering, and degree reverse lexicographic order (also called graded reverse lexicographic order or simply total degree order). Lexicographic order eliminates variables, however the resulting Gröbner bases are often very large and expensive to compute. Degree reverse lexicographic order typically provides for the fastest Gröbner basis computations. In this order monomials are compared first by total degree, with ties broken by taking the smallest monomial with respect to lexicographic ordering with the variables reversed.

In most cases (polynomials in finitely many variables with complex coefficients or, more generally, coefficients over any field, for example), Gröbner bases exist for any monomial ordering. Buchberger's algorithm is the oldest and most well-known method for computing them. Other methods are the Faugère's F4 and F5 algorithms, based on the same mathematics as the Buchberger algorithm, and involutive approaches, based on ideas from differential algebra. There are also three algorithms for converting a Gröbner basis with respect to one monomial order to a Gröbner basis with respect to a different monomial order: the FGLM algorithm, the Hilbert Driven Algorithm and the Gröbner walk algorithm. These algorithms are often employed to compute (difficult) lexicographic Gröbner bases from (easier) total degree Gröbner bases.

A Gröbner basis is termed reduced if the leading coefficient of each element of the basis is 1 and no monomial in any element of the basis is in the ideal generated by the leading terms of the other elements of the basis. In the worst case, computation of a Gröbner basis may require time that is exponential or even doubly exponential in the number of solutions of the polynomial system (for degree reverse lexicographic order and lexicographic order, respectively). Despite these complexity bounds, both standard and reduced Gröbner bases are often computable in practice, and most computer algebra systems contain routines to do so.

The concept and algorithms of Gröbner bases have been generalized to submodules of free modules over a polynomial ring. In fact, if L is a free module over a ring R, then one may consider R•"L as a ring by defining the product of two elements of L to be 0. This ring may be identified with R [e 1 , … , e l] / è' { e i e j | 1 d" i d" j d" l } é' , {\displaystyle R[e_{1},\ldots ,e_{l}]/\left\langle \{e_{i}e_{j} | 1\leq i\leq j\leq l\}\right\rangle ,} where e 1 , … , e l {\displaystyle e_{1},\ldots ,e_{l}} is a basis of L. This allows to identify a submodule of L generated by g 1 , … , g k {\displaystyle g_{1},\ldots ,g_{k}} with the ideal of R [e 1 , … , e l] {\displaystyle R[e_{1},\ldots ,e_{l}]} generated by g 1 , … , g k {\displaystyle g_{1},\ldots ,g_{k}} and the products e i e j {\displaystyle e_{i}e_{j}} , 1 d" i d" j d" l {\displaystyle 1\leq i\leq j\leq l} . If R is a polynomial ring, this reduces the theory and the algorithms of Gröbner bases of modules to the theory and the algorithms of Gröbner bases of ideals.

The concept and algorithms of Gröbner bases have also been generalized to ideals over various rings, commutative or not, like polynomial rings over a principal ideal ring or Weyl algebras.

Example and counterexample

The zeroes of f(x,y) form the red parabola; the zeroes of g(x,y) form the three blue vertical lines. Their intersection consists of three points. Let R = Q[x,y] be the ring of bivariate polynomials with rational coefficients and consider the ideal I = <f,g> generated by the polynomials

$$f(x,y) = x^2 - y$$
$$g(x,y) = x^3 - x$$

Two other elements of I are the polynomials

$$k(x,y) = -xf(x,y) + g(x,y) = xy - x$$
$$h(x,y) = xk(x,y)\text{''}\ (y - 1)f(x,y) = y^2 - y$$

Under lexicographic ordering with x > y we have

$$lt(f) = x^2$$
$$lt(g) = x^3$$
$$lt(h) = y^2$$

The ideal generated by {lt(f),lt(g)} only contains polynomials that are divisible by x^2 which excludes lt(h) = y^2; it follows that {f, g} is not a Gröbner basis for I. On the other hand, we can show that {f, k, h} is indeed a Gröbner basis for I. First note that f and g, and therefore also h, k and all the other polynomials in the ideal I have the following three zeroes in the (x,y) plane in common, as indicated in the figure: {(1,1),(-1,1),(0,0)}. Those three points are not collinear, so I does not contain any polynomial of the first degree. Neither can I contain any polynomials of the special form

$$m(x,y) = cx + p(y)$$

with c a nonzero rational number and p a polynomial in the variable y only; the reason being that such an m can never have two distinct zeroes with the same value for y (in this case, the points (1,1) and (-1,1)).

From the above it follows that I, apart from the zero polynomial, only contains polynomials whose leading term has degree greater than or equal to 2; therefore their leading terms are divisible by at least one of the three monomials

$$\{x^2, xy, y^2\} = \{lt(f), lt(k), lt(h)\}.$$

This means that {f, k, h} is a Gröbner basis for I with respect to lexicographic ordering with x > y.

Properties and applications of Gröbner bases

Unless explicitly stated, all the results that follow are true for any monomial ordering (see that article for the definitions of the different orders that are mentioned below). It is a common misconception that the lexicographical order is needed for some of these results. On the contrary, the lexicographical order is, almost always, the most difficult to compute, and using it makes impractical many computations that are relatively easy with graded reverse lexicographic order (grevlex), or, when elimination is needed, the elimination order (lexdeg) which restricts to grevlex on each block of variables.

Equality of ideals

Reduced Gröbner bases are unique for any given ideal and any monomial ordering. Thus two ideals are equal if and only if they have the same (reduced) Gröbner basis (usually a Gröbner basis software always produces reduced Gröbner bases).

Membership and inclusion of ideals

The reduction of a polynomial f by the Gröbner basis G of an ideal I yields 0 if and only if f is in I. This allows to test the membership of an element in an ideal. Another method consists in verifying that the Gröbner basis of G*"{f} is equal to G. To test if the ideal I generated by $f_1, ..., f_k$ is contained in the ideal J, it suffices to test that every f_i is in J. One may also test the equality of the reduced Gröbner bases of J and J*"$\{f_1, ..., f_k\}$.

Solutions of a system of algebraic equations

Any set of polynomials may be viewed as a system of polynomial equations by equating the polynomials to zero. The set of the solutions of such a system depends only on the generated ideal, and, therefore does not change when the given generating set is replaced by the Gröbner basis, for any ordering, of the generated ideal. Such a solution, with coordinates in an algebraically closed field containing the coefficients of the polynomials, is called a zero of the ideal. In the usual case of rational coefficients, this algebraically closed field is chosen as the complex field. An ideal does not have any zero (the system of equations is inconsistent) if and only if 1 belongs to the ideal (this is Hilbert's Nullstellensatz), or, equivalently, if its Gröbner basis (for any monomial ordering) contains 1, or, also, if the corresponding reduced Gröbner basis is .

Given the Gröbner basis G of an ideal I, it has only a finite number of zeros, if and only if, for each variable x, G contains a polynomial with a leading monomial that is a power of x (without any other variable appearing in the leading term). If it is the case the number of zeros, counted with multiplicity, is equal to the number of monomials that are not multiple of any leading monomial of G. This number is called the degree of the ideal. When the number of zeros is finite, the Gröbner basis for a lexicographical monomial ordering provides, theoretically, a solution: the first coordinates of a solution is a root of the greatest common divisor of polynomials of the basis that depends only of the first variable. After substituting this root in the basis, the second coordinates of this solution is a root of the greatest common divisor of the resulting polynomials that depends only on this second variable, and so on. This solving process is only theoretical, because it implies GCD computation and root-finding of polynomials with approximate coefficients, which are not practicable because of numeric instability. Therefore, other methods have been developed to solve polynomial systems through Gröbner bases (see System of polynomial equations for more details).

Dimension, degree and Hilbert series

The dimension of an ideal I in a polynomial ring R is the Krull dimension of the ring R/I and is equal to the dimension of the algebraic set of the zeros of I. It is also equal to number of hyperplanes in general position which are needed to have an intersection with the algebraic set, which is a finite number of points. The degree of the ideal and of its associated algebraic set is the number of points of this finite intersection, counted with multiplicity. In particular, the degree of a hypersurface is equal to the degree of its definition polynomial.

Both degree and dimension depends only on the set of the leading monomials of the Gröbner basis of the ideal for any monomial ordering. The dimension is the maximal size of a subset S of the variables such that there is no leading monomial depending only on the variables in S. Thus, if the ideal has dimension 0, then for each variable x there is a leading monomial in the Gröbner basis that is a power of x. Both dimension and degree may be deduced from the Hilbert series of the ideal, which is the series " i = 0 " d i t i {\displaystyle \sum _{i=0}^{\infty }d_{i}t^{i}} , where d i {\displaystyle d_{i}} is the number of monomials of degree i that are not multiple of any leading monomial in the Gröbner basis. The Hilbert series may be summed into a rational fraction

where d is the dimension of the ideal and P (t) {\displaystyle P(t)} is a polynomial such that P (1) {\displaystyle P(1)} is the degree of the ideal. Although the dimension and the degree do not depend on the choice of the monomial ordering, the Hilbert series and the polynomial P (t) {\displaystyle P(t)} change when one changes of monomial ordering. Most computer algebra systems that provide functions to compute Gröbner bases provide also functions for computing the Hilbert series, and thus also the dimension and the degree.

Elimination

The computation of Gröbner bases for an elimination monomial ordering allows computational elimination theory. This is based on the following theorem. The computation of Gröbner bases for an elimination monomial ordering allows computational elimination theory. This is based on the following theorem. Consider a polynomial ring K [x 1 , … , x n , y 1 , … , y m] = K [X , Y] , {\displaystyle K[x_{1},\ldots ,x_{n},y_{1},\ldots ,y_{m}]=K[X,Y],} in which the variables are split into two subsets X and Y. Let us also choose an elimination monomial ordering "eliminating" X, that is a monomial ordering for which two monomials are compared by comparing first the X-parts, and, in case of equality only, considering the Y-parts. This implies that a monomial containing an X-variable is greater than every monomial independent of X. If G is a Gröbner basis of an ideal I for this monomial ordering, then G)" K [Y] {\displaystyle G\cap K[Y]} is a Gröbner basis of I)" K [Y] {\displaystyle I\cap K[Y]} (this ideal is often called the elimination ideal). Moreover, G)" K [Y] {\displaystyle G\cap K[Y]} consists exactly of the polynomials of G whose leading terms belong to K[Y] (this makes very easy the computation of G)" K [Y] {\displaystyle G\cap K[Y]} , as only the leading monomials need to be checked).

This elimination property has many applications, some of them are reported in the next sections. Another application, in algebraic geometry, is that elimination realizes the geometric operation of projection of an affine algebraic set into a subspace of the ambient space: with above notation, the (Zariski closure of) the projection of the algebraic set defined by the ideal I into the Y-subspace is defined by the ideal I)" K [Y] . {\displaystyle I\cap K[Y].} The lexicographical ordering such that x 1 > ï" > x n {\displaystyle x_{1}>\cdots >x_{n}} is an elimination ordering for every partition { x 1 , … , x k } , { x k + 1 , … , x n } . {\displaystyle \{x_{1},\ldots ,x_{k}\},\{x_{k+1},\ldots ,x_{n}\}.} Thus a Gröbner basis for this ordering carries much more information than usually necessary. This may explain why Gröbner bases for the lexicographical ordering are usually the most difficult to compute.

Intersecting ideals

If I and J are two ideals generated respectively by $\{f_1, ..., f_m\}$ and $\{g_1, ..., g_k\}$, then a single Gröbner basis computation produces a Gröbner basis of their intersection I)" J. For this, one introduces a new indeterminate t, and one uses an elimination ordering such that the first block contains only t and the other block contains all the other variables (this means that a monomial containing t is greater than every monomial that do not contain t. With this monomial ordering, a Gröbner basis of I)" J consists in the polynomials that do not contain t, in the Gröbner basis of the ideal

In other words, I)" J is obtained by eliminating t in K. This may be proven by remarking that the ideal K consists in the polynomials (a " b) t + b {\displaystyle (a-b)t+b} such that a " I {\displaystyle a\in I} and b " J {\displaystyle b\in J} . Such a polynomial is independent of t if and only a=b, which means that b " I)" J . {\displaystyle b\in I\cap J.}

Implicitization of a rational curve

A rational curve is an algebraic curve that has a parametric equation of the form where f i (t) {\displaystyle f_{i}(t)} and g i (t) {\displaystyle g_{i}(t)} are univariate polynomials for 1 d" i d" n. One may (and will) suppose that f i (t) {\displaystyle f_{i}(t)} and g i (t) {\displaystyle g_{i}(t)} are coprime (they have no non-constant common factors). Implicitization consists in computing the implicit equations of such a curve. In case of n = 2, that is for plane curves, this may be computed with the resultant. The implicit equation is the following resultant: Elimination with Gröbner bases allow to implicitize for any value of n, simply by eliminating t in the ideal è' g 1 x 1 " f 1 , … , g n x 2 " f n é' . {\displaystyle \langle g_{1}x_{1}-f_{1},\ldots ,g_{n}x_{2}-f_{n}\rangle .} If n = 2, the result is the same as with the resultant, if the map t ¦! (x 1 , x 2) {\displaystyle t\mapsto (x_{1},x_{2})} is injective for almost every t. In the other case, the resultant is a power of the result of the elimination.

Saturation

When modeling a problem by polynomial equations, it is highly frequent that some quantities are supposed to be non-zero, because, if they are zero, the problem becomes very different. For example, when dealing with triangles, many properties become false if the triangle is degenerated, that is if the length of one side is equal to the sum of the lengths of the other sides. In such situations, there is no hope of deducing relevant information from the polynomial system if the degenerate solutions are not dropped out. More precisely, the system of equations defines an algebraic set which may have several irreducible components, and one has to remove the components on which the degeneracy conditions are everywhere zero. This is done by saturating the equations by the degeneracy conditions, which may be done by using the elimination property of Gröbner bases.

Definition of the saturation

The localization of a ring consists in adjoining to it the formal inverses of some elements. This section concerns only the case of a single element, or equivalently a finite number of elements (adjoining the inverses of several elements is equivalent to adjoin the inverse of their product. The localization of a ring R by an element f is the ring R f = R [t] / (1 " f t) , {\displaystyle R_{f}=R[t]/(1-ft),} where t is a new indeterminate representing the inverse of f. The localization of an ideal I of R is the ideal R f I {\displaystyle R_{f}I} of R f . {\displaystyle R_{f}.} When R is a polynomial ring, computing in R f {\displaystyle R_{f}} is not efficient, because of the need to manage the denominators. Therefore, the operation of localization is usually replaced by the operation of saturation.

The saturation with respect to f of an ideal I in R is the inverse image of R f I {\displaystyle R_{f}I} under the canonical map from R to R f . {\displaystyle R_{f}.} It is the ideal I : f " = { g " R | (" k " N) f k g " I } {\displaystyle I:f^{\infty }=\{g\in R|(\exists k\in \mathbb {N})f^{k}g\in I\}} consisting in all elements of R whose products by some power of f belongs to I. If J is the ideal generated by I and 1-ft in R[t], then I : f " = J)" R . {\displaystyle I:f^{\infty }=J\cap R.} It follows that, if R is a polynomial ring, a Gröbner basis computation eliminating t allows to compute a Gröbner basis of the saturation of an ideal by a polynomial. The important property of the saturation, which ensures that it removes from the algebraic set defined by the ideal I the irreducible components on which the polynomial f is zero is the following: The primary decomposition of I : f " {\displaystyle I:f^{\infty }} consists in the components of the primary decomposition of I that do not contain any power of f.

Computation of the saturation

A Gröbner basis of the saturation by f of a polynomial ideal generated by a finite set of polynomials F, may be obtained by eliminating t in F *" { 1 " t f } , {\displaystyle F\cup \{1-tf\},} that is by keeping the polynomials independent of t in the Gröbner basis of F *" { 1 " t f } {\displaystyle F\cup \{1-tf\}} for an elimination ordering eliminating t. Instead of using F, one may also start from a Gröbner basis of F. Which method is most efficient depends on the problem. However, if the saturation does not remove any component, that is if the ideal is equal to its saturated ideal, computing first the Gröbner basis of F is usually faster. On the other hand, if the saturation removes some components, the direct computation may be dramatically faster. If one wants to saturate with respect to several polynomials f 1 , … , f k {\displaystyle f_{1},\ldots ,f_{k}} or with respect to a single polynomial which is a product f = f 1 … f k , {\displaystyle f=f_{1}\ldots f_{k},} there are three ways to proceed which give the same result but may have very different computation times (it depends on the problem which is the most efficient).

- Saturating by f = f 1 … f k {\displaystyle f=f_{1}\ldots f_{k}} in a single Gröbner basis computation.
- Saturating by f 1 , {\displaystyle f_{1},} then saturating the result by f 2 , {\displaystyle f_{2},} and so on.
- Adding to F or to its Gröbner basis the polynomials 1 " t 1 f 1 , … , 1 " t k f k , {\displaystyle 1-t_{1}f_{1},\ldots ,1-t_{k}f_{k},} and eliminating the t i {\displaystyle t_{i}} in a single Gröbner basis computation.

Effective Nullstellensatz

Hilbert's Nullstellensatz has two versions. The first one asserts that a set of polynomials has an empty set of common zeros in an algebraic closure of the field of the coefficients if and only if 1 belongs to the generated ideal. This is easily tested with a Gröbner basis computation, because 1 belongs to an ideal if and only if 1 belongs to the Gröbner basis of the ideal, for any monomial ordering. The second version asserts that the set of common zeros (in an algebraic closure of the field of the coefficients) of an ideal is contained in the hypersurface of the zeros of a polynomial f, if and only if a power of f belongs to the ideal. This may be tested by a saturating the ideal by f; in fact, a power of f belongs to the ideal if and only if the saturation by f provides a Gröbner basis containing 1.

Implicitization in higher dimension

By definition, an affine rational variety of dimension k may be described by parametric equations of the form where p 0 , … , p n {\displaystyle p_{0},\ldots ,p_{n}} are n+1 polynomials in the k variables (parameters of the parameterization) t 1 , … , t k . {\displaystyle t_{1},\ldots ,t_{k}.} Thus the parameters t 1 , … , t k {\displaystyle t_{1},\ldots ,t_{k}} and the coordinates x 1 , … , x n {\displaystyle x_{1},\ldots ,x_{n}} of the points of the variety are zeros of the ideal

One could guess that it suffices to eliminate the parameters to obtain the implicit equations of the variety, as it has been done in the case of curves. Unfortunately this is not always the case. If the p i {\displaystyle p_{i}} have a common zero (sometimes called base point), every irreducible component of the non-empty algebraic set defined by the p i {\displaystyle p_{i}} is an irreducible component of the algebraic set defined by I. It follows that, in this case, the direct elimination of the t i {\displaystyle t_{i}} provides an empty set of polynomials. Therefore, if k>1, two Gröbner basis computations are needed to implicitize:

1. Saturate I {\displaystyle I} by p 0 {\displaystyle p_{0}} to get a Gröbner basis G {\displaystyle G}
2. Eliminate the t i {\displaystyle t_{i}} from G {\displaystyle G} to get a Gröbner basis of the ideal (of the implicit equations) of the variety.

Implementations

- CoCoA free computer algebra system for computing Gröbner bases.

- GAP free computer algebra system that can perform Gröbner bases calculations.
- FGb, Faugère's own implementation of his F4 algorithm, available as a Maple library. To the date, as of 2014, it is, with Magma, the fastest implementation for rational coefficients and coefficients in a finite field of prime order.
- Macaulay 2 free software for doing polynomial computations, particularly Gröbner bases calculations.
- Magma has a very fast implementation of the Faugère's F4 algorithm.
- Maple has implementations of the Buchberger and Faugère F4 algorithms, as well as Gröbner trace.
- Mathematica includes an implementation of the Buchberger algorithm, with performance-improving techniques such as the Gröbner walk, Gröbner trace, and an improvement for toric bases.
- SINGULAR free software for computing commutative and noncommutative Gröbner bases.
- SageMath provides a unified interface to several computer algebra systems (including SINGULAR and Macaulay), and includes a few Gröbner basis algorithms of its own.
- SymPy Python computer algebra system uses Gröbner bases to solve polynomial systems.

Cylindrical algebraic decomposition (CAD)

CAD is an algorithm which was introduced in 1973 by G. Collins to implement with an acceptable complexity the Tarski–Seidenberg theorem on quantifier elimination over the real numbers. This theorem concerns the formulas of the first-order logic whose atomic formulas are polynomial equalities or inequalities between polynomials with real coefficients. These formulas are thus the formulas which may be constructed from the atomic formulas by the logical operators and $(\wedge)$, or$(\vee)$, not$(\neg)$, for all$(\forall)$ andexists$(\exists)$. Tarski's theorem asserts that, from such a formula, one may compute an equivalent formula without quantifier $(\forall, \exists)$. The complexity of CAD is doubly exponential in the number of variables. This means that CAD allows, in theory, to solve every problem of real algebraic geometry which may be expressed by such a formula, that is almost every problem concerning explicitly given varieties and semi-algebraic sets.

While Gröbner basis computation has doubly exponential complexity only in rare cases, CAD has almost always this high complexity. This implies that, unless if most polynomials appearing in the input are linear, it may not solve problems with more than four variables. Since 1973, most of the research on this subject is devoted either to improve CAD or to find alternative algorithms in special cases of general interest. As an example of the state of art, there are efficient algorithms to find at least a point in every connected component of a semi-algebraic set, and thus to test if a semi-algebraic set is empty. On the other hand, CAD is yet, in practice, the best algorithm to count the number of connected components.

Asymptotic complexity vs. practical efficiency

The basic general algorithms of computational geometry have a double exponential worst case complexity. More precisely, if d is the maximal degree of the input polynomials and n the number of variables, their complexity is at most d 2 c n

{\displaystyle d^{2^{cn}}} $d^{2^{cn}}$ for some constant c, and, for some inputs, the complexity is at least d 2 c 2 n {\displaystyle d^{2^{c'n}}} $d^{2^{c'n}}$ for another constant c2 . During the last 20 years of 20th century, various algorithms have been introduced to solve specific subproblems with a better complexity. Most of these algorithms have a complexity d O (n 2) {\displaystyle d^{O(n^{2})}} $d^{O(n^{2})}$. Among these algorithms which solve a sub problem of the problems solved by Gröbner bases, one may cite testing if an affine variety is empty and solving nonhomogeneous polynomial systems which have a finite number of solutions. Such algorithms are rarely implemented because, on most entries Faugère's F4 and F5 algorithms have a better practical efficiency and probably a similar or better complexity (probably because the evaluation of the complexity of Gröbner basis algorithms on a particular class of entries is a difficult task which has been done only in a few special cases).

The main algorithms of real algebraic geometry which solve a problem solved by CAD are related to the topology of semi-algebraic sets. One may cite counting the number of connected components, testing if two points are in the same components or computing a Whitney stratification of a real algebraic set. They have a complexity of d O (n 2) {\displaystyle d^{O(n^{2})}} $d^{O(n^{2})}$, but the constant involved by O notation is so high that using them to solve any nontrivial problem effectively solved by CAD, is impossible even if one could use all the existing computing power in the world. Therefore, these algorithms have never been implemented and this is an active research area to search for algorithms with have together a good asymptotic complexity and a good practical efficiency.

Abstract modern viewpoint

The modern approaches to algebraic geometry redefine and effectively extend the range of basic objects in various levels of generality to schemes, formal schemes, ind-schemes, algebraic spaces, algebraic stacks and so on. The need for this arises already from the useful ideas within theory of varieties, e.g. the formal functions of Zariski can be accommodated by introducing nilpotent elements in structure rings; considering spaces of loops and arcs, constructing quotients by group actions and developing formal grounds for natural intersection theory and deformation theory lead to some of the further extensions. Most remarkably, in late 1950s, algebraic varieties were subsumed into Alexander Grothendieck's concept of a scheme. Their local objects are affine schemes or prime spectra which are locally ringed spaces which form a category which is antiequivalent to the category of commutative unital rings, extending the duality between the category of affine algebraic varieties over a field k, and the category of finitely generated reduced k-algebras. The gluing is along Zariski topology; one can glue within the category of locally ringed spaces, but also, using the Yoneda embedding, within the more abstract category of presheaves of sets over the category of affine schemes.

The Zariski topology in the set theoretic sense is then replaced by a Grothendieck topology. Grothendieck introduced Grothendieck topologies having in mind more exotic but geometrically finer and more sensitive examples than the crude Zariski topology, namely the étale topology, and the two flat Grothendieck topologies: fppf and fpqc; nowadays some other examples became prominent including Nisnevich topology. Sheaves can be furthermore generalized to stacks in the sense of Grothendieck, usually with some additional representability conditions leading to Artin stacks and, even finer, Deligne-Mumford stacks, both often called algebraic stacks. Sometimes other algebraic sites replace the category of affine schemes. For example, Nikolai Durov has introduced commutative algebraic monads as a generalization of local objects in a generalized algebraic geometry. Versions of a tropical geometry, of an absolute geometry over a field of one element and an algebraic analogue of Arakelov's geometry were realized in this setup.

Another formal generalization is possible to universal algebraic geometry in which every variety of algebras has its own algebraic geometry. The term variety of algebras should not be confused with algebraic variety. The language of schemes, stacks and generalizations has proved to be a valuable way of dealing with geometric concepts and became cornerstones of modern algebraic geometry. Algebraic stacks can be further generalized and for many practical questions like deformation theory and intersection theory, this is often the most natural approach. One can extend the Grothendieck site of affine schemes to a higher categorical site of derived affine schemes, by replacing the commutative rings with an infinity category of differential graded commutative algebras, or of simplicial commutative rings or a similar category with an appropriate variant of a Grothendieck topology. One can also replace presheaves of sets by presheaves of simplicial sets (or of infinity groupoids). Then, in presence of an appropriate homotopic machinery one can develop a notion of derived stack as such a presheaf on the infinity category of derived affine schemes, which is satisfying certain infinite categorical version of a sheaf axiom (and to be algebraic, inductively a sequence of representability conditions). Quillen model categories, Segal categories and quasicategories are some of the most often used tools to formalize this yielding the derived algebraic geometry, introduced by the school of Carlos Simpson, including Andre Hirschowitz, Bertrand Toën, Gabrielle Vezzosi, Michel Vaquié and others; and developed further by Jacob Lurie, Bertrand Toën, and Gabrielle Vezzosi. Another (noncommutative) version of derived algebraic geometry, using A-infinity categories has been developed from early 1990s by Maxim Kontsevich and followers.

History

Before the 16th century

Some of the roots of algebraic geometry date back to the work of the Hellenistic Greeks from the 5th century BC. The Delian problem, for instance, was to construct a length x so that the cube of side x contained the same volume as the rectangular box a^2b for given sides a and b. Menaechmus (circa 350 BC) considered the problem geometrically by intersecting the pair of plane conics $ay = x^2$ and $xy = ab$. The later work, in the 3rd century BC, of Archimedes and Apollonius studied more systematically problems on conic sections, and also involved the use of coordinates. The Arab mathematicians were able to solve by purely algebraic means certain cubic equations, and then to interpret the results geometrically. This was done, for instance, by Ibn al-Haytham in the 10th century AD. Subsequently, Persian mathematician Omar Khayyám (born 1048 A.D.) discovered a method for solving cubic equations by intersecting a parabola with a circle and seems to have been the first to conceive a general theory of cubic equations. A few years after Omar Khayyám, Sharaf al-Din al-Tusi's Treatise on equations has been described as "inaugurating the beginning of algebraic geometry".

Renaissance

Such techniques of applying geometrical constructions to algebraic problems were also adopted by a number of Renaissance mathematicians such as Gerolamo Cardano and Niccolò Fontana "Tartaglia" on their studies of the cubic equation. The geometrical approach to construction problems, rather than the algebraic one, was favored by most 16th and 17th century mathematicians, notably Blaise Pascal who argued against the use of algebraic and analytical methods in geometry. The French mathematicians Franciscus Vieta and later René Descartes and Pierre de Fermat revolutionized the conventional way of thinking about construction problems through the introduction of coordinate geometry. They were interested primarily in the properties of algebraic curves, such as those defined by Diophantine equations (in the case of Fermat), and the algebraic reformulation of the classical Greek works on conics and cubics (in the case of Descartes).

During the same period, Blaise Pascal and Gérard Desargues approached geometry from a different perspective, developing the synthetic notions of projective geometry. Pascal and Desargues also studied curves, but from the purely geometrical point of view: the analog of the Greek ruler and compass construction. Ultimately, the analytic geometry of Descartes and Fermat won out, for it supplied the 18th century mathematicians with concrete quantitative tools needed to study physical problems using the new calculus of Newton and Leibniz. However, by the end of the 18th century, most of the algebraic character of coordinate geometry was subsumed by the calculus of infinitesimals of Lagrange and Euler.

19th and early 20th century

It took the simultaneous 19th century developments of non-Euclidean geometry and Abelian integrals in order to bring the old algebraic ideas back into the geometrical fold. The first of these new developments was seized up by Edmond Laguerre and Arthur Cayley, who attempted to ascertain the generalized metric properties of projective space. Cayley introduced the idea of homogeneous polynomial forms, and more specifically quadratic forms, on projective space. Subsequently, Felix Klein studied projective geometry (along with other types of geometry) from the viewpoint that the geometry on a space is encoded in a certain class of transformations on the space. By the end of the 19th century, projective geometers were studying more general kinds of transformations on figures in projective space. Rather than the projective linear transformations which were normally regarded as giving the fundamental Kleinian geometry on projective space, they concerned themselves also with the higher degree birational transformations. This weaker notion of congruence would later lead members of the 20th century Italian school of algebraic geometry to classify algebraic surfaces up to birational isomorphism.

The second early 19th century development, that of Abelian integrals, would lead Bernhard Riemann to the development of Riemann surfaces. In the same period began the algebraization of the algebraic geometry through commutative algebra. The prominent results in this direction are Hilbert's basis theorem and Hilbert's Nullstellensatz, which are the basis of the connexion between algebraic geometry and commutative algebra, and Macaulay's multivariate resultant, which is the basis of elimination theory. Probably because of the size of the computation which is implied by multivariate resultants, elimination theory was forgotten during the middle of the 20th century until it was renewed by singularity theory and computational algebraic geometry.

20th century

B. L. van der Waerden, Oscar Zariski and André Weil developed a foundation for algebraic geometry based on contemporary commutative algebra, including valuation theory and the theory of ideals. One of the goals was to give a rigorous framework for proving the results of Italian school of algebraic geometry. In particular, this school used systematically the notion of generic point without any precise definition, which was first given by these authors during the 1930s. In the 1950s and 1960s, Jean-Pierre Serre and Alexander Grothendieck recast the foundations making use of sheaf theory. Later, from about 1960, and largely led by Grothendieck, the idea of schemes was worked out, in conjunction with a very refined apparatus of homological techniques. After a decade of rapid development the field stabilized in the 1970s, and new applications were made, both to number theory and to more classical geometric questions on algebraic varieties, singularities, moduli, and formal moduli. An important class of varieties, not easily understood directly from their defining equations, are the abelian varieties, which are the projective varieties whose points form an abelian group. The prototypical examples are the elliptic curves, which have a rich theory. They were instrumental in the proof of Fermat's last theorem and are also used in elliptic-curve cryptography. In parallel with the abstract trend of the algebraic geometry, which is concerned with general statements about varieties, methods for effective computation with concretely-given varieties have also been developed, which lead to the new area of computational algebraic geometry. One of the founding methods of this area is the theory of Gröbner bases, introduced by Bruno Buchberger in 1965. Another founding method, more specially devoted to real algebraic geometry, is the cylindrical algebraic decomposition, introduced by George E. Collins in 1973.

Analytic geometry

An analytic variety is defined locally as the set of common solutions of several equations involving analytic functions. It is analogous to the included concept of real or complex algebraic variety. Any complex manifold is an analytic variety. Since analytic varieties may have singular points, not all analytic varieties are manifolds. Modern analytic geometry is essentially equivalent to real and complex algebraic geometry, as has been shown by Jean-Pierre Serre in his paper GAGA, the name of which is French for Algebraic geometry and analytic geometry. Nevertheless, the two fields remain distinct, as the methods of proof are quite different and algebraic geometry includes also geometry in finite characteristic.

Applications

Algebraic geometry now finds applications in statistics, control theory, robotics, error-correcting codes, phylogenetics and geometric modelling. There are also connections to string theory, game theory, graph matchings, solitons and integer programming.

Noncommutative rings

Noncommutative rings resemble rings of matrices in many respects. Following the model of algebraic geometry, attempts have been made recently at defining noncommutative geometry based on noncommutative rings. Noncommutative rings and associative algebras (rings that are also vector spaces) are often studied via their categories of modules. A module over a ring is an abelian group that the ring acts on as a ring of endomorphisms, very much akin to the way fields (integral domains in which every non-zero element is invertible) act on vector spaces. Examples of noncommutative rings are given by rings of square matrices or more generally by rings of endomorphisms of abelian groups or modules, and by monoid rings.

Noncommutative ring

In mathematics, more specifically abstract algebra and ring theory, a noncommutative ring is a ring whose multiplication is not commutative; that is, there exists a and b in R with a·b ′″ b·a. Many authors use the term noncommutative ring to refer to rings which are not necessarily commutative, and hence include commutative rings in their definition. Noncommutative algebra is the study of results applying to rings that are not required to be commutative. Many important results in the field of noncommutative algebra area apply to commutative rings as special cases. Although some authors do not assume that rings have a multiplicative identity, in this article we make that assumption unless stated otherwise.

History

Beginning with division rings arising from geometry, the study of noncommutative rings has grown into a major area of modern algebra. The theory and exposition of noncommutative rings was expanded and refined in the 19th and 20th centuries by numerous authors. An incomplete list of such contributors includes E. Artin, Richard Brauer, P. M. Cohn, W. R. Hamilton, I. N. Herstein, N. Jacobson, K. Morita, E. Noether, Ø. Ore and others. Differences between commutative and noncommutative algebra Because noncommutative rings are a much larger class of rings than the commutative rings, their structure and behavior is less well understood. A great deal of work has been done successfully generalizing some results from commutative rings to noncommutative rings. A major difference between rings which are and are not commutative is the necessity to separately consider right ideals and left ideals. It is common for noncommutative ring theorists to enforce a condition on one of these types of ideals while not requiring it to hold for the opposite side. For commutative rings, the left–right distinction does not exist.

Important classes

Division rings

A division ring, also called a skew field, is a ring in which division is possible. Specifically, it is a nonzero ring in which every nonzero element a has a multiplicative inverse, i.e., an element x with a·x = x·a = 1. Stated differently, a ring is a division ring if and only if the group of units equals the set of all nonzero elements. Division rings differ from fields only in that their multiplication is not required to be commutative. However, by Wedderburn's little theorem all finite division rings are commutative and therefore finite fields. Historically, division rings were sometimes referred to as fields, while fields were called "commutative fields".

Semisimple rings

A module over a (not necessarily commutative) ring with unity is said to be semisimple (or completely reducible) if it is the direct sum of simple (irreducible) submodules. A ring is said to be (left)-semisimple if it is semisimple as a left module over itself. Surprisingly, a left-semisimple ring is also right-semisimple and vice versa. The left/right distinction is therefore unnecessary.

Semisimple module

In mathematics, especially in the area of abstract algebra known as module theory, a semisimple module or completely reducible module is a type of module that can be understood easily from its parts. A ring that is a semisimple module over itself is known as an Artinian semisimple ring. Some important rings, such as group rings of finite groups over fields of characteristic zero, are semisimple rings. An Artinian ring is initially understood via its largest semisimple quotient. The structure of Artinian semisimple rings is well understood by the Artin–Wedderburn theorem, which exhibits these rings as finite direct products of matrix rings.

Definition

A module over a (not necessarily commutative) ring with unity is said to be semisimple (or completely reducible) if it is the direct sum of simple (irreducible) submodules.

For a module M, the following are equivalent:

1. M is semisimple; i.e., a direct sum of irreducible modules.
2. M is the sum of its irreducible submodules.
3. Every submodule of M is a direct summand: for every submodule N of M, there is a complement P such that M = N •" P.

For 3 Ò! 2 {\displaystyle 3\Rightarrow 2} , the starting idea is to find an irreducible submodule by picking any nonzero x " M {\displaystyle x\in M} and letting P {\displaystyle P} be a maximal submodule such that x " P {\displaystyle x\notin P} . It can be shown that the complement of P {\displaystyle P} is irreducible. The most basic example of a semisimple module is a module over a field; i.e., a vector space. On the other hand, the ring Z of integers is not a semisimple module over itself (because, for example, it is not an artinian ring.)

Semisimple is stronger than completely decomposable, which is a direct sum of indecomposable submodules. Let A be an algebra over a field k. Then a left module M over A is said to be absolutely semisimple if, for any field extension F of k, F —" k M {\displaystyle F\otimes _{k}M} is a semisimple module over F —" k A {\displaystyle F\otimes _{k}A} .

Properties

- If M is semisimple and N is a submodule, then N and M/N are also semisimple.
- If each M i {\displaystyle M_{i}} is a semisimple module, then so is * i M i {\displaystyle \bigoplus _{i}M_{i}} .
- A module M is finitely generated and semisimple if and only if it is Artinian and its radical is zero.
  Endomorphism rings
- A semisimple module M over a ring R can also be thought of as a ring homomorphism from R into the ring of abelian group endomorphisms of M. The image of this homomorphism is a semiprimitive ring, and every semiprimitive ring is isomorphic to such an image.
- The endomorphism ring of a semisimple module is not only semiprimitive, but also von Neumann regular, (Lam 2001, p. 62).

Semisimple rings

A ring is said to be (left)-semisimple if it is semisimple as a left module over itself. Surprisingly, a left-semisimple ring is also right-semisimple and vice versa. The left/right distinction is therefore unnecessary, and one can speak of semisimple rings without ambiguity. A semisimple ring may be characterized in terms of homological algebra: namely, a ring R is semisimple if and only if any short exact sequence of left (or right) R-modules splits. In particular, any module over a semisimple ring is injective and projective. Since "projective" implies "flat", a semisimple ring is a von Neumann regular ring. Semisimple rings are of particular interest to algebraists. For example, if the base ring R is semisimple, then all R-modules would automatically be semisimple. Furthermore, every simple (left) R-module is isomorphic to a minimal left ideal of R, that is, R is a left Kasch ring. Semisimple rings are both Artinian and Noetherian. From the above properties, a ring is semisimple if and only if it is Artinian and its Jacobson radical is zero. If an Artinian semisimple ring contains a field as a central subring, it is called a semisimple algebra.

Examples

- semisimple if and only if it is artinian and reduced.
- If k is a field and G is a finite group of order n, then the group ring k [G] {\displaystyle k[G]} is semisimple if and only if the characteristic of k does not divide n. This is Maschke's theorem, an important result in group representation theory.
- By the Artin–Wedderburn theorem, a unital Artinian ring R is semisimple if and only if it is (isomorphic to) M n 1 (D 1) × M n 2 (D 2) × ï" × M n r (D r) {\displaystyle M_{n_{1}}(D_{1})\times M_{n_{2}}(D_{2})\times \dots \times M_{n_{r}}(D_{r})} , where each D i {\displaystyle D_{i}} is a division ring and each n i {\displaystyle n_{i}} is a positive integer, and M n (D) {\displaystyle M_{n}(D)} denotes the ring of n-by-n matrices with entries in D.
- An example of a semisimple non-unital ring is M " (K) {\displaystyle M_{\infty }(K)} , the row-finite, column-finite, infinite matrices over a field K.

Simple rings

One should beware that despite the terminology, not all simple rings are semisimple. The problem is that the ring may be "too big", that is, not (left/right) Artinian. In fact, if R is a simple ring with a minimal left/right ideal, then R is semisimple. Classic examples of simple, but not semisimple, rings are the Weyl algebras, such as the Q {\displaystyle \mathbb {Q} } -algebra which is a simple noncommutative domain. These and many other nice examples are discussed in more detail in several noncommutative ring theory texts, including chapter 3 of Lam's text, in which they are described as nonartinian simple rings. The module theory for the Weyl algebras is well studied and differs significantly from that of semisimple rings.

Jacobson semisimple

A ring is called Jacobson semisimple (or J-semisimple or semiprimitive) if the intersection of the maximal left ideals is zero, that is, if the Jacobson radical is zero. Every ring that is semisimple as a module over itself has zero Jacobson radical, but not every ring with zero Jacobson radical is semisimple as a module over itself. A J-semisimple ring is semisimple if and only if it is an artinian ring, so semisimple rings are often called artinian semisimple rings to avoid confusion. For example, the ring of integers, Z, is J-semisimple, but not artinian semisimple.

Semiprimitive rings

A semiprimitive ring or Jacobson semisimple ring or J-semisimple ring is a ring whose Jacobson radical is zero. This is a type of ring more general than a semisimple ring, but where simple modules still provide enough information about the ring. Rings such as the ring of integers are semiprimitive, and an artinian semiprimitive ring is just a semisimple ring. Semiprimitive rings can be understood as subdirect products of primitive rings, which are described by the Jacobson density theorem

Simple rings

A simple ring is a non-zero ring that has no two-sided ideal besides the zero ideal and itself. A simple ring can always be considered as a simple algebra. Rings which are simple as rings but not as modules do exist: the full matrix ring over a field does not have any nontrivial ideals (since any ideal of M(n,R) is of the form M(n,I) with I an ideal of R), but has nontrivial left ideals (namely, the sets of matrices which have some fixed zero columns). According to the Artin–Wedderburn theorem, every simple ring that is left or right Artinian is a matrix ring over a division ring. In particular, the only simple rings that are a finite-dimensional vector space over the real numbers are rings of matrices over either the real numbers, the complex numbers, or the quaternions. Any quotient of a ring by a maximal ideal is a simple ring. In particular, a field is a simple ring. A ring R is simple if and only its opposite ring R^{o} is simple. An example of a simple ring that is not a matrix ring over a division ring is the Weyl algebra.

Important theorems

Wedderburn's little theorem

Wedderburn's little theorem states that every finite domain is a field. In other words, for finite rings, there is no distinction between domains, skew-fields and fields. The Artin–Zorn theorem generalizes the theorem to alternative rings: every finite simple alternative ring is a field.

Artin–Wedderburn theorem

The Artin–Wedderburn theorem is a classification theorem for semisimple rings and semisimple algebras. The theorem states that an (Artinian) semisimple ring R is isomorphic to a product of finitely many n_i-by-n_i matrix rings over division rings D_i, for some integers n_i, both of which are uniquely determined up to permutation of the index i. In particular, any simple left or right Artinian ring is isomorphic to an n-by-n matrix ring over a division ring D, where both n and D are uniquely determined. As a direct corollary, the Artin–Wedderburn theorem implies that every simple ring that is finite-dimensional over a division ring (a simple algebra) is a matrix ring. This is Joseph Wedderburn's original result. Emil Artin later generalized it to the case of Artinian rings.

Jacobson density theorem

The Jacobson density theorem is a theorem concerning simple modules over a ring R. The theorem can be applied to show that any primitive ring can be viewed as a "dense" subring of the ring of linear transformations of a vector space. This theorem first appeared in the literature in 1945, in the famous paper "Structure Theory of Simple Rings Without Finiteness Assumptions" by Nathan Jacobson. This can be viewed as a kind of generalization of the Artin-Wedderburn theorem's conclusion about the structure of simple Artinian rings. More formally, the theorem can be stated as follows: The Jacobson Density Theorem. Let U be a simple right R-module, D = End(U_R), and X ," U a finite and D-linearly independent set. If A is a D-linear transformation on U then there exists r " R such that A(x) = x • r for all x in X.

Nakayama's lemma

Let J(R) be the Jacobson radical of R. If U is a right module over a ring, R, and I is a right ideal in R, then define U·I to be the set of all (finite) sums of elements of the form u·i, where · is simply the action of R on U. Necessarily, U·I is a submodule of U. If V is a maximal submodule of U, then U/V is simple. So U·J(R) is necessarily a subset of V, by the definition of J(R) and the fact that U/V is simple. Thus, if U contains at least one (proper) maximal submodule, U·J(R) is a proper submodule of U. However, this need not hold for arbitrary modules U over R, for U need not contain any maximal submodules. Naturally, if U is a Noetherian module, this holds. If R is Noetherian, and U is finitely generated, then U is a Noetherian module over R, and the conclusion is satisfied. Somewhat remarkable is that the weaker assumption, namely that U is finitely generated as an R-module (and no finiteness assumption on R), is sufficient to guarantee the conclusion. This is essentially the statement of Nakayama's lemma.

Precisely, one has the following.

Nakayama's lemma: Let U be a finitely generated right module over a ring R. If U is a non-zero module, then U·J(R) is a proper submodule of U. A version of the lemma holds for right modules over non-commutative unitary rings R. The resulting theorem is sometimes known as the Jacobson–Azumaya theorem.

Noncommutative localization

Localization is a systematic method of adding multiplicative inverses to a ring, and is usually applied to commutative rings. Given a ring R and a subset S, one wants to construct some ring R* and ring homomorphism from R to R*, such that the image of S consists of units (invertible elements) in R*. Further one wants R* to be the 'best possible' or 'most general' way to do this – in the usual fashion this should be expressed by a universal property. The localization of R by S is usually denoted by $S^{-1}R$; however other notations are used in some important special cases. If S is the set of the non zero elements of an integral domain, then the localization is the field of fractions and thus usually denoted Frac(R). Localizing non-commutative rings is more difficult; the localization does not exist for every set S of prospective units. One condition which ensures that the localization exists is the Ore condition. One case for non-commutative rings where localization has a clear interest is for rings of differential operators. It has the interpretation, for example, of adjoining a formal inverse D^{-1} for a differentiation operator D. This is done in many contexts in methods for differential equations. There is now a large mathematical theory about it, named microlocalization, connecting with numerous other branches. The micro- tag is to do with connections with Fourier theory, in particular.

Morita equivalence

Morita equivalence is a relationship defined between rings that preserves many ring-theoretic properties. It is named after Japanese mathematician Kiiti Morita who defined equivalence and a similar notion of duality in 1958. Two rings R and S (associative, with 1) are said to be (Morita) equivalent if there is an equivalence of the category of (left) modules over R, R-Mod, and the category of (left) modules over S, S-Mod. It can be shown that the left module categories R-Mod and S-Mod are equivalent if and only if the right module categories Mod-R and Mod-S are equivalent. Further it can be shown that any functor from R-Mod to S-Mod that yields an equivalence is automatically additive.

Brauer group

The Brauer group of a field K is an abelian group whose elements are Morita equivalence classes of central simple algebras of finite rank over K and addition is induced by the tensor product of algebras. It arose out of attempts to classify division algebras over a field and is named after the algebraist Richard Brauer. The group may also be defined in terms of Galois cohomology. More generally, the Brauer group of a scheme is defined in terms of Azumaya algebras.

Ore conditions

The Ore condition is a condition introduced by Øystein Ore, in connection with the question of extending beyond commutative rings the construction of a field of fractions, or more generally localization of a ring. The right Ore condition for a multiplicative subset S of a ring R is that for a " R and s " S, the intersection aS)" sR `" ". A domain that satisfies the right Ore condition is called a right Ore domain. The left case is defined similarly.

Goldie's theorem

In mathematics, Goldie's theorem is a basic structural result in ring theory, proved by Alfred Goldie during the 1950s. What is now termed a right Goldie ring is a ring R that has finite uniform dimension (also called "finite rank") as a right module over itself, and satisfies the ascending chain condition on right annihilators of subsets of R. Goldie's theorem states that the semiprime right Goldie rings are precisely those that have a semisimple Artinian right classical ring of quotients. The structure of this ring of quotients is then completely determined by the Artin–Wedderburn theorem. In particular, Goldie's theorem applies to semiprime right Noetherian rings, since by definition right Noetherian rings have the ascending chain condition on all right ideals. This is sufficient to guarantee that a right-Noetherian ring is right Goldie. The converse does not hold: every right Ore domain is a right Goldie domain, and hence so is every commutative integral domain. A consequence of Goldie's theorem, again due to Goldie, is that every semiprime principal right ideal ring is isomorphic to a finite direct sum of prime principal right ideal rings. Every prime principal right ideal ring is isomorphic to a matrix ring over a right Ore domain.

Representation theory

Representation theory is a branch of mathematics that draws heavily on non-commutative rings. It studies abstractalgebraic structures by representing their elements as linear transformations of vector spaces, and studies modules over these abstract algebraic structures. In essence, a representation makes an abstract algebraic object more concrete by describing its elements by matrices and the algebraic operations in terms of matrix addition and matrix multiplication, which is non-commutative. The algebraic objects amenable to such a description include groups, associative algebras and Lie algebras. The most prominent of these (and historically the first) is the representation theory of groups, in which elements of a group are represented by invertible matrices in such a way that the group operation is matrix multiplication.

Some relevant theorems

General

- Isomorphism theorems for rings
- Nakayama's lemma

Structure theorems

- The Artin–Wedderburn theorem determines the structure of semisimple rings
- The Jacobson density theorem determines the structure of primitive rings
- Goldie's theorem determines the structure of semiprimeGoldie rings
- The Zariski–Samuel theorem determines the structure of a commutative principal ideal ring
- The Hopkins–Levitzki theorem gives necessary and sufficient conditions for a Noetherian ring to be an Artinian ring
- Morita theory consists of theorems determining when two rings have "equivalent" module categories
- Cartan–Brauer–Hua theorem gives insight on the structure of division rings
- Wedderburn's little theorem states that finite domains are fields

Other

- The Skolem–Noether theorem characterizes the automorphisms of simple rings Structures and invariants of rings Dimension of a commutative ring

The Krull dimension of a commutative ring R is the supremum of the lengths n of all the increasing chains of prime ideals p 0 Š″ p 1 Š″ï″Š″ p n {\displaystyle {\mathfrak {p}}_{0}\subsetneq {\mathfrak {p}}_{1}\subsetneq \cdots \subsetneq {\mathfrak {p}}_{n}} $\mathfrak{p}_0 \subsetneq \mathfrak{p}_1 \subsetneq \cdots \subsetneq \mathfrak{p}_n$. For example, the polynomial ring k [t 1 , ï″ , t n] {\displaystyle k[t_{1},\cdots ,t_{n}]} $k[t_1,\cdots,t_n]$ over a field k has dimension n. The fundamental theorem in the dimension theory states the following numbers coincide for a noetherian local ring (R , m) {\displaystyle (R,{\mathfrak {m}})} $(R,\mathfrak{m})$:

- The Krull dimension of R.
- The minimum number of the generators of the m {\displaystyle {\mathfrak {m}}} m-primary ideals.
- The dimension of the graded ring gr m a (R) = •″ k e″ 0 m k / m k + 1 {\displaystyle \operatorname {gr} _{\mathfrak {m}}(R)=\oplus _{k\geq 0}{\mathfrak {m}}^{k}/{{\mathfrak {m}}^{k+1}}} $gr_{\mathfrak{m}}(R) = \oplus_{k\geq 0} \mathfrak{m}^k / \mathfrak{m}^{k+1}$ (equivalently, one plus the degree of its Hilbert polynomial).

A commutative ring R is said to be catenary if any pair of prime ideals p ,″ p 2 {\displaystyle {\mathfrak {p}}\subset {\mathfrak {p}}'} $\mathfrak{p} = \mathfrak{p}_0 \subsetneq \cdots \subsetneq \mathfrak{p}_n = \mathfrak{p}'$ can be extended to a chain of prime ideals p = p 0 Š″ï″Š″ p n = p 2 {\displaystyle {\mathfrak {p}}={\mathfrak {p}}_{0}\subsetneq \cdots \subsetneq {\mathfrak {p}}_{n}={\mathfrak {p}}'} $\mathfrak{p} \subset \mathfrak{p}'$ of same finite length such that there is no prime ideal that is strictly contained in two consecutive terms. Practically all noetherian rings that appear in application are catenary. If (R , m) {\displaystyle (R,{\mathfrak {m}})} $(R,\mathfrak{m})$ is a catenary local integral domain, then, by definition,

$$\dim R = ht\mathfrak{p} + \dim R / \mathfrak{p}$$

whereht a p = dim a R p {\displaystyle \operatorname {ht} {\mathfrak {p}}=\operatorname {dim} R_{\mathfrak {p}}} ht a p = dim a R p {\displaystyle \operatorname {ht} {\mathfrak {p}}=\operatorname {dim} R_{\mathfrak {p}}} $ht\mathfrak{p} = \dim R_{\mathfrak{p}}$ is the height of p {\displaystyle {\mathfrak {p}}} p. It is a deep theorem of Ratliff that the converse is also true. If R is an integral domain that is a finitely generated k-algebra, then its dimension is the transcendence degree of its field of fractions over k. If S is an integral extension of a commutative ring R, then S and R have the same dimension. Closely related concepts are those of depth and global dimension. In general, if R is a noetherian local ring, then the depth of R is less than or equal to the dimension of R. When the equality holds, R is called a Cohen–Macaulay ring. A regular local ring is an example of a Cohen–Macaulay ring. It is a theorem of Serre that R is a regular local ring if and only if it has finite global dimension and in that case the global dimension is the Krull dimension of R. The significance of this is that a global dimension is a homological notion.

Morita equivalence

Two rings R, S are said to be Morita equivalent if the category of left modules over R is equivalent to the category of left modules over S. In fact, two commutative rings which are Morita equivalent must be isomorphic, so the notion does not add anything new to the category of commutative rings. However, commutative rings can be Morita equivalent to noncommutative rings, so Morita equivalence is coarser than isomorphism. Morita equivalence is especially important in algebraic topology and functional analysis.

Morita equivalence

In abstract algebra, Morita equivalence is a relationship defined between rings that preserves many ring-theoretic properties. It is named after Japanese mathematician Kiiti Morita who defined equivalence and a similar notion of duality in 1958.

Motivation

Rings are commonly studied in terms of their modules, as modules can be viewed as representations of rings. Every ring R has a natural R-module structure on itself where the module action is defined as the multiplication in the ring, so the approach via modules is more general and gives useful information. Because of this, one often studies a ring by studying the category of modules over that ring. Morita equivalence takes this viewpoint to a natural conclusion by defining rings to be Morita equivalent if their module categories are equivalent. This notion is of interest only when dealing with noncommutative rings, since it can be shown that two commutative rings are Morita equivalent if and only if they are isomorphic.

Definition

Two rings R and S (associative, with 1) are said to be (Morita) equivalent if there is an equivalence of the category of (left) modules over R, R-Mod, and the category of (left) modules over S, S-Mod. It can be shown that the left module categories R-Mod and S-Mod are equivalent if and only if the right module categories Mod-R and Mod-S are equivalent. Further it can be shown that any functor from R-Mod to S-Mod that yields an equivalence is automatically additive.

Examples

The ring of n-by-n matrices with elements in R, denoted $M_n(R)$, is Morita-equivalent to R for any $n > 0$. Notice that this generalizes the classification of simple artinian rings given by Artin–Wedderburn theory. To see the equivalence, notice that if X is a left R-module then X^n is an $M_n(R)$-module where the module structure is given by matrix multiplication on the left of column vectors from X. This allows the definition of a functor from the category of left R-modules to the category of left $M_n(R)$-modules. The inverse functor is defined by realizing that for any $M_n(R)$-module there is a left R-module X such that the $M_n(R)$-module is obtained from X as described above.

Criteria for equivalence

Equivalences can be characterized as follows: if F:R-Mod '! {\displaystyle \to } →S-Mod and G:S-Mod '! {\displaystyle \to } →R-Mod are additive (covariant) functors, then F and G are an equivalence if and only if there is a balanced (S,R)-bimodule P such that $_SP$ and P_R are finitely generated projective generators and there are natural isomorphisms of the functors F a (") E" P —" R " {\displaystyle \operatorname {F} (-)\cong P\otimes _{R}-} $F(-) \cong P \otimes_R -$, and of the functors G a (") E" Hom a (S P , ") . {\displaystyle \operatorname {G} (-)\cong \operatorname {Hom} (_{S}P,-).} $G(-) \cong \mathrm{Hom}(_S P, -)$. Finitely generated projective generators are also sometimes called progenerators for their module category. For every right-exact functor F from the category of left-R modules to the category of left-S modules that commutes with direct sums, a theorem of homological algebra shows that there is a (S,R)-bimodule E such that the functor F a (") {\displaystyle \operatorname {F} (-)} F(–) is naturally isomorphic to the functor E —" R " {\displaystyle E\otimes _{R}-} $E \otimes_R -$. Since equivalences are by necessity exact and commute with direct sums, this implies that R and S are Morita equivalent if and only if there are bimodules $_RM_S$ and $_SN_R$ such that M —" S N E" R {\displaystyle M\otimes _{S}N\cong R} $M \otimes_S N \cong R$ as (R,R) bimodules and N —" R M E" S {\displaystyle N\otimes _{R}M\cong S} $N \otimes_R M \cong S$ as (S,S) bimodules. Moreover, N and M are related via an (S,R) bimodule isomorphism: N E" Hom a (M S , S S) {\displaystyle N\cong \operatorname {Hom} (M_{S},S_{S})} $N \cong \mathrm{Hom}(M_S, S_S)$.

More concretely, two rings R and S are Morita equivalent if and only if S E" End a (P R) {\displaystyle S\cong \operatorname {End} (P_{R})} $S \cong \mathrm{End}(P_R)$ for a progenerator module P_R, which is the case if and only if

$$S \cong e\mathbb{M}_n(R)e$$

(isomorphism of rings) for some positive integer n and full idempotent e in the matrix ring $M_n(R)$. It is known that if R is Morita equivalent to S, then the ring C(R) is isomorphic to the ring C(S), where C(-) denotes the center of the ring, and furthermore R/J(R) is Morita equivalent to S/J(S), where J(-) denotes the Jacobson radical. While isomorphic rings are Morita equivalent, Morita equivalent rings can be nonisomorphic. An easy example is that a division ring D is Morita equivalent to all of its matrix rings $M_n(D)$, but cannot be isomorphic when n > 1. In the special case of commutative rings, Morita equivalent rings are actually isomorphic. This follows immediately from the comment above, for if R is Morita equivalent to S, R = C a (R) E" C a (S) = S {\displaystyle R=\operatorname {C} (R)\cong \operatorname {C} (S)=S} $R = \mathrm{C}(R) \cong \mathrm{C}(S) = S$. In fact, if R and S are isomorphic commutative rings, every equivalence between R-Mod and S-Mod arises up to natural isomorphism from an isomorphism between R and S.

Properties preserved by equivalence

Many properties are preserved by the equivalence functor for the objects in the module category. Generally speaking, any property of modules defined purely in terms of modules and their homomorphisms (and not to their underlying elements or ring) is a categorical property which will be preserved by the equivalence functor. For example, if F(-) is the equivalence functor from R-Mod to S-Mod, then the R module M has any of the following properties if and only if the S module F(M) does: injective, projective, flat, faithful, simple, semisimple, finitely generated, finitely presented, Artinian, and Noetherian. Examples of properties not necessarily preserved include being free, and being cyclic. Many ring theoretic properties are stated in terms of their modules, and so these properties are preserved between Morita equivalent rings. Properties shared between equivalent rings are called Morita invariant properties. For example, a ring R is semisimple if and only if all of its modules are semisimple, and since semisimple modules are preserved under Morita equivalence, an equivalent ring S must also have all of its modules semisimple, and therefore be a semisimple ring itself. Sometimes it is not immediately obvious why a property should be preserved. For example, using one standard definition of von Neumann regular ring (for all a in R, there exists x in R such that a = axa) it is not clear that an equivalent ring should also be von Neumann regular. However another formulation is: a ring is von Neumann regular if and only if all of its modules are flat. Since flatness is preserved across Morita equivalence, it is now clear that von Neumann regularity is Morita invariant.

The following properties are Morita invariant:

- simple, semisimple
- von Neumann regular
- right (or left) Noetherian, right (or left) Artinian
- right (or left) self-injective
- quasi-Frobenius
- prime, right (or left) primitive, semiprime, semiprimitive
- right (or left) (semi-)hereditary
- right (or left) nonsingular
- right (or left) coherent
- semiprimary, right (or left) perfect, semiperfect
- semilocal

Examples of properties which are not Morita invariant include commutative, local, reduced, domain, right (or left) Goldie, Frobenius, invariant basis number, and Dedekind finite. There are at least two other tests for determining whether or not a ring property P {\displaystyle {\mathcal {P}}} $\mathcal{P}$ is Morita invariant. An element e in a ring R is a full idempotent when $e^2 = e$ and ReR = R.

- P {\displaystyle {\mathcal {P}}} $\mathcal{P}$ is Morita invariant if and only if whenever a ring R satisfies P {\displaystyle {\mathcal {P}}} $\mathcal{P}$, then so does eRe for every full idempotent e and so does every matrix ring $M_n(R)$ for every positive integer n;

  or
- $\mathcal{P}$ is Morita invariant if and only if: for any ring R and full idempotent e in R, R satisfies P {\displaystyle {\mathcal {P}}} $\mathcal{P}$ if and only if the ring eRe satisfies P {\displaystyle {\mathcal {P}}} $\mathcal{P}$.

Further directions

Dual to the theory of equivalences is the theory of dualities between the module categories, where the functors used are contravariant rather than covariant. This theory, though similar in form, has significant differences because there is no duality between the categories of modules for any rings, although dualities may exist for subcategories. In other words, because infinite dimensional modules[clarification needed] are not generally reflexive, the theory of dualities applies more easily to finitely generated algebras over noetherian rings. Perhaps not surprisingly, the criterion above has an analogue for dualities, where the natural isomorphism is given in terms of the hom functor rather than the tensor functor. Morita equivalence can also be defined in more structured situations, such as for symplectic groupoids and C*-algebras. In the case of C*-algebras, a stronger type equivalence, called strong Morita equivalence, is needed to obtain results useful in applications, because of the additional structure of C*-algebras (coming from the involutive *-operation) and also because C*-algebras do not necessarily have an identity element.

Significance in K-theory

If two rings are Morita equivalent, there is an induced equivalence of the respective categories of projective modules since the Morita equivalences will preserve exact sequences (and hence projective modules). Since the algebraic K-theory of a ring is defined (in Quillen's approach) in terms of the homotopy groups of (roughly) the classifying space of the nerve of the (small) category of finitely generated projective modules over the ring, Morita equivalent rings must have isomorphic K-groups. Finitely generated projective module over a ring and Picard group Let R be a commutative ring and P (R) {\displaystyle \mathbf {P} (R)} $\mathbf{P}(R)$ the set of isomorphism classes of finitely generated projective modules over R; let also P n (R) {\displaystyle \mathbf {P} _{n}(R)} $\mathbf{P}_n(R)$ subsets consisting of those with constant rank n. (The rank of a module M is the continuous function Spec a R '! Z , p ¦! dim a M—" R k (p) {\displaystyle \operatorname {Spec} R\to \mathbb {Z} ,\,{\mathfrak {p}}\mapsto \dim M\otimes _{R}k({\mathfrak {p}})} $SpecR \to \mathbb{Z}, \mathfrak{p} \mapsto \dim M \otimes_R k(\mathfrak{p})$.) P 1 (R) {\displaystyle \mathbf {P} _{1}(R)}

$\mathbf{P}_1(R)$ is usually denoted by Pic(R). It is an abelian group called the Picard group of R. If R is an integral domain with the field of fractions F of R, then there is an exact sequence of groups:

$$1 \to R^* \to F^* \xrightarrow{f \mapsto fR} Cart(R) \to Pic(R) \to 1$$

whereCart a (R) {\displaystyle \operatorname {Cart} (R)} $Cart(R)$ is the set of fractional ideals of R. If R is a regular domain (i.e., regular at any prime ideal), then Pic(R) is precisely the divisor class group of R. For example, if R is a principal ideal domain, then Pic(R) vanishes. In algebraic number theory, R will be taken to be the ring of integers, which is Dedekind and thus regular. It follows that Pic(R) is a finite group (finiteness of class number) that measures the deviation of the ring of integers from being a PID. One can also consider the group completionof P (R) {\displaystyle

\mathbf {P} (R)} $\mathbf{P}(R)$; this results in a commutative ring $K_0(R)$. Note that $K_0(R) = K_0(S)$ if two commutative rings R, S are Morita equivalent.

Structure of noncommutative rings

The structure of a noncommutative ring is more complicated than that of a commutative ring. For example, there exist simple rings, containing no non-trivial proper (two-sided) ideals, which contain non-trivial proper left or right ideals. Various invariants exist for commutative rings, whereas invariants of noncommutative rings are difficult to find. As an example, the nilradical of a ring, the set of all nilpotent elements, need not be an ideal unless the ring is commutative. Specifically, the set of all nilpotent elements in the ring of all n x n matrices over a division ring never forms an ideal, irrespective of the division ring chosen. There are, however, analogues of the nilradical defined for noncommutative rings, that coincide with the nilradical when commutativity is assumed.

The concept of the Jacobson radical of a ring; that is, the intersection of all right/left annihilators of simple right/left modules over a ring, is one example. The fact that the Jacobson radical can be viewed as the intersection of all maximal right/left ideals in the ring, shows how the internal structure of the ring is reflected by its modules. It is also a fact that the intersection of all maximal right ideals in a ring is the same as the intersection of all maximal left ideals in the ring, in the context of all rings; whether commutative or noncommutative. Noncommutative rings serve as an active area of research due to their ubiquity in mathematics. For instance, the ring of n-by-nmatrices over a field is noncommutative despite its natural occurrence in geometry, physics and many parts of mathematics. More generally, endomorphism rings of abelian groups are rarely commutative, the simplest example being the endomorphism ring of the Klein four-group. One of the best known noncommutative rings is the division ring of quaternions.

The ring of integers of a number field

The coordinate ring of an algebraic variety

If X is an affine algebraic variety, then the set of all regular functions on X forms a ring called the coordinate ring of X. For a projective variety, there is an analogous ring called the homogeneous coordinate ring. Those rings are essentially the same things as varieties: they correspond in essentially a unique way. This may be seen via either Hilbert's Nullstellensatz or scheme-theoretic constructions (i.e., Spec and Proj).

Ring of invariants

A basic (and perhaps the most fundamental) question in the classical invariant theory is to find and study polynomials in the polynomial ring k [V] {\displaystyle k[V]} $k[V]$ that are invariant under the action of a finite group (or more generally reductive) G on V. The main example is the ring of symmetric polynomials: symmetric polynomials are polynomials that are invariant under permutation of variable. The fundamental theorem of symmetric polynomials states that this ring is R [ó 1 , … , ó n] $R[\sigma_1,\ldots,\sigma_n]$ where ó i σ_i are elementary symmetric polynomials.

History

Commutative ring theory originated in algebraic number theory, algebraic geometry, and invariant theory. Central to the development of these subjects were the rings of integers in algebraic number fields and algebraic function fields, and the rings of polynomials in two or more variables. Noncommutative ring theory began with attempts to extend the complex numbers to various hypercomplex number systems. The genesis of the theories of commutative and noncommutative rings dates back to the early 19th century, while their maturity was achieved only in the third decade of the 20th century. More precisely, William Rowan Hamilton put forth the quaternions and biquaternions; James Cockle presented tessarines and coquaternions; and William Kingdon Clifford was an enthusiast of split-biquaternions, which he called algebraic motors. These noncommutative algebras, and the non-associative Lie algebras, were studied within universal algebra before the subject was divided into particular mathematical structure types. One sign of re-organization was the use of direct sums to describe algebraic structure.

The various hypercomplex numbers were identified with matrix rings by Joseph Wedderburn (1908) and Emil Artin (1928). Wedderburn's structure theorems were formulated for finite-dimensional algebras over a field while Artin generalized them to Artinian rings. In 1920, Emmy Noether, in collaboration with W. Schmeidler, published a paper about the theory of ideals in which they defined left and right ideals in a ring. The following year she published a landmark paper called Idealtheorie in Ringbereichen, analyzing ascending chain conditions with regard to (mathematical) ideals. Noted algebraist Irving Kaplansky called this work "revolutionary"; the publication gave rise to the term "Noetherian ring", and several other mathematical objects being called Noetherian.

4

Galois Theory

In mathematics, Galois theory provides a connection between field theory and group theory. Using Galois theory, certain problems in field theory can be reduced to group theory, which is in some sense simpler and better understood. The subject is named after Évariste Galois, who introduced it for studying the roots of a polynomial and characterizing the polynomial equations that are solvable by radicals in terms of properties of the permutation group of their roots—an equation is solvable by radicals if its roots may be expressed by a formula involving only integers, nth roots and the four basic arithmetic operations. The theory has been popularized (among mathematicians) and developed by Richard Dedekind, Leopold Kronecker and Emil Artin, and others, who, in particular, interpreted the permutation group of the roots as the automorphism group of a field extension. Galois theory has been generalized to Galois connections and Grothendieck's Galois theory.

APPLICATION TO CLASSICAL PROBLEMS

The birth and development of Galois theory was caused by the following question, whose answer is known as the Abel–Ruffini theorem: Why is there no formula for the roots of a fifth (or higher) degree polynomial equation in terms of the coefficients of the polynomial, using only the usual algebraic operations (addition, subtraction, multiplication, division) and application of radicals (square roots, cube roots, etc)? Galois' theory not only provides a beautiful answer to this question, but also explains in detail why it is possible to solve equations of degree four or lower in the above manner, and why their solutions take the form that they do. Further, it gives a conceptually clear, and often practical, means of telling when some particular equation of higher degree can be solved in that manner. Galois' theory also gives a clear insight into questions concerning problems in compass and straightedge construction. It gives an elegant characterization of the ratios of lengths that can be constructed with this method. Using this, it becomes relatively easy to answer such classical problems of geometry as

1. Which regular polygons are constructible polygons?
2. Why is it not possible to trisect every angle using a compass and straightedge?

History of Galois theory

Pre-history

Galois' theory originated in the study of symmetric functions – the coefficients of a monic polynomial are (up to sign) the elementary symmetric polynomials in the roots. For instance, $(x - a)(x - b) = x^2 - (a + b)x + ab$, where 1, a + b and ab are the elementary polynomials of degree 0, 1 and 2 in two variables. This was first formalized by the 16th-century French mathematician François Viète, in Viète's formulas, for the case of positive real roots. In the opinion of the 18th-century British mathematician Charles Hutton, the expression of coefficients of a polynomial in terms of the roots (not only for positive roots) was first understood by the 17th-century French mathematician Albert Girard; Hutton writes: ...[Girard was] the first person who understood the general doctrine of the formation of the coefficients of the powers from the sum of the roots and their products. He was the first who discovered the rules for summing the powers of the roots of any equation. In this vein, the discriminant is a symmetric function in the roots that reflects properties of the roots – it is zero if and only if the polynomial has a multiple root, and for quadratic and cubic polynomials it is positive if and only if all roots are real and distinct, and negative if and only if there is a pair of distinct complex conjugate roots. See Discriminant:Nature of the roots for details. The cubic was first partly solved by the 15–16th-century Italian mathematician Scipione del Ferro, who did not however publish his results; this method, though, only solved one type of cubic equation. This solution was then rediscovered independently in 1535 by Niccolò Fontana Tartaglia, who shared it with Gerolamo Cardano, asking him to not publish it. Cardano then extended this to numerous other cases, using similar arguments; see more details at Cardano's method. After the discovery of del Ferro's work, he felt that Tartaglia's method was no longer secret, and thus he published his solution in his 1545 Ars Magna. His student Lodovico Ferrari solved the quartic polynomial; his solution was also included in Ars Magna. In this book, however, Cardano did not provide a "general formula" for the solution of a cubic equation, as he had neither complex numbers at his disposal, nor the algebraic notation to be able to describe a general cubic equation. With the benefit of modern notation and complex numbers, the formulae in this book do work in the general case, but Cardano did not know this. It was Rafael Bombelli who managed to understand how to work with complex numbers in order to solve all forms of cubic equation.

A further step was the 1770 paper Réflexions sur la résolution algébrique des équations by the French-Italian mathematician Joseph Louis Lagrange, in his method of Lagrange resolvents, where he analyzed Cardano's and Ferrari's solution of cubics and quartics by considering them in terms of permutations of the roots, which yielded an auxiliary polynomial of lower degree, providing a unified understanding of the solutions and laying the groundwork for group theory and Galois' theory. Crucially, however, he did not consider composition of permutations. Lagrange's method did not extend to quintic equations or higher, because the resolvent had higher degree.

The quintic was almost proven to have no general solutions by radicals by Paolo Ruffini in 1799, whose key insight was to use permutation groups, not just a single permutation. His solution contained a gap, which Cauchy considered minor, though this was not patched until the work of the Norwegian mathematician Niels Henrik Abel, who published a proof in 1824, thus establishing the Abel–Ruffini theorem. While Ruffini and Abel established that the general quintic could not be solved, some particular quintics can be solved, such as $(x " 1)^5 = 0$, and the precise criterion by which a given quintic or higher polynomial could be determined to be solvable or not was given by Évariste Galois, who showed that whether a polynomial was solvable or not was equivalent to whether or not the permutation group of its roots – in modern terms, its Galois group – had a certain structure – in modern terms, whether or not it was a solvable group. This group was always solvable for polynomials of degree four or less, but not always so for polynomials of degree five and greater, which explains why there is no general solution in higher degree.

Galois' writings

In 1830 Galois (at the age of 18) submitted to the Paris Academy of Sciences a memoir on his theory of solvability by radicals; Galois' paper was ultimately rejected in 1831 as being too sketchy and for giving a condition in terms of the roots of the equation instead of its coefficients. Galois then died in a duel in 1832, and his paper, "Mémoire sur les conditions de résolubilité des équations par radicaux", remained unpublished until 1846 when it was published by Joseph Liouville accompanied by some of his own explanations. Prior to this publication, Liouville announced Galois' result to the Academy in a speech he gave on 4 July 1843. According to Allan Clark, Galois's characterization "dramatically supersedes the work of Abel and Ruffini."

Aftermath

Galois' theory was notoriously difficult for his contemporaries to understand, especially to the level where they could expand on it. For example, in his 1846 commentary, Liouville completely missed the group-theoretic core of Galois' method.Joseph Alfred Serret who attended some of Liouville's talks, included Galois' theory in his 1866 (third edition) of his textbook Cours d'algèbre supérieure. Serret's pupil, Camille Jordan, had an even better understanding reflected in his 1870 book Traité des substitutions et des équations algébriques. Outside France, Galois' theory remained more obscure for a longer period. In Britain, Cayley failed to grasp its depth and popular British algebra textbooks did not even mention Galois' theory until well after the turn of the century. In Germany, Kronecker's writings focused more on Abel's result. Dedekind wrote little about Galois' theory, but lectured on it at Göttingen in 1858, showing a very good understanding.Eugen Netto's books of the 1880s, based on Jordan's Traité, made Galois theory accessible to a wider German and American audience as did Heinrich Martin Weber's 1895 algebra textbook.

Permutation group approach to Galois theory

Given a polynomial, it may be that some of the roots are connected by various algebraic equations. For example, it may be that for two of the roots, say A and B, that $A^2 + 5B^3 = 7$. The central idea of Galois' theory is to consider permutations (or rearrangements) of the roots such that any algebraic equation satisfied by the roots is still satisfied after the roots have been permuted. Originally, the theory has been developed for algebraic equations whose coefficients are rational numbers. It extends naturally to equations with coefficients in any field, but this will not be considered in the simple examples below. These permutations together form a permutation group, also called the Galois group of the polynomial, which is explicitly described in the following examples.

First example: a quadratic equation

$$x^2 - 4x + 1 = 0.$$

By using the quadratic formula, we find that the two roots are

$$A = 2 + \sqrt{3},$$
$$B = 2 - \sqrt{3}.$$

Examples of algebraic equations satisfied by A and B include

$$A + B = 4,$$

And

$$AB = 1.$$

Obviously, in either of these equations, if we exchange A andB, we obtain another true statement. For example, the equation A + B = 4 becomes simply B + A = 4. Furthermore, it is true, but less obvious, that this holds for every possible algebraic relation between A and B such that all coefficients are rational (in any such relation, swapping A and B yields another true relation). This results from the theory of symmetric polynomials, which, in this simple case, may be replaced by formula manipulations involving binomial theorem. (One might object that A and B are related by the algebraic equation $A - B - 2\sqrt{3} = 0$, which does not remain true when A andB are exchanged. However, this relation is not considered here, because it has the coefficient $-2\sqrt{3}$ which is not rational.)

We conclude that the Galois group of the polynomial x^2 " 4x + 1 consists of two permutations: the identity permutation which leaves A and B untouched, and the transposition permutation which exchanges A and B. It is a cyclic group of order two, and therefore isomorphic to Z/2Z. A similar discussion applies to any quadratic polynomial $ax^2 + bx + c$, where a, b and c are rational numbers.

- If the polynomial has rational roots, for examplex $x^2 - 4x + 4 = (x-2)^2$, or $x^2 - 3x + 2 = (x-2)(x-1)$, then the Galois group is trivial; that is, it contains only the identity permutation.
- If it has two irrational roots, for example $x^2 - 2$, then the Galois group contains two permutations, just as in the above example.

Second example

Consider the polynomial

$$x^4 - 10x^2 + 1,$$

which can also be written as

$$(x^2 - 5)^2 - 24.$$

We wish to describe the Galois group of this polynomial, again over the field of rational numbers. The polynomial has four roots:

$$\begin{aligned} A &= \sqrt{2} + \sqrt{3}, \\ B &= \sqrt{2} - \sqrt{3}, \\ C &= -\sqrt{2} + \sqrt{3}, \\ D &= -\sqrt{2} - \sqrt{3}. \end{aligned}$$

There are 24 possible ways to permute these four roots, but not all of these permutations are members of the Galois group. The members of the Galois group must preserve any algebraic equation with rational coefficients involving A, B, C and D.

Among these equations, we have:

$$\begin{aligned} AB &= -1 \\ AC &= 1 \\ A + D &= 0 \end{aligned}$$

It follows that, if ö is a permutation that belongs to the Galois group, we must have:

$$\begin{aligned} \varphi(B) &= \frac{-1}{\varphi(A)}, \\ \varphi(C) &= \frac{1}{\varphi(A)}, \\ \varphi(D) &= -\varphi(A). \end{aligned}$$

This implies that the permutation is well defined by the image of A, and that the Galois group has 4 elements, which are:

(A, B, C, D) '! (A, B, C, D)
(A, B, C, D) '! (B, A, D, C)
(A, B, C, D) '! (C, D, A, B)
(A, B, C, D) '! (D, C, B, A)

This implies that the Galois group is isomorphic to the Klein four-group.

Modern approach by field theory

In the modern approach, one starts with a field extensionL/K (read "L over K"), and examines the group of field automorphisms of L/K (these are bijectivering homomorphismsá : L $\rightarrow$ L such that á(x) = x for all x"K). See the article on Galois groups for further explanation and examples The connection between the two approaches is as follows. The coefficients of the polynomial in question should be chosen from the base field K. The top field L should be the field obtained by adjoining the roots of the polynomial in question to the base field. Any permutation of the roots which respects algebraic equations as described above gives rise to an automorphism of L/K, and vice versa. In the first example above, we were studying the extension

$\mathbf{Q}(\sqrt{3}) / \mathbf{Q}$, where Q is the field of rational numbers, and

$\mathbf{Q}(\sqrt{3})$ is the field obtained from Q by adjoining $\sqrt{3}$. In the second example, we were studying the extension Q(A,B,C,D)/Q. There are several advantages to the modern approach over the permutation group approach.

- It permits a far simpler statement of the fundamental theorem of Galois theory.
- The use of base fields other than Q is crucial in many areas of mathematics. For example, in algebraic number theory, one often does Galois theory using number fields, finite fields or local fields as the base field.
- It allows one to more easily study infinite extensions. Again this is important in algebraic number theory, where for example one often discusses the absolute Galois group of Q, defined to be the Galois group of K/Q where K is an algebraic closure of Q.
- It allows for consideration of inseparable extensions. This issue does not arise in the classical framework, since it was always implicitly assumed that arithmetic took place in characteristic zero, but nonzero characteristic arises frequently in number theory and in algebraic geometry.
- It removes the rather artificial reliance on chasing roots of polynomials. That is, different polynomials may yield the same extension fields, and the modern approach recognizes the connection between these polynomials.

Solvable groups and solution by radicals

The notion of a solvable group in group theory allows one to determine whether a polynomial is solvable in radicals, depending on whether its Galois group has the property of solvability. In essence, each field extension L/K corresponds to a factor group in a composition series of the Galois group. If a factor group in the composition series is cyclic of order n, and if in the corresponding field extension L/K the field K already contains a primitive nth root of unity, then it is a radical extension and the elements of L can then be expressed using the nth root of some element of K.

If all the factor groups in its composition series are cyclic, the Galois group is called solvable, and all of the elements of the corresponding field can be found by repeatedly taking roots, products, and sums of elements from the base field (usually Q). One of the great triumphs of Galois Theory was the proof that for every n> 4, there exist polynomials of degree n which are not solvable by radicals (this was proven independently, using a similar method, by Niels Henrik Abel a few years before, and is the Abel–Ruffini theorem), and a systematic way for testing whether a specific polynomial is solvable by radicals. The Abel–Ruffini theorem results from the fact that for n> 4 the symmetric groupS_n contains a simple, noncyclic, normal subgroup, namely the alternating groupA_n.

A non-solvable quintic example

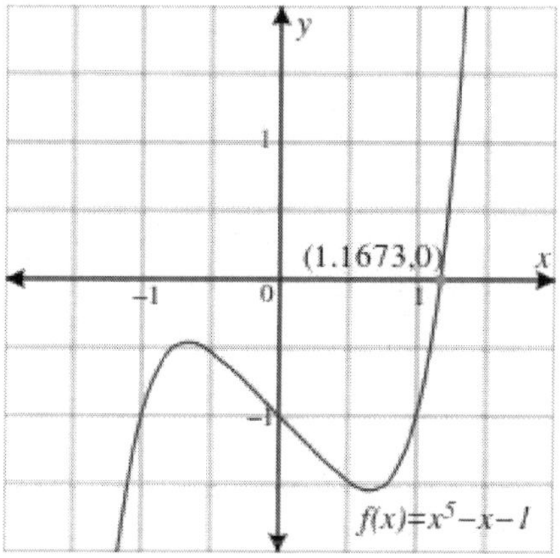

For the polynomial f(x) = x^5 Š x Š 1, the lone real root x = 1.1673... is algebraic, but not expressible in terms of radicals. The other four roots are complex numbers. Van der Waerden cites the polynomial $f(x) = x^5 - x - 1$. By the rational root theorem this has no rational zeroes. Neither does it have linear factors modulo 2 or 3. The Galois group of f(x) modulo 2 is cyclic of order 6, because f(x) modulo 2 factors into polynomials of orders 2 and 3, $(x^2 + x + 1)(x^3 + x^2 + 1)$. f(x) modulo 3 has no linear or quadratic factor, and hence is irreducible. Thus its modulo 3 Galois group contains an element of order 5. It is known that a Galois group modulo a prime is isomorphic to a subgroup of the Galois group over the rationals. A permutation group on 5 objects with elements of orders 6 and 5 must be the symmetric group S_5, which is therefore the Galois group of f(x). This is one of the simplest examples of a non-solvable quintic polynomial. According to Serge Lang, Emil Artin found this example.

Inverse Galois problem

The inverse Galois problem is to find a field extension with a given Galois group As long as one does not also specify the ground field, the problem is not very difficult, and all finite groups do occur as Galois groups. For showing this, one may proceed as follows. Choose a field K and a finite group G. Cayley's theorem says that G is (up to isomorphism) a subgroup of the symmetric groupS on the elements of G. Choose indeterminates $\{x_{á}\}$, one for each element á of G, and adjoin them to K to get the field $F = K(\{x_{á}\})$. Contained within F is the field L of symmetric rational functions in the $\{x_{á}\}$. The Galois group of F/L is S, by a basic result of Emil Artin. G acts on F by restriction of action of S. If the fixed field of this action is M, then, by the fundamental theorem of Galois theory, the Galois group of F/M is G. On the other hand, it is an open problem whether every finite group is the Galois group of a field extension of the field Q of the rational numbers. Igor Shafarevich proved that every solvable finite group is the Galois group of some extension of Q. Various people have solved the inverse Galois problem for selected non-Abelian simple groups. Existence of solutions has been shown for all but possibly one (Mathieu groupM_{23}) of the 26 sporadic simple groups. There is even a polynomial with integral coefficients whose Galois group is the Monster group.

Inverse Galois problem

In Galois theory, the inverse Galois problem concerns whether or not every finite group appears as the Galois group of some Galois extension of the rational numbers Q. This problem, first posed in the early 19th century, is unsolved. There are some permutation groups for which generic polynomials are known, which define all algebraic extensions of Q having a particular group as Galois group. These groups include all of degree no greater than 5. There also are groups known not to have generic polynomials, such as the cyclic group of order 8. More generally, let G be a given finite group, and let K be a field. Then the question is this: is there a Galois extension field L/K such that the Galois group of the extension is isomorphic to G? One says that G is realizable over K if such a field L exists.

Partial results

There is a great deal of detailed information in particular cases. It is known that every finite group is realizable over any function field in one variable over the complex numbers C, and more generally over function fields in one variable over any algebraically closed field of characteristic zero. Igor Shafarevich showed that every finite solvable group is realizable over Q. It is also known that every sporadic group, except possibly the Mathieu group M_{23}, is realizable over Q.

David Hilbert had shown that this question is related to a rationality question for G: If K is any extension of Q, on which G acts as an automorphism group and the invariant field K^G is rational over Q, then G is realizable over Q. Here rational means that it is a purely transcendental extension of Q, generated by an algebraically independent set. This criterion can for example be used to show that all the symmetric groups are realizable. Much detailed work has been carried out on the question, which is in no sense solved in general. Some of this is based on constructing G geometrically as a Galois covering of the projective line: in algebraic terms, starting with an extension of the field Q(t) of rational functions in an indeterminate t. After that, one applies Hilbert's irreducibility theorem to specialise t, in such a way as to preserve the Galois group. All permutation groups of degree 16 or less are known to be realizable over Q ; the group PSL(2,16):2 of degree 17 may not be . All 13 non-Abelian simple groups smaller than PSL(2,25) (order 7800) are known to be realizable over Q.

A simple example: cyclic groups

It is possible, using classical results, to construct explicitly a polynomial whose Galois group over Q is the cyclic group Z/nZ for any positive integer n. To do this, choose a prime p such that p a ≡ 1 (mod n); this is possible by Dirichlet's theorem. Let Q(ì) be the cyclotomic extension of Q generated by ì, where ì is a primitive p^{th} root of unity; the Galois group of Q(ì)/Q is cyclic of order p - 1. Since n divides p Š 1, the Galois group has a cyclic subgroup H of order (p - 1)/n. The fundamental theorem of Galois theory implies that the corresponding fixed field, F = $Q(ì)^H$, has Galois group Z/nZ over Q. By taking appropriate sums of conjugates of ì, following the construction of Gaussian periods, one can find an element á of F that generates F over Q, and compute its minimal polynomial. This method can be extended to cover all finite abelian groups, since every such group appears in fact as a quotient of the Galois group of some cyclotomic extension of Q. (This statement should not though be confused with the Kronecker–Weber theorem, which lies significantly deeper.)

2Worked example: the cyclic group of order three

For n = 3, we may take p = 7. Then Gal(Q(ì) / Q) is cyclic of order six. Let us take the generator ç of this group which sends ì to $ì^3$. We are interested in the subgroup H = {1, $ç^3$} of order two. Consider the element á = ì + $ç^3$(ì). By construction, á is fixed by H, and only has three conjugates over Q:

$$á = ç^0(á) = ì + ì^6,$$
$$â = ç^1(á) = ì^3 + ì^4,$$
$$ã = ç^2(á) = ì^2 + ì^5.$$

Using the identity:

$$1 + ì + ì^2 + ... + ì^6 = 0,$$

one finds that

$$á + â + ã = “1,$$
$$áâ + âã + ãá = “2,$$
$$áâã = 1.$$

Therefore á is a root of the polynomial

$$(x “ á)(x “ â)(x “ ã) = x^3 + x^2 “ 2x “ 1,$$

which consequently has Galois group Z/3Z over Q.

Symmetric and alternating groups

Hilbert showed that all symmetric and alternating groups are represented as Galois groups of polynomials with rational coefficients. The polynomial x^n + ax + b has discriminant

We take the special case

$$f(x, s) = x^n “ sx “ s.$$

Substituting a prime integer for s in f(x, s) gives a polynomial (called a specialization of f(x, s)) that by Eisenstein's criterion is irreducible. Then f(x, s) must be irreducible over Q(s). Furthermore, f(x, s) can be written and f(x, 1/2) can be factored to: whose second factor is irreducible by Eisenstein's criterion. We have now shown that the group Gal(f(x, s)/Q(s)) is doubly transitive. We can then find that this Galois group has a transposition. Use the scaling (1 Š n)x = ny to get and with we arrive at:

$$g(y, t) = y^n “ nty + (n “ 1)t$$

which can be arranged to

$$y^n “ y “ (n “ 1)(y “ 1) + (t “ 1)(“ny + n “ 1).$$

Gal(f(x, s)/Q(s)) is implied. Any finite doubly transitive permutation group containing a transposition is a full symmetric group. Hilbert's irreducibility theorem then implies that an infinite set of rational numbers give specializations of f(x, t) whose Galois groups are S_n over the rational field Q. In fact this set of rational numbers is dense in Q.

The discriminant of g(y, t) equals and this is not in general a perfect square. Alternating groupsSolutions for alternating groups must be handled differently for odd and even degrees. Odd DegreeLet Under this substitution the discriminant of g(y, t) equals which is a perfect square when n is odd. Even DegreeLet: Under this substitution the discriminant of g(y, t) equals: which is a perfect square when n is even. Again, Hilbert's irreducibility theorem implies the existence of infinitely many specializations whose Galois groups are alternating groups.

Rigid groups

Suppose that $C_1, ..., C_n$ are conjugacy classes of a finite group G, and A be the set of n-tuples $(g_1, ..., g_n)$ of G such that g_i is in C_i and the product $g_1...g_n$ is trivial. Then A is called rigid if it is nonempty, G acts transitively on it by conjugation, and each element of A generates G. Thompson (1984) showed that if a finite group G has a rigid set then it can often be realized as a Galois group over a cyclotomic extension of the rationals. (More precisely, over the cyclotomic extension of the rationals generated by the values of the irreducible characters of G on the conjugacy classes C_i.)

This can be used to show that many finite simple groups, including the monster group, are Galois groups of extensions of the rationals. The monster group is generated by a triad of elements of orders 2, 3, and 29. All such triads are conjugate. The prototype for rigidity is the symmetric group S_n, which is generated by an n-cycle and a transposition whose product is an (n Š 1)-cycle. The construction in the preceding section used these generators to establish a polynomial's Galois group.

A construction with an elliptic modular function

Let n > 1 be any integer. A lattice Ë in the complex plane with period ratio ô has a sublattice Ë2 with period ratio nô. The latter lattice is one of a finite set of sublattices permuted by the modular group PSL(2, Z), which is based on changes of basis for Ë. Let j denote the elliptic modular function of Felix Klein. Define the polynomial $ö_n$ as the product of the differences (X Š j($Ë_i$)) over the conjugate sublattices. As a polynomial in X, $ö_n$ has coefficients that are polynomials over Q in j(ô). On the conjugate lattices, the modular group acts as PGL(2, Z/nZ). It follows that $ö_n$ has Galois group isomorphic to PGL(2, Z/nZ) over Q(J(ô)).

Use of Hilbert's irreducibility theorem gives an infinite (and dense) set of rational numbers specializing $ö_n$ to polynomials with Galois group PGL(2, Z/nZ) over Q. The groups PGL(2, Z/nZ) include infinitely many non-solvable groups.

5

Topics in Group Theory

LIST OF GROUP THEORY TOPICS

In mathematics and abstract algebra, group theory studies the algebraic structures known as groups. The concept of a group is central to abstract algebra: other well-known algebraic structures, such as rings, fields, and vector spaces, can all be seen as groups endowed with additional operations and axioms. Groups recur throughout mathematics, and the methods of group theory have influenced many parts of algebra. Linear algebraic groups and Lie groups are two branches of group theory that have experienced advances and have become subject areas in their own right. Various physical systems, such as crystals and the hydrogen atom, may be modelled by symmetry groups. Thus group theory and the closely related representation theory have many important applications in physics, chemistry, and materials science. Group theory is also central to public key cryptography.

Category theory

Schematic representation of a category with objects X, Y, Z and morphisms f, g, g'''f. (The category's three identity morphisms 1_X, 1_Y and 1_Z, if explicitly represented, would appear as three arrows, from the letters X, Y, and Z to themselves, respectively.) Category theory formalizes mathematical structure and its concepts in terms of a labeleddirected graph called a category, whose nodes are called objects, and whose labelled directed edges are called arrows (or morphisms). A category has two basic properties: the ability to compose the arrows associatively, and the existence of an identity arrow for each object. The language of category theory has been used to formalize concepts of other high-level abstractions such as sets, rings, and groups. Informally, category theory is a general theory of functions. Several terms used in category theory, including the term "morphism", are used differently from their uses in the rest of mathematics. In category theory, morphisms obey conditions specific to category theory itself. Samuel Eilenberg and Saunders Mac Lane introduced the concepts of categories, functors, and natural transformations in 1942–45 in their study of algebraic topology, with the goal of understanding the processes that preserve mathematical structure. Category theory has practical applications in programming language theory, for example the usage of monads in functional programming. It may also be used as an axiomatic foundation for mathematics, as an alternative to set theory and other proposed foundations.

Basic concepts

Categories represent abstractions of other mathematical concepts. Many areas of mathematics can be formalised by category theory as categories. Hence category theory uses abstraction to make it possible to state and prove many intricate and subtle mathematical results in these fields in a much simpler way. A basic example of a category is the category of sets, where the objects are sets and the arrows are functions from one set to another. However, the objects of a category need not be sets, and the arrows need not be functions. Any way of formalising a mathematical concept such that it meets the basic conditions on the behaviour of objects and arrows is a valid category—and all the results of category theory apply to it. The "arrows" of category theory are often said to represent a process connecting two objects, or in many cases a "structure-preserving" transformation connecting two objects. There are, however, many applications where much more abstract concepts are represented by objects and morphisms. The most important property of the arrows is that they can be "composed", in other words, arranged in a sequence to form a new arrow.

Applications of categories

Categories now appear in many branches of mathematics, some areas of theoretical computer science where they can correspond to types or to database schemas, and mathematical physics where they can be used to describe vector spaces. Probably the first application of category theory outside pure mathematics was the "metabolism-repair" model of autonomous living organisms by Robert Rosen.

Utility

Categories, objects, and morphisms

The study of categories is an attempt to axiomatically capture what is commonly found in various classes of related mathematical structures by relating them to the structure-preserving functions between them. A systematic study of category theory then allows us to prove general results about any of these types of mathematical structures from the axioms of a category. Consider the following example. The classGrp of groups consists of all objects having a "group structure". One can proceed to provetheorems about groups by making logical deductions from the set of axioms defining groups. For example, it is immediately proven from the axioms that the identity element of a group is unique.

Instead of focusing merely on the individual objects (e.g., groups) possessing a given structure, category theory emphasizes the morphisms – the structure-preserving mappings – between these objects; by studying these morphisms, one is able to learn more about the structure of the objects. In the case of groups, the morphisms are the group homomorphisms. A group homomorphism between two groups "preserves the group structure" in a precise sense; informally it is a "process" taking one group to another, in a way that carries along information about the structure of the first group into the second group. The study of group homomorphisms then provides a tool for studying general properties of groups and consequences of the group axioms. A similar type of investigation occurs in many mathematical theories, such as the study of continuous maps (morphisms) between topological spaces in topology (the associated category is called Top), and the study of smooth functions (morphisms) in manifold theory. Not all categories arise as "structure preserving (set) functions", however; the standard example is the category of homotopies between pointed topological spaces. If one axiomatizes relations instead of functions, one obtains the theory of allegories.

Functors

A category is itself a type of mathematical structure, so we can look for "processes" which preserve this structure in some sense; such a process is called a functor. Diagram chasing is a visual method of arguing with abstract "arrows" joined in diagrams. Functors are represented by arrows between categories, subject to specific defining commutativity conditions. Functors can define (construct) categorical diagrams and sequences (viz. Mitchell, 1965)[citation needed]. A functor associates to every object of one category an object of another category, and to every morphism in the first category a morphism in the second. As a result, this defines a category of categories and functors – the objects are categories, and the morphisms (between categories) are functors. Studying categories and functors is not just studying a class of mathematical structures and the morphisms between them but rather the relationships between various classes of mathematical structures. This fundamental idea first surfaced in algebraic topology. Difficult topological questions can be translated into algebraic questions which are often easier to solve. Basic constructions, such as the fundamental group or the fundamental groupoid of a topological space, can be expressed as functors to the category of groupoids in this way, and the concept is pervasive in algebra and its applications.

Natural transformations

Abstracting yet again, some diagrammatic and/or sequential constructions are often "naturally related" – a vague notion, at first sight. This leads to the clarifying concept of natural transformation, a way to "map" one functor to another. Many important constructions in mathematics can be studied in this context. "Naturality" is a principle, like general covariance in physics, that cuts deeper than is initially apparent. An arrow between two functors is a natural transformation when it is subject to certain naturality or commutativity conditions. Functors and natural transformations ('naturality') are the key concepts in category theory.

Categories

A categoryC consists of the following three mathematical entities:

- A class ob(C), whose elements are called objects;
- A class hom(C), whose elements are called morphisms or maps or arrows. Each morphism f has a source object a and target object b.

The expression f :a $\rightarrow$ b, would be verbally stated as "f is a morphism from a to b". The expression hom(a, b) – alternatively expressed as $hom_C(a, b)$, mor(a, b), or C(a, b) –

- denotes the hom-class of all morphisms from a to b.
- A binary operation'", called composition of morphisms, such that for any three objects a, b, and c, we have $\circ$: hom(b, c) × hom(a, b) → hom(a, c). The composition of f : a → b and g : b → c is written as $g \circ f$ or gf,[a] governed by two axioms:
  o

Associativity: If $f : a \rightarrow b, g : b \rightarrow c$ and $h : c \rightarrow d$ then $h \circ (g \circ f) = (h \circ g) \circ f$, and

o Identity: For every object x, there exists a morphism 1_x :x → x called the identity morphism for x, such that for every morphism f : a → b, we have $1_b \circ f = f = f \circ 1_a$.

From the axioms, it can be proved that there is exactly one identity morphism for every object. Some authors deviate from the definition just given by identifying each object with its identity morphism.

Morphisms

Relations among morphisms (such as fg = h) are often depicted using commutative diagrams, with "points" (corners) representing objects and "arrows" representing morphisms. Morphisms can have any of the following properties. A morphism $f : a \rightarrow b$ is a:

- monomorphism (or monic) if $f \circ g_1 = f \circ g_2$ implies $g_1 = g_2$ for all morphisms $g_1, g_2 : x \rightarrow a$.
- epimorphism (or epic) if $g_1 \circ f = g_2 \circ f$ implies $g_1 = g_2$ for all morphisms $g_1, g_2 : b \rightarrow x$.
- bimorphism if f is both epic and monic.
- isomorphism if there exists a morphism g : b → a such that $f \circ g = 1_b$ and $g \circ f = 1_a$.[b]
- endomorphism if a = b. end(a) denotes the class of endomorphisms of a.
- automorphism if f is both an endomorphism and an isomorphism. aut(a) denotes the class of automorphisms of a.
- retraction if a right inverse of f exists, i.e. if there exists a morphism g : b → a with $f \circ g = 1_b$.
- section if a left inverse of f exists, i.e. if there exists a morphism g : b → a with $g \circ f = 1_a$.

Every retraction is an epimorphism, and every section is a monomorphism. Furthermore, the following three statements are equivalent:

- f is a monomorphism and a retraction;
- f is an epimorphism and a section;
- f is an isomorphism.

Functors

Functors are structure-preserving maps between categories. They can be thought of as morphisms in the category of all (small) categories. A (covariant) functor F from a category C to a category D, written F :C → D, consists of:

- for each object x in C, an object F(x) in D; and
- for each morphism f : x → y in C, a morphism F(f) : F(x) → F(y),

such that the following two properties hold:

- For every object x in C, $F(1_x) = 1_{F(x)}$;
- For all morphisms f :x → y and g : y → z, F(g ∘ f) = F(g) ∘ F(f).

A contravariant functor F: C → D is like a covariant functor, except that it "turns morphisms around" ("reverses all the arrows"). More specifically, every morphism f :x → y in C must be assigned to a morphism F(f) : F(y) → F(x) in D. In other words, a contravariant functor acts as a covariant functor from the opposite categoryC^{op} to D.

Natural transformations

A natural transformation is a relation between two functors. Functors often describe "natural constructions" and natural transformations then describe "natural homomorphisms" between two such constructions. Sometimes two quite different constructions yield "the same" result; this is expressed by a natural isomorphism between the two functors. If F and G are (covariant) functors between the categories C and D, then a natural transformation ç from F to G associates to every object X in C a morphism ç$_X$:F(X) → G(X) in D such that for every morphism f : X → Y in C, we have ç$_Y$ ∘ F(f) = G(f) ∘ ç$_X$; this means that the following diagram is commutative:

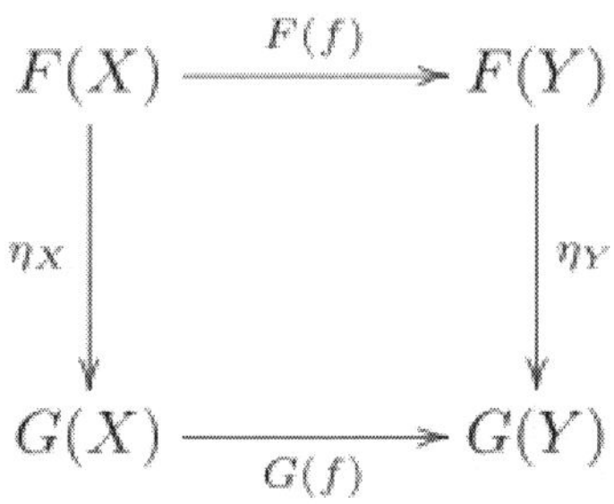

The two functors F and G are called naturally isomorphic if there exists a natural transformation from F to G such that ç$_X$ is an isomorphism for every object X in C.

Other concepts

Universal constructions, limits, and colimits

Using the language of category theory, many areas of mathematical study can be categorized. Categories include sets, groups and topologies. Each category is distinguished by properties that all its objects have in common, such as the empty set or the product of two topologies, yet in the definition of a category, objects are considered atomic, i.e., we do not know whether an object A is a set, a topology, or any other abstract concept. Hence, the challenge is to define special objects without referring to the internal structure of those objects. To define the empty set without referring to elements, or the product topology without referring to open sets, one can characterize these objects in terms of their relations to other objects, as given by the morphisms of the respective categories. Thus, the task is to find universal properties that uniquely determine the objects of interest. Numerous important constructions can be described in a purely categorical way if the category limit can be developed and dualized to yield the notion of a colimit.

Equivalent categories

It is a natural question to ask: under which conditions can two categories be considered essentially the same, in the sense that theorems about one category can readily be transformed into theorems about the other category? The major tool one employs to describe such a situation is called equivalence of categories, which is given by appropriate functors between two categories. Categorical equivalence has found numerous applications in mathematics.

Further concepts and results

The definitions of categories and functors provide only the very basics of categorical algebra; additional important topics are listed below. Although there are strong interrelations between all of these topics, the given order can be considered as a guideline for further reading.

- The functor category D^C has as objects the functors from C to D and as morphisms the natural transformations of such functors. The Yoneda lemma is one of the most famous basic results of category theory; it describes representable functors in functor categories.
- Duality: Every statement, theorem, or definition in category theory has a dual which is essentially obtained by "reversing all the arrows". If one statement is true in a category C then its dual is true in the dual category C^{op}. This duality, which is transparent at the level of category theory, is often obscured in applications and can lead to surprising relationships.
- Adjoint functors: A functor can be left (or right) adjoint to another functor that maps in the opposite direction. Such a pair of adjoint functors typically arises from a construction defined by a universal property; this can be seen as a more abstract and powerful view on universal properties.

Higher-dimensional categories

Many of the above concepts, especially equivalence of categories, adjoint functor pairs, and functor categories, can be situated into the context of higher-dimensional categories. Briefly, if we consider a morphism between two objects as a "process taking us from one object to another", then higher-dimensional categories allow us to profitably generalize this by considering "higher-dimensional processes". For example, a (strict) 2-category is a category together with "morphisms between morphisms", i.e., processes which allow us to transform one morphism into another. We can then "compose" these "bimorphisms" both horizontally and vertically, and we require a 2-dimensional "exchange law" to hold, relating the two composition laws. In this context, the standard example is Cat, the 2-category of all (small) categories, and in this example, bimorphisms of morphisms are simply natural transformations of morphisms in the usual sense. Another basic example is to consider a 2-category with a single object; these are essentially monoidal categories. Bicategories are a weaker notion of 2-dimensional categories in which the composition of morphisms is not strictly associative, but only associative "up to" an isomorphism. This process can be extended for all natural numbers n, and these are called n-categories. There is even a notion of ù-category corresponding to the ordinal number ù. Higher-dimensional categories are part of the broader mathematical field of higher-dimensional algebra, a concept introduced by Ronald Brown. For a conversational introduction to these ideas, see John Baez, 'A Tale of n-categories' (1996).

Historical notes

In 1942–45, Samuel Eilenberg and Saunders Mac Lane introduced categories, functors, and natural transformations as part of their work in topology, especially algebraic topology. Their work was an important part of the transition from intuitive and geometric homology to axiomatic homology theory. Eilenberg and Mac Lane later wrote that their goal was to understand natural transformations. That required defining functors, which required categories. Stanislaw Ulam, and some writing on his behalf, have claimed that related ideas were current in the late 1930s in Poland. Eilenberg was Polish, and studied mathematics in Poland in the 1930s. Category theory is also, in some sense, a continuation of the work of Emmy Noether (one of Mac Lane's teachers) in formalizing abstract processes; Noether realized that understanding a type of mathematical structure requires understanding the processes that preserve that structure. To achieve this understanding, Eilenberg and Mac Lane proposed an axiomatic formalization of the relation between structures and the processes that preserve them.

The subsequent development of category theory was powered first by the computational needs of homological algebra, and later by the axiomatic needs of algebraic geometry. General category theory, an extension of universal algebra having many new features allowing for semantic flexibility and higher-order logic, came later; it is now applied throughout mathematics. Certain categories called topoi (singular topos) can even serve as an alternative to axiomatic set theory as a foundation of mathematics. A topos can also be considered as a specific type of category with two additional topos axioms. These foundational applications of category theory have been worked out in fair detail as a basis for, and justification of, constructive mathematics. Topos theory is a form of abstract sheaf theory, with geometric origins, and leads to ideas such as pointless topology.

Categorical logic is now a well-defined field based on type theory for intuitionistic logics, with applications in functional programming and domain theory, where a cartesian closed category is taken as a non-syntactic description of a lambda calculus. At the very least, category theoretic language clarifies what exactly these related areas have in common (in some abstract sense). Category theory has been applied in other fields as well. For example, John Baez has shown a link between Feynman diagrams in Physics and monoidal categories. Another application of category theory, more specifically: topos theory, has been made in mathematical music theory, see for example the book The Topos of Music, Geometric Logic of Concepts, Theory, and Performance by Guerino Mazzola. More recent efforts to introduce undergraduates to categories as a foundation for mathematics include those of William Lawvere and Rosebrugh (2003) and Lawvere and Stephen Schanuel (1997) and Mirroslav Yotov (2012).

6

Module Theory

MODULE (MATHEMATICS)

In mathematics, a module is one of the fundamental algebraic structures used in abstract algebra. A module over a ring is a generalization of the notion of vector space over a field, wherein the corresponding scalars are the elements of an arbitrary given ring (with identity) and a multiplication (on the left and/or on the right) is defined between elements of the ring and elements of the module. Thus, a module, like a vector space, is an additive abelian group; a product is defined between elements of the ring and elements of the module that is distributive over the addition operation of each parameter and is compatible with the ring multiplication. Modules are very closely related to the representation theory of groups. They are also one of the central notions of commutative algebra and homological algebra, and are used widely in algebraic geometry and algebraic topology.

Motivation

In a vector space, the set of scalars is a field and acts on the vectors by scalar multiplication, subject to certain axioms such as the distributive law. In a module, the scalars need only be a ring, so the module concept represents a significant generalization. In commutative algebra, both ideals and quotient rings are modules, so that many arguments about ideals or quotient rings can be combined into a single argument about modules. In non-commutative algebra the distinction between left ideals, ideals, and modules becomes more pronounced, though some ring-theoretic conditions can be expressed either about left ideals or left modules. Much of the theory of modules consists of extending as many of the desirable properties of vector spaces as possible to the realm of modules over a "well-behaved" ring, such as a principal ideal domain. However, modules can be quite a bit more complicated than vector spaces; for instance, not all modules have a basis, and even those that do, free modules, need not have a unique rank if the underlying ring does not satisfy the invariant basis number condition, unlike vector spaces, which always have a (possibly infinite) basis whose cardinality is then unique. (These last two assertions require the axiom of choice in general, but not in the case of finite-dimensional spaces, or certain well-behaved infinite-dimensional spaces such as L^p spaces.)

Formal definition

Suppose that R is a ring and 1_R is its multiplicative identity. A left R-module M consists of an abelian group (M, +) and an operation . : R × M '! M such that for all r, s in R and x, y in M, we have:

$$r\cdot(x+y)=r\cdot x+r\cdot y$$

$$(r+s)\cdot x=r\cdot x+s\cdot x$$

$$(rs)\cdot x=r\cdot(s\cdot x)$$

The operation of the ring on M is called scalar multiplication, and is usually written by juxtaposition, i.e. as rx for r in R and x in M, though here it is denoted as r ÅŠ x to distinguish it from the ring multiplication operation, denoted here by juxtaposition. The notation $_R$M indicates a left R-module M. A right R-module M or M_R is defined similarly, except that the ring acts on the right; i.e., scalar multiplication takes the form ÅŠ : M × R '! M, and the above axioms are written with scalars r and s on the right of x and y. Authors who do not require rings to be unital omit condition 4 above in the definition of an R-module, and so would call the structures defined above "unital left R-modules". In this article, consistent with the glossary of ring theory, all rings and modules are assumed to be unital.

If one writes the scalar action as f_r so that f_r(x) = r ÅŠ x, and f for the map that takes each r to its corresponding map f_r, then the first axiom states that every f_r is a group endomorphism of M, and the other three axioms assert that the map f : R '! End(M) given by r ¦! f_r is a ring homomorphism from R to the endomorphism ring End(M). Thus a module is a ring action on an abelian group (cf. group action. Also consider monoid action of multiplicative structure of R). In this sense, module theory generalizes representation theory, which deals with group actions on vector spaces, or equivalently group ring actions. A bimodule is a module that is a left module and a right module such that the two multiplications are compatible. If R is commutative, then left R-modules are the same as right R-modules and are simply called R-modules.

Examples

- If K is a field, then K-vector spaces (vector spaces over K) and K-modules are identical.
- If K is a field, and K[x] a univariate polynomial ring, then a K[x]-module M is a K-module with an additional action of x on M that commutes with the action of K on M. In other words, a K[x]-module is a K-vector space M combined with a linear map from M to M. Applying the Structure theorem for finitely generated modules over a principal ideal domain to this example shows the existence of the rational and Jordan canonical forms.
- The concept of a Z-module agrees with the notion of an abelian group. That is, every abelian group is a module over the ring of integers Z in a unique way. For n > 0, let n Å" x = x + x + ... + x (n summands), 0 Å" x = 0, and ("n) Å" x = "(n Å" x). Such a module need not have a basis—groups containing torsion elements do not. (For example, in the group of integers modulo 3, one cannot find even

one element which satisfies the definition of a linearly independent set since when an integer such as 3 or 6 multiplies an element, the result is 0. However, if a finite field is considered as a module over the same finite field taken as a ring, it is a vector space and does have a basis.)

- The decimal fractions (including negative ones) form a module over the integers. Only singletons are linearly independent sets, but there is no singleton that can serve as a basis, so the module has no basis and no rank.
- If R is any ring and n a natural number, then the cartesian product R^n is both a left and right R-module over if we use the component-wise operations. Hence when n = 1, R is an R-module, where the scalar multiplication is just ring multiplication. The case n = 0 yields the trivial R-module {0} consisting only of its identity element. Modules of this type are called free and if R has invariant basis number (e.g. any commutative ring or field) the number n is then the rank of the free module.
- If $M_n(R)$ is the ring of n × n matrices over a ring R, M is an $M_n(R)$-module, and e_i is the n × n matrix with 1 in the (i, i)-entry (and zeros elsewhere), then e_iM is an R-module, since $re_im = e_irm$ Š e_iM. So M breaks up as the direct sum of R-modules, $M = e_1M$ •Š ... •Š e_nM. Conversely, given an R-module M_0, then $M_0^{\bullet Šn}$ is an $M_n(R)$-module. In fact, the category of R-modules and the category of $M_n(R)$-modules are equivalent. The special case is that the module M is just R as a module over itself, then R^n is an $M_n(R)$-module.
- If S is a nonempty set, M is a left R-module, and M^S is the collection of all functions f : S '! M, then with addition and scalar multiplication in M^S defined by (f + g)(s) = f(s) + g(s) and (rf)(s) = rf(s), M^S is a left R-module. The right R-module case is analogous. In particular, if R is commutative then the collection of R-module homomorphisms h : M '! N (see below) is an R-module (and in fact a submodule of N^M).
- If X is a smooth manifold, then the smooth functions from X to the real numbers form a ring C″(X). The set of all smooth vector fields defined on X form a module over C″(X), and so do the tensor fields and the differential forms on X. More generally, the sections of any vector bundle form a projective module over C″(X), and by Swan's theorem, every projective module is isomorphic to the module of sections of some bundle; the category of C″(X)-modules and the category of vector bundles over X are equivalent.
- If R is any ring and I is any left ideal in R, then I is a left R-module, and analogously right ideals in R are right R-modules.
- If R is a ring, we can define the ring R^{op} which has the same underlying set and the same addition operation, but the opposite multiplication: if ab = c in R, then ba = c in R^{op}. Any left R-module M can then be seen to be a right module over R^{op}, and any right module over R can be considered a left module over R^{op}.
- There are modules of a Lie algebra as well.

Submodules and homomorphisms

Suppose M is a left R-module and N is a subgroup of M. Then N is a submodule (or more explicitly an R-submodule) if for any n in N and any r in R, the product r Å″ n is in N (or n Å″ r for a right R-module). The set of submodules of a given module M, together with the two binary operations + and)″, forms a lattice which satisfies the modular law: Given submodules U, N_1, N_2 of M such that N_1 ,″ N_2, then the following two submodules are equal: (N_1 + U))″ N_2 = N_1 + (U)″ N_2). If M and N are left R-modules, then a map f : M '! N is a homomorphism of R-modules if for any m, n in M and r, s in R,

$$f(r \cdot m + s \cdot n) = r \cdot f(m) + s \cdot f(n)$$

This, like any homomorphism of mathematical objects, is just a mapping which preserves the structure of the objects. Another name for a homomorphism of R-modules is an R-linear map. A bijective module homomorphism is an isomorphism of modules, and the two modules are called isomorphic. Two isomorphic modules are identical for all practical purposes, differing solely in the notation for their elements. The kernel of a module homomorphism f : M '! N is the submodule of M consisting of all elements that are sent to zero by f. The isomorphism theorems familiar from groups and vector spaces are also valid for R-modules. Given a ring R, the set of all left R-modules together with their module homomorphisms forms an abelian category, denoted by R-Mod (see category of modules).

Types of modules

Finitely generated. An R-module M is finitely generated if there exist finitely many elements x_1, ..., x_n in M such that every element of M is a linear combination of those elements with coefficients from the ring R. Cyclic. A module is called a cyclic module if it is generated by one element. Free. A free R-module is a module that has a basis, or equivalently, one that is isomorphic to a direct sum of copies of the ring R. These are the modules that behave very much like vector spaces. Projective. Projective modules are direct summands of free modules and share many of their desirable properties. Injective. Injective modules are defined dually to projective modules. Flat. A module is called flat if taking the tensor product of it with any exact sequence of R-modules preserves exactness. Torsionless module. A module is called torsionless if it embeds into its algebraic dual. Simple. A simple module S is a module that is not {0} and whose only submodules are {0} and S. Simple modules are sometimes called irreducible. Semisimple. A semisimple module is a direct sum (finite or not) of simple modules. Historically these modules are also called completely reducible. Indecomposable. An indecomposable module is a non-zero module that cannot be written as a direct sum of two non-zero submodules. Every simple module is indecomposable, but there are indecomposable modules which are not simple (e.g. uniform modules). Faithful. A faithful module M is one where the action of each r ''' 0 in R on M is nontrivial (i.e. r Å″ x ''' 0 for some x in M). Equivalently, the annihilator of M is the zero ideal.

Torsion-free. A torsion-free module is a module over a ring such that 0 is the only element annihilated by a regular element (non zero-divisor) of the ring. Noetherian. A Noetherian module is a module which satisfies the ascending chain condition on submodules, that is, every increasing chain of submodules becomes stationary after finitely many steps. Equivalently, every submodule is finitely generated. Artinian. An Artinian module is a module which satisfies the descending chain condition on submodules, that is, every decreasing chain of submodules becomes stationary after finitely many steps. Graded. A graded module is a module with a decomposition as a direct sum M = $*_x M_x$ over a graded ring R = $*_x R_x$ such that

R_xM_y ," M_{x+y} for all x and y. Uniform. A uniform module is a module in which all pairs of nonzero submodules have nonzero intersection.

Further notions

Relation to representation theory

If M is a left R-module, then the action of an element r in R is defined to be the map M '! M that sends each x to rx (or xr in the case of a right module), and is necessarily a group endomorphism of the abelian group (M, +). The set of all group endomorphisms of M is denoted $End_Z(M)$ and forms a ring under addition and composition, and sending a ring element r of R to its action actually defines a ring homomorphism from R to $End_Z(M)$. Such a ring homomorphism R '! $End_Z(M)$ is called a representation of R over the abelian group M; an alternative and equivalent way of defining left R-modules is to say that a left R-module is an abelian group M together with a representation of R over it. A representation is called faithful if and only if the map R '! $End_Z(M)$ is injective. In terms of modules, this means that if r is an element of R such that rx = 0 for all x in M, then r = 0. Every abelian group is a faithful module over the integers or over some modular arithmetic Z/nZ.

Generalizations

Any ring R can be viewed as a preadditive category with a single object. With this understanding, a left R-module is just a covariant additive functor from R to the category Ab of abelian groups, and right R-modules are contravariant additive functors. This suggests that, if C is any preadditive category, a covariant additive functor from C to Ab should be considered a generalized left module over C. These functors form a functor category C-Mod which is the natural generalization of the module category R-Mod. Modules over commutative rings can be generalized in a different direction: take a ringed space (X, O_X) and consider the sheaves of O_X-modules (see sheaf of modules). These form a category O_X-Mod, and play an important role in modern algebraic geometry. If X has only a single point, then this is a module category in the old sense over the commutative ring $O_X(X)$. One can also consider modules over a semiring. Modules over rings are abelian groups, but modules over semirings are only commutative monoids. Most applications of modules are still possible. In particular, for any semiring S, the matrices over S form a semiring over which the tuples of elements from S are a module (in this generalized sense only). This allows a further generalization of the concept of vector space incorporating the semirings from theoretical computer science. Over near-rings, one can consider near-ring modules, a nonabelian generalization of modules.[citation needed]

7

Commutative Algebra

A 1915 postcard from one of the pioneers of commutative algebra, Emmy Noether, to E. Fischer, discussing her work in commutative algebra. Commutative algebra is the branch of algebra that studies commutative rings, their ideals, and modules over such rings. Both algebraic geometry and algebraic number theory build on commutative algebra. Prominent examples of commutative rings include polynomial rings; rings of algebraic integers, including the ordinary integers ℤ Z {\displaystyle \mathbb {Z} } ℤ ; and p-adic integers. Commutative algebra is the main technical tool in the local study of schemes. The study of rings that are not necessarily commutative is known as noncommutative algebra; it includes ring theory, representation theory, and the theory of Banach algebras. Commutative algebra is essentially the study of the rings occurring in algebraic number theory and algebraic geometry.

In algebraic number theory, the rings of algebraic integers are Dedekind rings, which constitute therefore an important class of commutative rings. Considerations related to modular arithmetic have led to the notion of a valuation ring. The restriction of algebraic field extensions to subrings has led to the notions of integral extensions and integrally closed domains as well as the notion of ramification of an extension of valuation rings. The notion of localization of a ring (in particular the localization with respect to a prime ideal, the localization consisting in inverting a single element and the total quotient ring) is one of the main differences between commutative algebra and the theory of non-commutative rings. It leads to an important class of commutative rings, the local rings that have only one maximal ideal. The set of the prime ideals of a commutative ring is naturally equipped with a topology, the Zariski topology. All these notions are widely used in algebraic geometry and are the basic technical tools for the definition of scheme theory, a generalization of algebraic geometry introduced by Grothendieck. Many other notions of commutative algebra are counterparts of geometrical notions occurring in algebraic geometry. This is the case of Krull dimension, primary decomposition, regular rings, Cohen–Macaulay rings, Gorenstein rings and many other notions.

History

The subject, first known as ideal theory, began with Richard Dedekind's work on ideals, itself based on the earlier work of Ernst Kummer and Leopold Kronecker. Later, David Hilbert introduced the term ring to generalize the earlier term number ring. Hilbert introduced a more abstract approach to replace the more concrete and computationally oriented methods grounded in such things as complex analysis and classical invariant theory. In turn, Hilbert strongly influenced Emmy Noether, who recast many earlier results in terms of an ascending chain condition, now known as the Noetherian condition. Another important milestone was the work of Hilbert's student Emanuel Lasker, who introduced primary ideals and proved the first version of the Lasker–Noether theorem.

The main figure responsible for the birth of commutative algebra as a mature subject was Wolfgang Krull, who introduced the fundamental notions of localization and completion of a ring, as well as that of regular local rings. He established the concept of the Krull dimension of a ring, first for Noetherian rings before moving on to expand his theory to cover general valuation rings and Krull rings. To this day, Krull's principal ideal theorem is widely considered the single most important foundational theorem in commutative algebra. These results paved the way for the introduction of commutative algebra into algebraic geometry, an idea which would revolutionize the latter subject. Much of the modern development of commutative algebra emphasizes modules. Both ideals of a ring R and R-algebras are special cases of R-modules, so module theory encompasses both ideal theory and the theory of ring extensions. Though it was already incipient in Kronecker's work, the modern approach to commutative algebra using module theory is usually credited to Krull and Noether.

Noetherian rings

In mathematics, more specifically in the area of modern algebra known as ring theory, a Noetherian ring, named after Emmy Noether, is a ring in which every non-empty set of ideals has a maximal element. Equivalently, a ring is Noetherian if it satisfies the ascending chain condition on ideals; that is, given any chain:

$$I_1 \subseteq \cdots I_{k-1} \subseteq I_k \subseteq I_{k+1} \subseteq \cdots$$

there exists an n such that:

$$I_n = I_{n+1} = \cdots$$

For a commutative ring to be Noetherian it suffices that every prime ideal of the ring is finitely generated. (The result is due to I. S. Cohen.) The notion of a Noetherian ring is of fundamental importance in both commutative and noncommutative ring theory, due to the role it plays in simplifying the ideal structure of a ring. For instance, the ring of integers and the polynomial ring over a field are both Noetherian rings, and consequently, such theorems as the Lasker–Noether theorem, the Krull intersection theorem, and the Hilbert's basis theorem hold for them. Furthermore, if a ring is Noetherian, then it satisfies the descending chain condition on prime ideals. This property suggests a deep theory of dimension for Noetherian rings beginning with the notion of the Krull dimension.

Primary decomposition

An ideal Q of a ring is said to be primary if Q is proper and whenever xy " Q, either x " Q or y^n " Q for some positive integer n. In Z, the primary ideals are precisely the ideals of the form (p^e) where p is prime and e is a positive integer. Thus, a primary decomposition of (n) corresponds to representing (n) as the intersection of finitely many primary ideals. The Lasker–Noether theorem, given here, may be seen as a certain generalization of the fundamental theorem of arithmetic: Lasker-Noether Theorem. Let R be a commutative Noetherian ring and let I be an ideal of R. Then I may be written as the intersection of finitely many primary ideals with distinct radicals; that is:

$$I = \bigcap_{i=1}^{t} Q_i$$

with Q_i primary for all i and Rad(Q_i) ''' Rad(Q_j) for i ''' j. Furthermore, if:

$$I = \bigcap_{i=1}^{k} P_i$$

is decomposition of I with Rad(P_i) ''' Rad(P_j) for i ''' j, and both decompositions of I are irredundant (meaning that no proper subset of either {Q_1, ..., Q_t} or {P_1, ..., P_k} yields an intersection equal to I), t = k and (after possibly renumbering the Q_i) Rad(Q_i) = Rad(P_i) for all i. For any primary decomposition of I, the set of all radicals, that is, the set {Rad(Q_1), ..., Rad(Q_t)} remains the same by the Lasker–Noether theorem. In fact, it turns out that (for a Noetherian ring) the set is precisely the assassinator of the module R/I; that is, the set of all annihilators of R/I (viewed as a module over R) that are prime.

Localization

The localization is a formal way to introduce the "denominators" to a given ring or a module. That is, it introduces a new ring/module out of an existing one so that it consists of fractions

$$\frac{m}{s}$$

where the denominators s range in a given subset S of R. The archetypal example is the construction of the ring Q of rational numbers from the ring Z of integers.

Completion

A completion is any of several related functors on rings and modules that result in complete topological rings and modules. Completion is similar to localization, and together they are among the most basic tools in analysing commutative rings. Complete commutative rings have simpler structure than the general ones and Hensel's lemma applies to them. Zariski topology on prime ideals The Zariski topology defines a topology on the spectrum of a ring (the set of prime ideals). In this formulation, the Zariski-closed sets are taken to be the sets

$$V(I) = \{P \in \mathrm{Spec}(A) \mid I \subseteq P\}$$

where A is a fixed commutative ring and I is an ideal. This is defined in analogy with the classical Zariski topology, where closed sets in affine space are those defined by polynomial equations . To see the connection with the classical picture, note that for any set S of polynomials (over an algebraically closed field), it follows from Hilbert's Nullstellensatz that the points of V(S) (in the old sense) are exactly the tuples $(a_1, ..., a_n)$ such that $(x_1 - a_1, ..., x_n - a_n)$ contains S; moreover, these are maximal ideals and by the "weak" Nullstellensatz, an ideal of any affine coordinate ring is maximal if and only if it is of this form. Thus, V(S) is "the same as" the maximal ideals containing S. Grothendieck's innovation in defining Spec was to replace maximal ideals with all prime ideals; in this formulation it is natural to simply generalize this observation to the definition of a closed set in the spectrum of a ring.

Examples

The fundamental example in commutative algebra is the ring of integers Z

{\displaystyle \mathbb {Z} } . The existence of primes and the unique factorization theorem laid the foundations for concepts such as Noetherian rings and the primary decomposition.

Other important examples are:

Polynomial rings $R[x_1,...,x_n]$

- The p-adic integers
- Rings of algebraic integers.

onnections with algebraic geometry

Commutative algebra (in the form of polynomial rings and their quotients, used in the definition of algebraic varieties) has always been a part of algebraic geometry. However, in the late 1950s, algebraic varieties were subsumed into Alexander Grothendieck's concept of a scheme. Their local objects are affine schemes or prime spectra, which are locally ringed spaces, which form a category that is antiequivalent (dual) to the category of commutative unital rings, extending the duality between the category of affine algebraic varieties over a field k, and the category of finitely generated reduced k-algebras. The gluing is along the Zariski topology; one can glue within the category of locally ringed spaces, but also, using the Yoneda embedding, within the more abstract category of presheaves of sets over the category of affine schemes. The Zariski topology in the set-theoretic sense is then replaced by a Zariski topology in the sense of Grothendieck topology. Grothendieck introduced Grothendieck topologies having in mind more exotic but geometrically finer and more sensitive examples than the crude Zariski topology, namely the étale topology, and the two flat Grothendieck topologies: fppf and fpqc. Nowadays some other examples have become prominent, including the Nisnevich topology. Sheaves can be furthermore generalized to stacks in the sense of Grothendieck, usually with some additional representability conditions, leading to Artin stacks and, even finer, Deligne-Mumford stacks, both often called algebraic stacks.

8

Homological Algebra

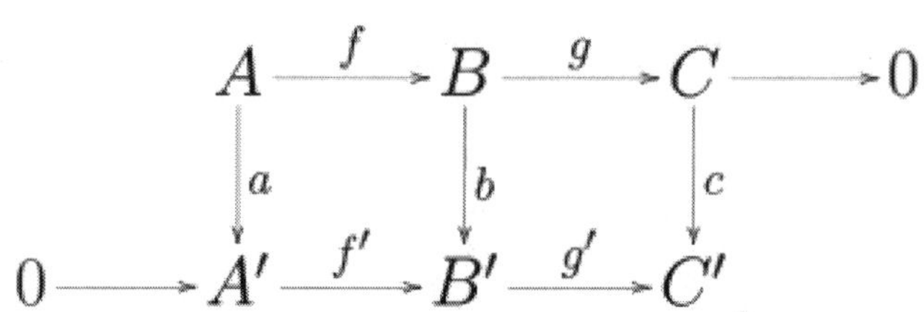

A diagram used in the snake lemma, a basic result in homological algebra. Homological algebra is the branch of mathematics that studies homology in a general algebraic setting. It is a relatively young discipline, whose origins can be traced to investigations in combinatorial topology (a precursor to algebraic topology) and abstract algebra (theory of modules and syzygies) at the end of the 19th century, chiefly by Henri Poincaré and David Hilbert. The development of homological algebra was closely intertwined with the emergence of category theory. By and large, homological algebra is the study of homological functors and the intricate algebraic structures that they entail. One quite useful and ubiquitous concept in mathematics is that of chain complexes, which can be studied through both their homology and cohomology. Homological algebra affords the means to extract information contained in these complexes and present it in the form of homological invariants of rings, modules,

objects. A powerful tool for doing this is provided by spectral sequences. From its very origins, homological algebra has played an enormous role in algebraic topology. Its influence has gradually expanded and presently includes commutative algebra, algebraic geometry, algebraic number theory, representation theory, mathematical physics, operator algebras, complex analysis, and the theory of partial differential equations. K-theory is an independent discipline which draws upon methods of homological algebra, as does the noncommutative geometry of Alain Connes.

HISTORY OF HOMOLOGICAL ALGEBRA

Homological algebra began to be studied in its most basic form in the 1800s as a branch of topology, but it wasn't until the 1940s that it became an independent subject with the study of objects such as the ext functor and the tor functor, among others.

Chain complexes and homology

The notion of chain complex is central in homological algebra. An abstract chain complex is a sequence $(C_\bullet, d_\bullet)$ of abelian groups and group homomorphisms, with the property that the composition of any two consecutive maps is zero: C ''' : ï" ö' C n + 1 ö' d n + 1 C n ö' d n C n " 1 ö' d n " 1 ï" , d n ''' d n + 1 = 0.

The elements of C_n are called n-chains and the homomorphisms d_n are called the boundary maps or differentials. The chain groups C_n may be endowed with extra structure; for example, they may be vector spaces or modules over a fixed ring R. The differentials must preserve the extra structure if it exists; for example, they must be linear maps or homomorphisms of R-modules. For notational convenience, restrict attention to abelian groups (more correctly, to the category Ab of abelian groups); a celebrated theorem by Barry Mitchell implies the results will generalize to any abelian category. Every chain complex defines two further sequences of abelian groups, the cycles Z_n = Ker d_n and the boundaries B_n = Im d_{n+1}, where Ker d and Im d denote the kernel and the image of d. Since the composition of two consecutive boundary maps is zero, these groups are embedded into each other as

$$B_n \subseteq Z_n \subseteq C_n.$$

Subgroups of abelian groups are automatically normal; therefore we can define the nth homology group $H_n(C)$ as the factor group of the n-cycles by the n-boundaries,

$$H_n(C) = Z_n / B_n = \mathrm{Ker}\, d_n / \mathrm{Im}\, d_{n+1}.$$

A chain complex is called acyclic or an exact sequence if all its homology groups are zero. Chain complexes arise in abundance in algebra and algebraic topology. For example, if X is a topological space then the singular chains $C_n(X)$ are formal linear combinations of continuous maps from the standard n-simplex into X; if K is a simplicial complex then the simplicial chains $C_n(K)$ are formal linear combinations of the n-simplices of K; if A = F/R is a presentation of an abelian group A by generators and relations, where F is a free abelian group spanned by the generators and R is the subgroup of relations, then letting $C_1(A)$ = R, $C_0(A)$ = F, and $C_n(A)$ = 0 for all other n defines a sequence of abelian groups. In all these cases, there are natural differentials d_n making C_n into a chain complex, whose homology reflects the structure of the topological space X, the simplicial complex K, or the abelian group A. In the case of topological spaces, we arrive at the notion of singular homology, which plays a fundamental role in

investigating the properties of such spaces, for example, manifolds. On a philosophical level, homological algebra teaches us that certain chain complexes associated with algebraic or geometric objects (topological spaces, simplicial complexes, R-modules) contain a lot of valuable algebraic information about them, with the homology being only the most readily available part. On a technical level, homological algebra provides the tools for manipulating complexes and extracting this information. Here are two general illustrations.

- Two objects X and Y are connected by a map f between them. Homological algebra studies the relation, induced by the map f, between chain complexes associated with X and Y and their homology. This is generalized to the case of several objects and maps connecting them. Phrased in the language of category theory, homological algebra studies the functorial properties of various constructions of chain complexes and of the homology of these complexes. An object X admits multiple descriptions (for example, as a topological space and as a simplicial complex) or the complex C ‴ (X) {\displaystyle C_{\bullet }(X)} $C_\bullet(X)$ is constructed using some
- 'presentation' of X, which involves non-canonical choices. It is important to know the effect of change in the description of X on chain complexes associated with X. Typically, the complex and its homology H ‴ (C) {\displaystyle H_{\bullet }(C)} $H_\bullet(C)$ are functorial with respect to the presentation; and the homology (although not the complex itself) is actually independent of the presentation chosen, thus it is an invariant of X.

Standard tools

Exact sequences

In the context of group theory, a sequence

$$G_0 \xrightarrow{f} G_1 \xrightarrow{f} G_2 \xrightarrow{f} \cdots \xrightarrow{f} G_n$$

of groups and group homomorphisms is called exact if the image (or range) of each homomorphism is equal to the kernel of the next:

$$\mathrm{im}(f_k) = \ker(f_{k+1}).$$

Note that the sequence of groups and homomorphisms may be either finite or infinite. A similar definition can be made for certain other algebraic structures. For example, one could have an exact sequence of vector spaces and linear maps, or of modules and module homomorphisms. More generally, the notion of an exact sequence makes sense in any category with kernels and cokernels.

Short exact sequence

The most common type of exact sequence is the short exact sequence. This is an exact sequence of the form where *f* is a monomorphism and g is an epimorphism. In this case, A is a subobject of B, and the corresponding quotient is isomorphic to C:

$$C \cong B / f(A).$$

(where f(A) = im(f)).

A short exact sequence of abelian groups may also be written as an exact sequence with five terms:

$$0 \rightarrow A \xrightarrow{f} B \xrightarrow{g} C \rightarrow 0$$

where 0 represents the zero object, such as the trivial group or a zero-dimensional vector space. The placement of the 0's forces *f* to be a monomorphism and g to be an epimorphism (see below).

Long exact sequence

A long exact sequence is an exact sequence indexed by the natural numbers.

The five lemma

Consider the following commutative diagram in any abelian category (such as the category of abelian groups or the category of vector spaces over a given field) or in the category of groups.

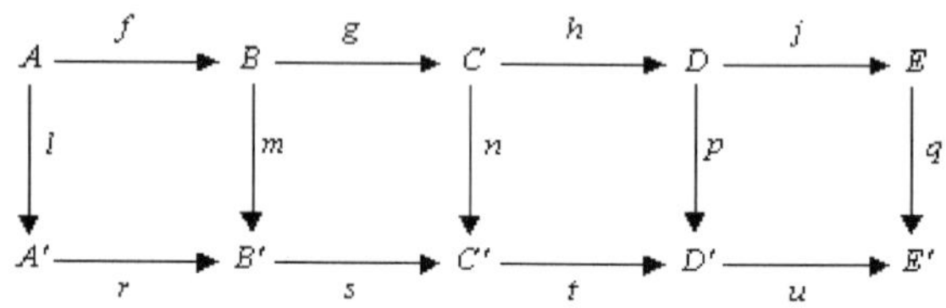

The five lemma states that, if the rows are exact, m and p are isomorphisms, l is an epimorphism, and q is a monomorphism, then n is also an isomorphism.

Five lemma

In mathematics, especially homological algebra and other applications of abelian category theory, the five lemma is an important and widely used lemma about commutative diagrams. The five lemma is not only valid for abelian categories but also works in the category of groups, for example. The five lemma can be thought of as a combination of two other theorems, the four lemmas, which are dual to each other.

Statements

Consider the following commutative diagram in any abelian category (such as the category of abelian groups or the category of vector spaces over a given field) or in the category of groups. The five lemma states that, if the rows are exact, m and p are isomorphisms, l is an epimorphism, and q is a monomorphism, then n is also an isomorphism.

The two four-lemmas state:

(1) If the rows in the commutative diagram
are exact and m and p are epimorphisms and q is a monomorphism, then n is an epimorphism.

(2) If the rows in the commutative diagram
are exact and m and p are monomorphisms and l is an epimorphism, then n is a monomorphism.

Proof

The method of proof we shall use is commonly referred to as diagram chasing. We shall prove the five lemma by individually proving each of the two four lemmas. To perform diagram chasing, we assume that we are in a category of modules over some ring, so that we may speak of elements of the objects in the diagram and think of the morphisms of the diagram as functions (in fact, homomorphisms) acting on those elements. Then a morphism is a monomorphism if and only if it is injective, and it is an epimorphism if and only if it is surjective. Similarly, to deal with exactness, we can think of kernels and images in a function-theoretic sense. The proof will still apply to any (small) abelian category because of Mitchell's embedding theorem, which states that any small abelian category can be represented as a category of modules over some ring. For the category of groups, just turn all additive notation below into multiplicative notation, and note that commutativity of abelian group is never used. So, to prove (1), assume that m and p are surjective and q is injective.

A proof of (1) in the case where t (c 2) = 0

A proof of (1) in the case where t (c 2) ‘‘‘ 0

- Let c2 be an element of C2 .
- Since p is surjective, there exists an element d in D with p(d) = t(c2).
- By commutativity of the diagram, u(p(d)) = q(j(d)).
- Since im t = ker u by exactness, 0 = u(t(c2)) = u(p(d)) = q(j(d)).
- Since q is injective, j(d) = 0, so d is in ker j = im h.
- Therefore, there exists c in C with h(c) = d.
- Then t(n(c)) = p(h(c)) = t(c2). Since t is a homomorphism, it follows that t(c2 “ n(c)) = 0.
- By exactness, c2 “ n(c) is in the image of s, so there exists b2 in B2 with s(b2) = c2 “ n(c).
- Since m is surjective, we can find b in B such that b2 = m(b).
- By commutativity, n(g(b)) = s(m(b)) = c2 “ n(c).
- Since n is a homomorphism, n(g(b) + c) = n(g(b)) + n(c) = c2 “ n(c) + n(c) = c2 .
- Therefore, n is surjective.

Then, to prove (2), assume that m and p are injective and l is surjective.

A proof of (2).

- Let c in C be such that n(c) = 0.
- t(n(c)) is then 0.
- By commutativity, p(h(c)) = 0.
- Since p is injective, h(c) = 0.
- By exactness, there is an element b of B such that g(b) = c.
- By commutativity, s(m(b)) = n(g(b)) = n(c) = 0.
- By exactness, there is then an element a2 of A2 such that r(a2) = m(b).
- Since l is surjective, there is a in A such that l(a) = a2 .
- By commutativity, m(f(a)) = r(l(a)) = m(b).
- Since m is injective, f(a) = b.
- So c = g(f(a)).
- Since the composition of g and f is trivial, c = 0.
- Therefore, n is injective.

Combining the two four lemmas now proves the entire five lemma.

Applications

The five lemma is often applied to long exact sequences: when computing homology or cohomology of a given object, one typically employs a simpler subobject whose homology/cohomology is known, and arrives at a long exact sequence which involves the unknown homology groups of the original object. This alone is often not sufficient to determine the unknown homology groups, but if one can compare the original object and sub object to well-understood ones via morphisms, then a morphism between the respective long exact sequences is induced, and the five lemma can then be used to determine the unknown homology groups.

Short five lemma

In mathematics, especially homological algebra and other applications of abelian category theory, the short five lemma is a special case of the five lemma. It states that for the following commutative diagram (in any abelian category, or in the category of groups), if the rows are short exact sequences, and if g and h are isomorphisms, then f is an isomorphism as well.

It follows immediately from the five lemma.

The essence of the lemma can be summarized as follows: if you have a homomorphism f from an object B to an object B2, and this homomorphism induces an isomorphism from a subobject A of B to a subobject A2 of B2 and also an isomorphism from the factor object B/A to B2 / A2, then f itself is an isomorphism. Note however that the existence of f (such that the diagram commutes) has to be assumed from the start; two objects B and B2 that simply have isomorphic sub- and factor objects need not themselves be isomorphic (for example, in the category of abelian groups, B could be the cyclic group of order four and B2 the Klein four-group).

The snake lemma

In an abelian category (such as the category of abelian groups or the category of vector spaces over a given field), consider a commutative diagram:

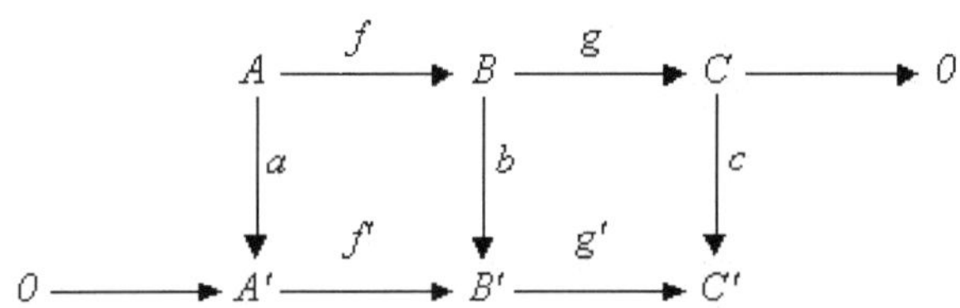

where the rows are exact sequences and 0 is the zero object. Then there is an exact sequence relating the kernels and cokernels of a, b, and c:

$$\ker a \to \ker b \to \ker c \to \operatorname{coker} a \to \operatorname{coker} b \to \operatorname{coker} c$$

Furthermore, if the morphism f is a monomorphism, then so is the morphism ker a '! ker b, and if g' is an epimorphism, then so is coker b '! coker c.

Snake lemma

The snake lemma is a tool used in mathematics, particularly homological algebra, to construct long exact sequences. The snake lemma is valid in every abelian category and is a crucial tool in homological algebra and its applications, for instance in algebraic topology. Homomorphisms constructed with its help are generally called connecting homomorphisms.

Statement

In an abelian category (such as the category of abelian groups or the category of vector spaces over a given field), consider a commutative diagram: where the rows are exact sequences and 0 is the zero object. Then there is an exact sequence relating the kernels and cokernels of a, b, and c: where d is a homomorphism, known as the connecting homomorphism. Furthermore, if the morphism f is a monomorphism, then so is the morphism ker a a ö′ ker a b {\displaystyle \operatorname {ker} a~{\color {Gray}\longrightarrow }~\operatorname {ker} b} , and if g′ is an epimorphism, then so is coker a b ö′ coker a c {\displaystyle \operatorname {coker} b~{\color {Gray}\longrightarrow }~\operatorname {coker} c} .

The cokernels here are: coker a a = A 2 / im a a , coker a b = B 2 / im a b , coker a c = C 2 / im a c . {\displaystyle \operatorname {coker} a=A′/\operatorname {im} a,\operatorname {coker} b=B′/\operatorname {im} b,\operatorname {coker} c=C′/ \operatorname {im} c.}

Explanation of the name

To see where the snake lemma gets its name, expand the diagram above as follows: and then note that the exact sequence that is the conclusion of the lemma can be drawn on this expanded diagram in the reversed "S" shape of a slithering snake.

Construction of the maps

The maps between the kernels and the maps between the cokernels are induced in a natural manner by the given (horizontal) maps because of the diagram's commutativity. The exactness of the two induced sequences follows in a straightforward way from the exactness of the rows of the original diagram. The important statement of the lemma is that a connecting homomorphism d exists which completes the exact sequence. In the case of abelian groups or modules over some ring, the map d can be constructed as follows: Pick an element x in ker c and view it as an element of C; since g is surjective, there exists y in B with g(y) = x. Because of the commutativity of the diagram, we have g′(b(y)) = c(g(y)) = c(x) = 0 (since x is in the kernel of c), and therefore b(y) is in the kernel of g′ . Since the bottom row is exact, we find an element z in A′ with f ′(z) = b(y). z is unique by injectivity of f ′. We then define d(x) = z + im(a). Now one has to check that d is well-defined (i.e., d(x) only depends on x and not on the choice of y), that it is a homomorphism, and that the resulting long sequence is indeed exact. One may routinely verify the exactness by diagram chasing (see the proof of Lemma 9.1 in). Once that is done, the theorem is proven for abelian groups or modules over a ring. For the general case, the argument may be rephrased in terms of properties of arrows and cancellation instead of elements. Alternatively, one may invoke Mitchell's embedding theorem.

Naturality

In the applications, one often needs to show that long exact sequences are "natural" (in the sense of natural transformations). This follows from the naturality of the sequence produced by the snake lemma. is a commutative diagram with exact rows, then the snake lemma can be applied twice, to the "front" and to the "back", yielding two long exact sequences; these are related by a commutative diagram of the form

In popular culture

The proof of the snake lemma is taught by Jill Clayburgh's character at the very beginning of the 1980 film It's My Turn.

Abelian categories

In mathematics, an abelian category is a category in which morphisms and objects can be added and in which kernels and cokernels exist and have desirable properties. The motivating prototype example of an abelian category is the category of abelian groups, Ab. The theory originated in a tentative attempt to unify several cohomology theories by Alexander Grothendieck. Abelian categories are very stable categories, for example they are regular and they satisfy the snake lemma. The class of Abelian categories is closed under several categorical constructions, for example, the category of chain complexes of an Abelian category, or the category of functors from a small category to an Abelian category are Abelian as well. These stability properties make them inevitable in homological algebra and beyond; the theory has major applications in algebraic geometry, cohomology and pure category theory. Abelian categories are named after Niels Henrik Abel.

More concretely, a category is abelian if

- it has a zero object,
- it has all binary products and binary coproducts, and
- it has all kernels and cokernels.
- all monomorphisms and epimorphisms are normal.

Spectral sequence

Fix an abelian category, such as a category of modules over a ring. A spectral sequence is a choice of a nonnegative integer r_0 and a collection of three sequences:

1. For all integers r e" r_0, an object E_r, called a sheet (as in a sheet of paper), or sometimes a page or a term,
2. Endomorphisms d_r : E_r '! E_r satisfying d_r o d_r = 0, called boundary maps or differentials,
3. Isomorphisms of E_{r+1} with $H(E_r)$, the homology of E_r with respect to d_r.

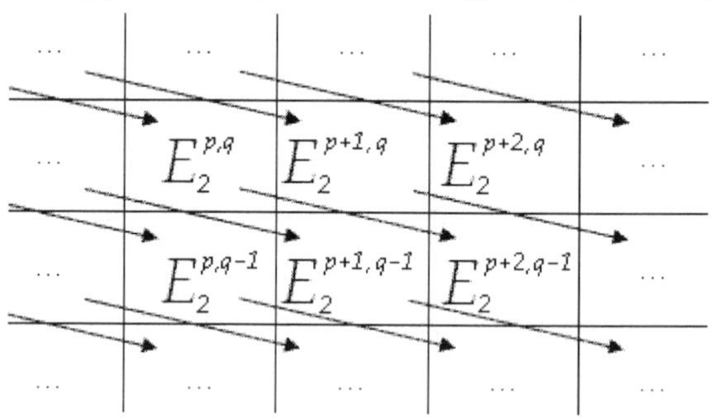

The E_2 sheet of a cohomological spectral sequence

A doubly graded spectral sequence has a tremendous amount of data to keep track of, but there is a common visualization technique which makes the structure of the spectral sequence clearer. We have three indices, r, p, and q. For each r, imagine that we have a sheet of graph paper. On this sheet, we will take p to be the horizontal direction and q to be the vertical direction. At each lattice point we have the object E r p , q {\displaystyle E_{r}^{p,q}} .

It is very common for n = p + q to be another natural index in the spectral sequence. n runs diagonally, northwest to southeast, across each sheet. In the homological case, the differentials have bidegree (“r, r ” 1), so they decrease n by one. In the cohomological case, n is increased by one. When r is zero, the differential moves objects one space down or up. This is similar to the differential on a chain complex. When r is one, the differential moves objects one space to the left or right. When r is two, the differential moves objects just like a knight’s move in chess. For higher r, the differential acts like a generalized knight’s move.

Derived functor

Suppose we are given a covariant left exact functor F : A '! B between two abelian categories A and B. If 0 '! A '! B '! C '! 0 is a short exact sequence in A, then applying F yields the exact sequence 0 '! F(A) '! F(B) '! F(C) and one could ask how to continue this sequence to the right to form a long exact sequence. Strictly speaking, this question is ill-posed, since there are always numerous different ways to continue a given exact sequence to the right. But it turns out that (if A is “nice” enough) there is one canonical way of doing so, given by the right derived functors of F. For every ie”1, there is a functor R^iF: A '! B, and the above sequence continues like so: 0 '! F(A) '! F(B) '! F(C) '! R^1F(A) '! R^1F(B) '! R^1F(C) '! R^2F(A) '! R^2F(B) '! .... From this we see that F is an exact functor if and only if R^1F = 0; so in a sense the right derived functors of F measure “how far” F is from being exact.

Foundational aspects

Cohomology theories have been defined for many different objects such as topological spaces, sheaves, groups, rings, Lie algebras, and C*-algebras. The study of modern algebraic geometry would be almost unthinkable without sheaf cohomology. Central to homological algebra is the notion of exact sequence; these can be used to perform actual calculations. A classical tool of homological algebra is that of derived functor; the most basic examples are functors Ext and Tor. With a diverse set of applications in mind, it was natural to try to put the whole subject on a uniform basis. There were several attempts before the subject settled down. An approximate history can be stated as follows:

- Cartan-Eilenberg: In their 1956 book “Homological Algebra”, these authors used projective and injective module resolutions.
- ‘Tohoku’: The approach in a celebrated paper by Alexander Grothendieck which appeared in the Second Series of the Tohoku Mathematical Journal in 1957, using the abelian category concept (to include sheaves of abelian groups).
- The derived category of Grothendieck and Verdier. Derived categories date back to Verdier’s 1967 thesis. They are examples of triangulated categories used in a number of modern theories.

These move from computability to generality.

The computational sledgehammer par excellence is the spectral sequence; these are essential in the Cartan-Eilenberg and Tohoku approaches where they are needed, for instance, to compute the derived functors of a composition of two functors. Spectral sequences are less essential in the derived category approach, but still play a role whenever concrete computations are necessary. There have been attempts at ‘non-commutative’ theories which extend first cohomology as torsors (important in Galois cohomology).

9

Representation Theory

Representation theory is a branch of mathematics that studies abstract algebraic structures by representing their elements as linear transformations of vector spaces, and studies modules over these abstract algebraic structures. In essence, a representation makes an abstract algebraic object more concrete by describing its elements by matrices and the algebraic operations in terms of matrix addition and matrix multiplication. The algebraic objects amenable to such a description include groups, associative algebras and Lie algebras. The most prominent of these (and historically the first) is the representation theory of groups, in which elements of a group are represented by invertible matrices in such a way that the group operation is matrix multiplication.

Representation theory is a useful method because it reduces problems in abstract algebra to problems in linear algebra, a subject that is well understood. Furthermore, the vector space on which a group (for example) is represented can be infinite-dimensional, and by allowing it to be, for instance, a Hilbert space, methods of analysis can be applied to the theory of groups. Representation theory is also important in physics because, for example, it describes how the symmetry group of a physical system affects the solutions of equations describing that system. Representation theory is pervasive across fields of mathematics, for two reasons. First, the applications of representation theory are diverse: in addition to its impact on algebra, representation theory:

- illuminates and generalizes Fourier analysis via harmonic analysis,
- is connected to geometry via invariant theory and the Erlangen program,
- has an impact in number theory via automorphic forms and the Langlands program.

Secondly, there are diverse approaches to representation theory. The same objects can be studied using methods from algebraic geometry, module theory, analytic number theory, differential geometry, operator theory, algebraic combinatorics and topology. The success of representation theory has led to numerous generalizations. One of the most general is in category theory. The algebraic objects to which representation theory applies can be viewed as particular kinds of categories, and the representations as functors from the object category to the category of vector spaces. This description points to two obvious generalizations: first, the algebraic objects can be replaced by more general categories; second, the target category of vector spaces can be replaced by other well-understood categories.

Definitions and concepts

Let V be a vector space over a field F. For instance, suppose V is R^n or C^n, the standard n-dimensional space of column vectors over the real or complex numbers respectively. In this case, the idea of representation theory is to do abstract algebra concretely by using n × n matrices of real or complex numbers. There are three main sorts of algebraic objects for which this can be done: groups, associative algebras and Lie algebras.

- The set of all invertible n × n matrices is a group under matrix multiplication and the representation theory of groups analyzes a group by describing ("representing") its elements in terms of invertible matrices.
- Matrix addition and multiplication make the set of all n × n matrices into an associative algebra and hence there is a corresponding representation theory of associative algebras.
- If we replace matrix multiplication MN by the matrix commutator MN " NM, then the n × n matrices become instead a Lie algebra, leading to a representation theory of Lie algebras.

This generalizes to any field F and any vector space V over F, with linear maps replacing matrices and composition replacing matrix multiplication: there is a group GL(V,F) of automorphisms of V, an associative algebra $End_F(V)$ of all endomorphisms of V, and a corresponding Lie algebra gl(V,F).

Definition3

There are two ways to say what a representation is. The first uses the idea of an action, generalizing the way that matrices act on column vectors by matrix multiplication. A representation of a group G or (associative or Lie) algebra A on a vector space V is a map

$$\Phi : G \times V \rightarrow V \quad \text{or} \quad \Phi : A \times V \rightarrow V$$

with two properties. First, for any g in G (or a in A), the map

$$\begin{aligned} \Phi(g) : V & \rightarrow V \\ v & \mapsto \Phi(g, v) \end{aligned}$$

is linear (over F). Second, if we introduce the notation g · v for Ö Φ (g, v),

then for any g_1, g_2 in G and v in V:

$$(1) \quad e \cdot v = v$$

where e is the identity element of G and g_1g_2 is the product in G. The requirement for associative algebras is analogous, except that associative algebras do not always have an identity element, in which case equation (1) is ignored. Equation (2) is an abstract expression of the associativity of matrix multiplication. This doesn't hold for the matrix commutator and also there is no identity element for the commutator. Hence for Lie algebras, the only requirement is that for any x_1, x_2 in A and v in V: where $[x_1, x_2]$ is the Lie bracket, which generalizes the matrix commutator MN " NM.

The second way to define a representation focuses on the map ö sending g in G to a linear map ö(g): V ′! V, which satisfies

$$\varphi(g_1 g_2) = \varphi(g_1) \circ \varphi(g_2) \quad \text{for all } g_1, g_2 \in G$$

and similarly in the other cases. This approach is both more concise and more abstract. From this point of view:

- a representation of a group G on a vector space V is a group homomorphism ö: G ′! GL(V,F);
- a representation of an associative algebra A on a vector space V is an algebra homomorphism ö: A ′! $End_F(V)$;
- a representation of a Lie algebra a on a vector space V is a Lie algebra homomorphism ö: a ′! gl(V,F).

Terminology

The vector space V is called the representation space of ö and its dimension (if finite) is called the dimension of the representation (sometimes degree, as in). It is also common practice to refer to V itself as the representation when the homomorphism ö is clear from the context; otherwise the notation (V,ö) can be used to denote a representation. When V is of finite dimension n, one can choose a basis for V to identify V with F^n and hence recover a matrix representation with entries in the field F. An effective or faithful representation is a representation (V,ö) for which the homomorphism ö is injective.

Equivariant maps and isomorphisms

If V and W are vector spaces over F, equipped with representations ö and ø of a group G, then an equivariant map from V to W is a linear map á: V ′! W such that

$$\alpha(g \cdot v) = g \cdot \alpha(v)$$

for all g in G and v in V. In terms of ö: G ′! GL(V) and ø: G ′! GL(W), this means

$$\alpha \circ \varphi(g) = \psi(g) \circ \alpha$$

for all g in G, that is, the following diagram commutes:

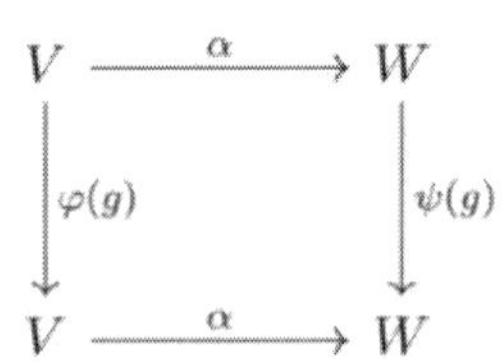

Equivariant maps for representations of an associative or Lie algebra are defined similarly. If á is invertible, then it is said to be an isomorphism, in which case V and W (or, more precisely, ö and ø) are isomorphic representations, also phrased as equivalent representations. An equivariant map is often called an intertwining map of representations. Also, in the case of a group G, it is on occasion called a G-map.

Isomorphic representations are, for practical purposes, "the same"; they provide the same information about the group or algebra being represented. Representation theory therefore seeks to classify representations up to isomorphism. Subrepresentations, quotients, and irreducible representations If (V,ø) is a representation of (say) a group G, and W is a linear subspace of V that is preserved by the action of G in the sense that g · w " W for all w " W (Serre calls these W stable under G), then W is called a subrepresentation: by defining ö(g) to be the restriction of ø(g) to W, (W, ö) is a representation of G and the inclusion of W into V is an equivariant map. The quotient space V/W can also be made into a representation of G.

If V has exactly two subrepresentations, namely the trivial subspace {0} and V itself, then the representation is said to be irreducible; if V has a proper nontrivial subrepresentation, the representation is said to be reducible. The definition of an irreducible representation implies Schur's lemma: an equivariant map á: V '! W between irreducible representations is either the zero map or an isomorphism, since its kernel and image are subrepresentations. In particular, when V = W, this shows that the equivariant endomorphisms of V form an associative division algebra over the underlying field F. If F is algebraically closed, the only equivariant endomorphisms of an irreducible representation are the scalar multiples of the identity. Irreducible representations are the building blocks of representation theory: if a representation V is not irreducible then it is built from a subrepresentation and a quotient that are both "simpler" in some sense; for instance, if V is finite-dimensional, then both the subrepresentation and the quotient have smaller dimension.

Direct sums and indecomposable representations

If (V,ö) and (W,ø) are representations of (say) a group G, then the direct sum of V and W is a representation, in a canonical way, via the equation

$$g \cdot (v, w) = (g \cdot v, g \cdot w).$$

The direct sum of two representations carries no more information about the group G than the two representations do individually. If a representation is the direct sum of two proper nontrivial subrepresentations, it is said to be decomposable. Otherwise, it is said to be indecomposable.

Complete reducibility

In favorable circumstances, every finite-dimensional representation is a direct sum of irreducible representations: such representations are said to be semisimple. In this case, it suffices to understand only the irreducible representations. Examples where this "complete reducibility" phenomenon occur include finite and compact groups, and semisimple Lie algebras. In cases where complete reducibility does not hold, one must understand how indecomposable representations can be built from irreducible representations as extensions of a quotient by a subrepresentation. Representation theory is notable for the number of branches it has, and the diversity of the approaches to studying representations of groups and algebras. Although, all the theories have in common the basic concepts discussed already, they differ considerably in detail. The differences are at least 3-fold:

1. Representation theory depends upon the type of algebraic object being represented. There are several different classes of groups, associative algebras and Lie algebras, and their representation theories all have an individual flavour.
2 Representation theory depends upon the nature of the vector space on which the algebraic object is represented. The most important distinction is between

finite-dimensional representations and infinite-dimensional ones. In the infinite-dimensional case, additional structures are important (for example,
3. whether or not the space is a Hilbert space, Banach space, etc.). Additional algebraic structures can also be imposed in the finite-dimensional case.
4. Representation theory depends upon the type of field over which the vector space is defined. The most important cases are the field of complex numbers, the field of real numbers, finite fields, and fields of p-adic numbers. Additional difficulties arise for fields of positive characteristic and for fields that are not algebraically closed.

Finite groups

Group representations are a very important tool in the study of finite groups. They also arise in the applications of finite group theory to geometry and crystallography. Representations of finite groups exhibit many of the features of the general theory and point the way to other branches and topics in representation theory.

Over a field of characteristic zero, the representation of a finite group G has a number of convenient properties. First, the representations of G are semisimple (completely reducible). This is a consequence of Maschke's theorem, which states that any subrepresentation V of a G-representation W has a G-invariant complement. One proof is to choose any projection δ from W to V and replace it by its average δ_G defined by δ_G is equivariant, and its kernel is the required complement. The finite-dimensional G-representations can be understood using character theory: the character of a representation ö: G '! GL(V) is the class function $\div_{ö}$: G '! F defined by where T r {\displaystyle \mathrm {Tr} } is the trace. An irreducible representation of G is completely determined by its character. Maschke's theorem holds more generally for fields of positive characteristic p, such as the finite fields, as long as the prime p is coprime to the order of G. When p and | G | have a common factor, there are G-representations that are not semisimple, which are studied in a subbranch called modular representation theory.

Inner product on V in the sense that for all g in G and v, w in W. Hence any G-representation is unitary. Unitary representations are automatically semisimple, since Maschke's result can be proven by taking the orthogonal complement of a subrepresentation. When studying representations of groups that are not finite, the unitary representations provide a good generalization of the real and complex representations of a finite group. Results such as Maschke's theorem and the unitary property that rely on averaging can be generalized to more general groups by replacing the average with an integral, provided that a suitable notion of integral can be defined. This can be done for compact topological groups (including compact Lie groups), using Haar measure, and the resulting theory is known as abstract harmonic analysis. Over arbitrary fields, another class of finite groups that have a good representation theory are the finite groups of Lie type. Important examples are linear algebraic groups over finite fields. The representation theory of linear algebraic groups and Lie groups extends these examples to infinite-dimensional groups, the latter being intimately related to Lie algebra representations. The importance of character theory for finite groups has an analogue in the theory of weights for representations of Lie groups and Lie algebras. Representations of a finite group G are also linked directly to algebra representations via the group algebra F[G], which is a vector space over F with the elements of G as a basis, equipped with the multiplication operation defined by the group operation, linearity, and the requirement that the group operation and scalar multiplication commute.

Modular representations

Modular representations of a finite group G are representations over a field whose characteristic is not coprime to |G|, so that Maschke's theorem no longer holds (because |G| is not invertible in F and so one cannot divide by it). Nevertheless, Richard Brauer extended much of character theory to modular representations, and this theory played an important role in early progress towards the classification of finite simple groups, especially for simple groups whose characterization was not amenable to purely group-theoretic methods because their Sylow 2-subgroups were "too small". As well as having applications to group theory, modular representations arise naturally in other branches of mathematics, such as algebraic geometry, coding theory, combinatorics and number theory.

Modular representation theory

Modular representation theory is a branch of mathematics, and that part of representation theory that studies linear representations of finite groups over a field K of positive characteristic p, necessarily a prime number. As well as having applications to group theory, modular representations arise naturally in other branches of mathematics, such as algebraic geometry, coding theory[citation needed], combinatorics and number theory. Within finite group theory, character-theoretic results proved by Richard Brauer using modular representation theory played an important role in early progress towards the classification of finite simple groups, especially for simple groups whose characterization was not amenable to purely group-theoretic methods because their Sylow 2 subgroups were too small in an appropriate sense. Also, a general result on embedding of elements of order 2 in finite groups called the Z* theorem, proved by George Glauberman using the theory developed by Brauer, was particularly useful in the classification program. If the characteristic p of K does not divide the order |G|, then modular representations are completely reducible, as with ordinary (characteristic 0) representations, by virtue of Maschke's theorem. In the other case, when |G| a" 0 mod p, the process of averaging over the group needed to prove Maschke's theorem breaks down, and representations need not be completely reducible. Much of the discussion below implicitly assumes that the field K is sufficiently large (for example, K algebraically closed suffices), otherwise some statements need refinement.

History

The earliest work on representation theory over finite fields is by Dickson (1902) who showed that when p does not divide the order of the group, the representation theory is similar to that in characteristic 0. He also investigated modular invariants of some finite groups. The systematic study of modular representations, when the characteristic p divides the order of the group, was started by Brauer (1935) and was continued by him for the next few decades.

Example

Finding a representation of the cyclic group of two elements over F_2 is equivalent to the problem of finding matrices whose square is the identity matrix. Over every field of characteristic other than 2, there is always a basis such that the matrix can be written as a diagonal matrix with only 1 or "1 occurring on the diagonal, such as Over F_2, there are many other possible matrices, such as Over an algebraically closed field of positive characteristic, the representation theory of a finite cyclic group is fully explained by the theory of the Jordan normal form. Non-diagonal Jordan forms occur when the characteristic divides the order of the group.

Ring theory interpretation

Given a field K and a finite group G, the group algebra K[G] (which is the K-vector space with K-basis consisting of the elements of G, endowed with algebra multiplication by extending the multiplication of G by linearity) is an Artinian ring. When the order of G is divisible by the characteristic of K, the group algebra is not semisimple, hence has non-zero Jacobson radical. In that case, there are finite-dimensional modules for the group algebra that are not projective modules. By contrast, in the characteristic 0 case every irreducible representation is a direct summand of the regular representation, hence is projective.

Brauer characters

Modular representation theory was developed by Richard Brauer from about 1940 onwards to study in greater depth the relationships between the characteristic p representation theory, ordinary character theory and structure of G, especially as the latter relates to the embedding of, and relationships between, its p-subgroups. Such results can be applied in group theory to problems not directly phrased in terms of representations. Brauer introduced the notion now known as the Brauer character. When K is algebraically closed of positive characteristic p, there is a bijection between roots of unity in K and complex roots of unity of order prime to p. Once a choice of such a bijection is fixed, the Brauer character of a representation assigns to each group element of order coprime to p the sum of complex roots of unity corresponding to the eigenvalues (including multiplicities) of that element in the given representation.

The Brauer character of a representation determines its composition factors but not, in general, its equivalence type. The irreducible Brauer characters are those afforded by the simple modules. These are integral (though not necessarily non-negative) combinations of the restrictions to elements of order coprime to p of the ordinary irreducible characters. Conversely, the restriction to the elements of order prime to p of each ordinary irreducible character is uniquely expressible as a non-negative integer combination of irreducible Brauer characters.

Reduction (mod p)

In the theory initially developed by Brauer, the link between ordinary representation theory and modular representation theory is best exemplified by considering the group algebra of the group G over a complete discrete valuation ring R with residue field K of positive characteristic p and field of fractions F of characteristic 0, such as the p-adic integers. The structure of R[G] is closely related both to the structure of the group algebra K[G] and to the structure of the semisimple group algebra F[G], and there is much interplay between the module theory of the three algebras. Each R[G]-module naturally gives rise to an F[G]-module, and, by a process often known informally as reduction (mod p), to a K[G]-module. On the other hand, since R is a principal ideal domain, each finite-dimensional F[G]-module arises by extension of scalars from an R[G]-module. In general, however, not all K[G]-modules arise as reductions (mod p) of R[G]-modules. Those that do are liftable.

Number of simple modules

In ordinary representation theory, the number of simple modules k(G) is equal to the number of conjugacy classes of G. In the modular case, the number l(G) of simple modules is equal to the number of conjugacy classes whose elements have order coprime to the relevant prime p, the so-called p-regular classes.

Blocks and the structure of the group algebra

In modular representation theory, while Maschke's theorem does not hold when the characteristic divides the group order, the group algebra may be decomposed as the direct sum of a maximal collection of two-sided ideals known as blocks. When the field F has characteristic 0, or characteristic coprime to the group order, there is still such a decomposition of the group algebra F[G] as a sum of blocks (one for each isomorphism type of simple module), but the situation is relatively transparent when F is sufficiently large: each block is a full matrix algebra over F, the endomorphism ring of the vector space underlying the associated simple module.

To obtain the blocks, the identity element of the group G is decomposed as a sum of primitive idempotents in Z(R[G]), the center of the group algebra over the maximal order R of F. The block corresponding to the primitive idempotent e is the two-sided ideal e R[G]. For each indecomposable R[G]-module, there is only one such primitive idempotent that does not annihilate it, and the module is said to belong to (or to be in) the corresponding block (in which case, all its composition factors also belong to that block). In particular, each simple module belongs to a unique block. Each ordinary irreducible character may also be assigned to a unique block according to its decomposition as a sum of irreducible Brauer characters. The block containing the trivial module is known as the principal block.

Projective modules

In ordinary representation theory, every indecomposable module is irreducible, and so every module is projective. However, the simple modules with characteristic dividing the group order are rarely projective. Indeed, if a simple module is projective, then it is the only simple module in its block, which is then isomorphic to the endomorphism algebra of the underlying vector space, a full matrix algebra. In that case, the block is said to have 'defect 0'. Generally, the structure of projective modules is difficult to determine.

For the group algebra of a finite group, the (isomorphism types of) projective indecomposable modules are in a one-to-one correspondence with the (isomorphism types of) simple modules: the socle of each projective indecomposable is simple (and isomorphic to the top), and this affords the bijection, as non-isomorphic projective indecomposables have non-isomorphic socles. The multiplicity of a projective indecomposable module as a summand of the group algebra (viewed as the regular module) is the dimension of its socle (for large enough fields of characteristic zero, this recovers the fact that each simple module occurs with multiplicity equal to its dimension as a direct summand of the regular module).

Each projective indecomposable module (and hence each projective module) in positive characteristic p may be lifted to a module in characteristic 0. Using the ring R as above, with residue field K, the identity element of G may be decomposed as a sum of mutually orthogonal primitive idempotents (not necessarily central) of K[G]. Each projective indecomposable K[G]-module is isomorphic to e.K[G] for a primitive idempotent e that occurs in this decomposition. The idempotent e lifts to a primitive idempotent, say E, of R[G], and the left module E.R[G] has reduction (mod p) isomorphic to e.K[G].

Some orthogonality relations for Brauer characters

When a projective module is lifted, the associated character vanishes on all elements of order divisible by p, and (with consistent choice of roots of unity), agrees with the Brauer character of the original characteristic p module on p-regular elements. The (usual character-ring) inner product of the Brauer character of a projective indecomposable with any other Brauer character can thus be defined: this is 0 if the second Brauer character is that of the socle of a non-isomorphic projective indecomposable, and 1 if the second Brauer character is that of its own socle. The multiplicity of an ordinary irreducible character in the character of the lift of a projective indecomposable is equal to the number of occurrences of the Brauer character of the socle of the projective indecomposable when the restriction of the ordinary character to p-regular elements is expressed as a sum of irreducible Brauer characters.

Decomposition matrix and Cartan matrix

The composition factors of the projective indecomposable modules may be calculated as follows: Given the ordinary irreducible and irreducible Brauer characters of a particular finite group, the irreducible ordinary characters may be decomposed as non-negative integer combinations of the irreducible Brauer characters. The integers involved can be placed in a matrix, with the ordinary irreducible characters assigned rows and the irreducible Brauer characters assigned columns. This is referred to as the decomposition matrix, and is frequently labelled D. It is customary to place the trivial ordinary and Brauer characters in the first row and column respectively. The product of the transpose of D with D itself results in the Cartan matrix, usually denoted C; this is a symmetric matrix such that the entries in its j-th row are the multiplicities of the respective simple modules as composition factors of the j-th projective indecomposable module. The Cartan matrix is non-singular; in fact, its determinant is a power of the characteristic of K. Since a projective indecomposable module in a given block has all its composition factors in that same block, each block has its own Cartan matrix.

Defect groups

To each block B of the group algebra K[G], Brauer associated a certain p-subgroup, known as its defect group (where p is the characteristic of K). Formally, it is the largest p-subgroup D of G for which there is a Brauer correspondent of B for the subgroup D C G (D) {\displaystyle DC_{G}(D)} , where C G (D) {\displaystyle C_{G}(D)} is the centralizer of D in G. The defect group of a block is unique up to conjugacy and has a strong influence on the structure of the block. For example, if the defect group is trivial, then the block contains just one simple module, just one ordinary character, the ordinary and Brauer irreducible characters agree on elements of order prime to the relevant characteristic p, and the simple module is projective. At the other extreme, when K has characteristic p, the Sylow p-subgroup of the finite group G is a defect group for the principal block of K[G].

The order of the defect group of a block has many arithmetical characterizations related to representation theory. It is the largest invariant factor of the Cartan matrix of the block, and occurs with multiplicity one. Also, the power of p dividing the index of the defect group of a block is the greatest common divisor of the powers of p dividing the dimensions of the simple modules in that block, and this coincides with the greatest common divisor of the powers of p dividing the degrees of the ordinary irreducible characters in that block.

Other relationships between the defect group of a block and character theory include Brauer's result that if no conjugate of the p-part of a group element g is in the defect group of a given block, then each irreducible character in that block vanishes at g. This is one of many consequences of Brauer's second main theorem. The defect group of a block also has several characterizations in the more module-theoretic approach to block theory, building on the work of J. A. Green, which associates a p-subgroup known as the vertex to an indecomposable module, defined in terms of relative projectivity of the module. For example, the vertex of each indecomposable module in a block is contained (up to conjugacy) in the defect group of the block, and no proper subgroup of the defect group has that property. Brauer's first main theorem states that the number of blocks of a finite group that have a given p-subgroup as defect group is the same as the corresponding number for the normalizer in the group of that p-subgroup.

The easiest block structure to analyse with non-trivial defect group is when the latter is cyclic. Then there are only finitely many isomorphism types of indecomposable modules in the block, and the structure of the block is by now well understood, by virtue of work of Brauer, E.C. Dade, J.A.Green and J.G.Thompson, among others. In all other cases, there are infinitely many isomorphism types of indecomposable modules in the block. Blocks whose defect groups are not cyclic can be divided into two types: tame and wild. The tame blocks (which only occur for the prime 2) have as a defect group a dihedral group, semidihedral group or (generalized) quaternion group, and their structure has been broadly determined in a series of papers by Karin Erdmann. The indecomposable modules in wild blocks are extremely difficult to classify, even in principle.

Unitary representations

A unitary representation of a group G is a linear representation ö of G on a real or (usually) complex Hilbert space V such that ö(g) is a unitary operator for every g " G. Such representations have been widely applied in quantum mechanics since the 1920s, thanks in particular to the influence of Hermann Weyl, and this has inspired the development of the theory, most notably through the analysis of representations of the Poincaré group by Eugene Wigner. One of the pioneers in constructing a general theory of unitary representations (for any group G rather than just for particular groups useful in applications) was George Mackey, and an extensive theory was developed by Harish-Chandra and others in the 1950s and 1960s.

A major goal is to describe the "unitary dual", the space of irreducible unitary representations of G. The theory is most well-developed in the case that G is a locally compact (Hausdorff) topological group and the representations are strongly continuous. For G abelian, the unitary dual is just the space of characters, while for G compact, the Peter–Weyl theorem shows that the irreducible unitary representations are finite-dimensional and the unitary dual is discrete. For example, if G is the circle group S^1, then the characters are given by integers, and the unitary dual is Z. For non-compact G, the question of which representations are unitary is a subtle one. Although irreducible unitary representations must be "admissible" (as Harish-Chandra modules) and it is easy to detect which admissible representations have a nondegenerate invariant sesquilinear form, it is hard to determine when this form is positive definite. An effective description of the unitary dual, even for relatively well-behaved groups such as real reductive Lie groups (discussed below), remains an important open problem in representation theory. It has been solved for many particular groups, such as SL(2,R) and the Lorentz group.

Harmonic analysis

The duality between the circle group S^1 and the integers Z, or more generally, between a torus T^n and Z^n is well known in analysis as the theory of Fourier series, and the Fourier transform similarly expresses the fact that the space of characters on a real vector space is the dual vector space. Thus unitary representation theory and harmonic analysis are intimately related, and abstract harmonic analysis exploits this relationship, by developing the analysis of functions on locally compact topological groups and related spaces. A major goal is to provide a general form of the Fourier transform and the Plancherel theorem. This is done by constructing a measure on the unitary dual and an isomorphism between the regular representation of G on the space $L^2(G)$ of square integrable functions on G and its representation on the space of L^2 functions on the unitary dual. Pontrjagin duality and the Peter–Weyl theorem achieve this for abelian and compact G respectively.

Another approach involves considering all unitary representations, not just the irreducible ones. These form a category, and Tannaka–Krein duality provides a way to recover a compact group from its category of unitary representations. If the group is neither abelian nor compact, no general theory is known with an analogue of the Plancherel theorem or Fourier inversion, although Alexander Grothendieck extended Tannaka–Krein duality to a relationship between linear algebraic groups and tannakian categories. Harmonic analysis has also been extended from the analysis of functions on a group G to functions on homogeneous spaces for G. The theory is particularly well developed for symmetric spaces and provides a theory of automorphic forms (discussed below).

Harmonic analysis

The harmonics of color. The harmonic-analysis chart shows how the different wavelengths interact with red light. At a difference of ë/2 (half wavelength), red is perfectly in sync with its second harmonic in the ultraviolet. All other wavelengths in the visual spectrum have less than a ë/2 difference between them, forming harmonic oscillations in the combined waves. At ë/14, the oscillations cycle every 14th wave, while at ë/8 they will cycle every 8th. The oscillations are most rapid at ë/4, cycling every 4th wave, while at ë/3 they cycle every 7th wave, and at ë/2.5 they cycle every 13th. The lower section shows how the ë/4 harmonic interacts in visible light (green and red), as photographed in an optical flat. Harmonic analysis is a branch of mathematics concerned with the representation of functions or signals as the superposition of basic waves, and the study of and generalization of the notions of Fourier series and Fourier transforms (i.e. an extended form of Fourier analysis). In the past two centuries, it has become a vast subject with applications in areas as diverse as number theory, representation theory, signal processing, quantum mechanics, tidal analysis and neuroscience. The term "harmonics" originated as the Ancient Greek word harmonikos, meaning "skilled in music". In physical eigenvalue problems, it began to mean waves whose frequencies are integer multiples of one another, as are the frequencies of the harmonics of music notes, but the term has been generalized beyond its original meaning. The classical Fourier transform on R^n is still an area of ongoing research, particularly concerning Fourier transformation on more general objects such as tempered distributions. For instance, if we impose some requirements on a distribution f, we can attempt to translate these requirements in terms of the Fourier transform of f. The Paley–Wiener theorem is an example of this. The Paley–Wiener theorem immediately implies that if f is a nonzero distribution of compact support (these include functions of compact support), then its Fourier transform is never compactly supported. This is a very elementary form of an uncertainty principle in a harmonic-analysis setting. See also: Convergence of Fourier series. Fourier series can be conveniently studied in the context of Hilbert spaces, which provides a connection between harmonic analysis and functional analysis.

Applied harmonic analysis

Bass-guitar time signal of open-string A note (55 Hz) Fourier transform of bass-guitar time signal of open-string A note (55 Hz) Many applications of harmonic analysis in science and engineering begin with the idea or hypothesis that a phenomenon or signal is composed of a sum of individual oscillatory components. Ocean tides and vibrating strings are common and simple examples. The theoretical approach is often to try to describe the system by a differential equation or system of equations to predict the essential features, including the amplitude, frequency, and phases of the oscillatory components. The specific equations depend on the field, but theories generally try to select equations that represent major principles that are applicable. The experimental approach is usually to acquire data that accurately quantifies the phenomenon. For example, in a study of tides, the experimentalist would acquire samples of water depth as a function of time at closely enough spaced intervals to see each oscillation and over a long enough duration that multiple oscillatory periods are likely included. In a study on vibrating strings, it is common for the experimentalist to acquire a sound waveform sampled at a rate at least twice that of the highest frequency expected and for a duration many times the period of the lowest frequency expected.

For example, the top signal at the right is a sound waveform of a bass guitar playing an open string corresponding to an A note with a fundamental frequency of 55 Hz. The waveform appears oscillatory, but it is more complex than a simple sine wave, indicating the presence of additional waves. The different wave components contributing to the sound can be revealed by applying a mathematical analysis technique known as the Fourier transform, the result of which is shown in the lower figure. Note that there is a prominent peak at 55 Hz, but that there are other peaks at 110 Hz, 165 Hz, and at other frequencies corresponding to integer multiples of 55 Hz. In this case, 55 Hz is identified as the fundamental frequency of the string vibration, and the integer multiples are known as harmonics.

Abstract harmonic analysis

Lord Kelvin's harmonic analyser 1876, Hunterian Museum, Glasgow One of the most modern branches of harmonic analysis, having its roots in the mid-20th century, is analysis on topological groups. The core motivating ideas are the various Fourier transforms, which can be generalized to a transform of functions defined on Hausdorff locally compact topological groups. The theory for abelian locally compact groups is called Pontryagin duality. Harmonic analysis studies the properties of that duality and Fourier transform and attempts to extend those features to different settings, for instance, to the case of non-abelian Lie groups. For general non-abelian locally compact groups, harmonic analysis is closely related to the theory of unitary group representations. For compact groups, the Peter–Weyl theorem explains how one may get harmonics by choosing one irreducible representation out of each equivalence class of representations. This choice of harmonics enjoys some of the useful properties of the classical Fourier transform in terms of carrying convolutions to pointwise products, or otherwise showing a certain understanding of the underlying group structure. See also: Non-commutative harmonic analysis. If the group is neither abelian nor compact, no general satisfactory theory is currently known ("satisfactory" means at least as strong as the Plancherel theorem). However, many specific cases have been analyzed, for example SL_n. In this

case, representations in infinite dimensions play a crucial role.

Other branches

- Study of the eigenvalues and eigenvectors of the Laplacian on domains, manifolds, and (to a lesser extent) graphs is also considered a branch of harmonic analysis. See e.g., hearing the shape of a drum.
- Harmonic analysis on Euclidean spaces deals with properties of the Fourier transform on R^n that have no analog on general groups. For example, the fact that the Fourier transform is rotation-invariant. Decomposing the Fourier transform into its radial and spherical components leads to topics such as Bessel functions and spherical harmonics.
- Harmonic analysis on tube domains is concerned with generalizing properties of Hardy spaces to higher dimensions.

Lie groups

A Lie group is a group that is also a smooth manifold. Many classical groups of matrices over the real or complex numbers are Lie groups. Many of the groups important in physics and chemistry are Lie groups, and their representation theory is crucial to the application of group theory in those fields. The representation theory of Lie groups can be developed first by considering the compact groups, to which results of compact representation theory apply. This theory can be extended to finite-dimensional representations of semisimple Lie groups using Weyl's unitary trick: each semisimple real Lie group G has a complexification, which is a complex Lie group G^c, and this complex Lie group has a maximal compact subgroup K. The finite-dimensional representations of G closely correspond to those of K. A general Lie group is a semidirect product of a solvable Lie group and a semisimple Lie group (the Levi decomposition). The classification of representations of solvable Lie groups is intractable in general, but often easy in practical cases. Representations of semidirect products can then be analysed by means of general results called Mackey theory, which is a generalization of the methods used in Wigner's classification of representations of the Poincaré group.

Lie algebras

A Lie algebra over a field F is a vector space over F equipped with a skew-symmetric bilinear operation called the Lie bracket, which satisfies the Jacobi identity. Lie algebras arise in particular as tangent spaces to Lie groups at the identity element, leading to their interpretation as "infinitesimal symmetries". An important approach to the representation theory of Lie groups is to study the corresponding representation theory of Lie algebras, but representations of Lie algebras also have an intrinsic interest. Lie algebras, like Lie groups, have a Levi decomposition into semisimple and solvable parts, with the representation theory of solvable Lie algebras being intractable in general. In contrast, the finite-dimensional representations of semisimple Lie algebras are completely understood, after work of Élie Cartan. A representation of a semisimple Lie algebra g is analysed by choosing a Cartan subalgebra, which is essentially a generic maximal subalgebra h of g on which the Lie bracket is zero ("abelian"). The representation of g can be decomposed into weight spaces that are eigenspaces for the action of h and the infinitesimal analogue of characters. The structure of semisimple Lie algebras then reduces the analysis of representations to easily understood combinatorics of the possible weights that can occur.

Infinite-dimensional Lie algebras

There are many classes of infinite-dimensional Lie algebras whose representations have been studied. Among these, an important class are the Kac–Moody algebras. They are named after Victor Kac and Robert Moody, who independently discovered them. These algebras form a generalization of finite-dimensional semisimple Lie algebras, and share many of their combinatorial properties. This means that they have a class of representations that can be understood in the same way as representations of semisimple Lie algebras. Affine Lie algebras are a special case of Kac–Moody algebras, which have particular importance in mathematics and theoretical physics, especially conformal field theory and the theory of exactly solvable models. Kac discovered an elegant proof of certain combinatorial identities, Macdonald identities, which is based on the representation theory of affine Kac–Moody algebras.

Lie superalgebras

Lie superalgebras are generalizations of Lie algebras in which the underlying vector space has a Z_2-grading, and skew-symmetry and Jacobi identity properties of the Lie bracket are modified by signs. Their representation theory is similar to the representation theory of Lie algebras.

Linear algebraic groups

Linear algebraic groups (or more generally, affine group schemes) are analogues in algebraic geometry of Lie groups, but over more general fields than just R or C. In particular, over finite fields, they give rise to finite groups of Lie type. Although linear algebraic groups have a classification that is very similar to that of Lie groups, their representation theory is rather different (and much less well understood) and requires different techniques, since the Zariski topology is relatively weak, and techniques from analysis are no longer available.

Invariant theory

Invariant theory studies actions on algebraic varieties from the point of view of their effect on functions, which form representations of the group. Classically, the theory dealt with the question of explicit description of polynomial functions that do not change, or are invariant, under the transformations from a given linear group. The modern approach analyses the decomposition of these representations into irreducibles. Invariant theory of infinite groups is inextricably linked with the development of linear algebra, especially, the theories of quadratic forms and determinants. Another subject with strong mutual influence is projective geometry, where invariant theory can be used to organize the subject, and during the 1960s, new life was breathed into the subject by David Mumford in the form of his geometric invariant theory. The representation theory of semisimple Lie groups has its roots in invariant theory and the strong links between representation theory and algebraic geometry have many parallels in differential geometry, beginning with Felix Klein's Erlangen program and Élie Cartan's connections, which place groups and symmetry at the heart of geometry. Modern developments link representation theory and invariant theory to areas as diverse as holonomy, differential operators and the theory of several complex variables.

Automorphic forms and number theory

Automorphic forms are a generalization of modular forms to more general analytic functions, perhaps of several complex variables, with similar transformation properties. The generalization involves replacing the modular group PSL_2 (R) and a chosen congruence subgroup by a semisimple Lie group G and a discrete subgroup Ã. Just as modular forms can be viewed as differential forms on a quotient of the upper half space H = PSL_2 (R)/SO(2), automorphic forms can be viewed as differential forms (or similar objects) on Ã\G/K, where K is (typically) a maximal compact subgroup of G. Some care is required, however, as the quotient typically has singularities. The quotient of a semisimple Lie group by a compact subgroup is a symmetric space and so the theory of automorphic forms is intimately related to harmonic analysis on symmetric spaces. Before the development of the general theory, many important special cases were worked out in detail, including the Hilbert modular forms and Siegel modular forms. Important results in the theory include the Selberg trace formula and the realization by Robert Langlands that the Riemann-Roch theorem could be applied to calculate the dimension of the space of automorphic forms. The subsequent notion of "automorphic representation" has proved of great technical value for dealing with the case that G is an algebraic group, treated as an adelic algebraic group. As a result, an entire philosophy, the Langlands program has developed around the relation between representation and number theoretic properties of automorphic forms.

Associative algebras

In one sense, associative algebra representations generalize both representations of groups and Lie algebras. A representation of a group induces a representation of a corresponding group ring or group algebra, while representations of a Lie algebra correspond bijectively to representations of its universal enveloping algebra. However, the representation theory of general associative algebras does not have all of the nice properties of the representation theory of groups and Lie algebras.

Module theory

When considering representations of an associative algebra, one can forget the underlying field, and simply regard the associative algebra as a ring, and its representations as modules. This approach is surprisingly fruitful: many results in representation theory can be interpreted as special cases of results about modules over a ring.

Hopf algebras and quantum groups

Hopf algebras provide a way to improve the representation theory of associative algebras, while retaining the representation theory of groups and Lie algebras as special cases. In particular, the tensor product of two representations is a representation, as is the dual vector space. The Hopf algebras associated to groups have a commutative algebra structure, and so general Hopf algebras are known as quantum groups, although this term is often restricted to certain Hopf algebras arising as deformations of groups or their universal enveloping algebras. The representation theory of quantum groups has added surprising insights to the representation theory of Lie groups and Lie algebras, for instance through the crystal basis of Kashiwara.

Representation theory of Hopf algebras

In abstract algebra, a representation of a Hopf algebra is a representation of its underlying associative algebra. That is, a representation of a Hopf algebra H over a field K is a K-vector space V with an action H × V '! V usually denoted by juxtaposition (that is, the image of (h,v) is written hv). The vector space V is called an H-module.

Properties

The module structure of a representation of a Hopf algebra H is simply its structure as a module for the underlying associative algebra. The main use of considering the additional structure of a Hopf algebra is when considering all H-modules as a category. The additional structure is also used to define invariant elements of an H-module V. An element v in V is invariant under H if for all h in H, hv = å(h)v, where å is the counit of H. The subset of all invariant elements of V forms a submodule of V. Categories of representations as a motivation for Hopf algebras For an associative algebra H, the tensor product V_1 —" V_2 of two H-modules V_1 and V_2 is a vector space, but not necessarily an H-module. For the tensor product to be a functorial product operation on H-modules, there must be a linear binary operation $\Delta : H \to H \otimes H$ H such that for any v in V_1 —" V_2 and any h in H,

$$hv = \Delta h(v_{(1)} \otimes v_{(2)}) = h_{(1)} v_{(1)} \otimes h_{(2)} v_{(2)},$$

and for any v in V_1 —" V_2 and a and b in H,

$$\Delta(ab)(v_{(1)} \otimes v_{(2)}) = (ab)v = a[b[v]] = \Delta a[\Delta b(v_{(1)} \otimes v_{(2)})] = (\Delta a)(\Delta b)(v_{(1)} \otimes v_{(2)}).$$

using sumless Sweedler's notation, which is somewhat like an index free form of Einstein's summation convention. This is satisfied if there is a Ä such that Ä(ab) = Ä(a)Ä(b) for all a, b in H. For the category of H-modules to be a strict monoidal category with respect to $\otimes$ ", $V_1 \otimes (V_2 \otimes V_3)$ and $(V_1 \otimes V_2) \otimes V_3$ must be equivalent and there must be unit object $å_H$, called the trivial module, such that $å_H$ —" V, V and V—" $å_H$ are equivalent.

This means that for any v in

$$V_1 \otimes (V_2 \otimes V_3) = (V_1 \otimes V_2) \otimes V_3$$

and for h in H,

$$((\mathrm{id} \otimes \Delta)\Delta h)(v_{(1)} \otimes v_{(2)} \otimes v_{(3)}) = h_{(1)} v_{(1)} \otimes h_{(2)(1)} v_{(2)} \otimes h_{(2)(2)} v_{(3)} = hv = ((\Delta \otimes \mathrm{id})\Delta h)(v_{(1)} \otimes v_{(2)} \otimes v_{(3)}).$$

This will hold for any three H-modules if Ä satisfies

$$(\mathrm{id} \otimes \Delta)\Delta A = (\Delta \otimes \mathrm{id})\Delta A.$$

The trivial module must be one-dimensional, and so an algebra homomorphism å : H '! F may be defined such that hv = å(h)v for all v in $å_H$. The trivial module may be identified with F, with 1 being the element such that 1 —" v = v = v —" 1 for all v. It follows that for any v in any H-module V, any c in $å_H$ and any h in H,

$$(\varepsilon(h_{(1)})h_{(2)})cv = h_{(1)} c \otimes h_{(2)} v = h(c \otimes v) = h(cv) = (h_{(1)} \varepsilon(h_{(2)}))cv.$$

The existence of an algebra homomorphism å satisfying

$$\varepsilon(h_{(1)})h_{(2)} = h = h_{(1)} \varepsilon(h_{(2)})$$

is a sufficient condition for the existence of the trivial module. It follows that in order for the category of H-modules to be a monoidal category with respect to the tensor product, it is sufficient for H to have maps Ä and å satisfying these conditions. This is the motivation for the definition of a bialgebra, where Ä is called the comultiplication and å is called the counit. In order for each H-module V to have a dual representation V such that the underlying vector spaces are dual and the operation * is functorial over the monoidal category of H- modules, there must be a linear map S : H '! H such that for any h in H, x in V and y in V*, where è' Å" , Å" é'

$(\cdot,\cdot)$ is the usual pairing of dual vector spaces. If the map ö : V —" V " '! å H {\displaystyle \varphi :V\otimes V^{*}\rightarrow \varepsilon _{H}} $\varphi : V \otimes V^{*} \rightarrow \varepsilon_{H}$ induced by the pairing is to be an H-homomorphism, then for any h in H, x in V and y in V*,

$$\varphi(h(x \otimes y)) = \varphi(x \otimes S(h_{(1)})h_{(2)}y) = \varphi(S(h_{(2)})h_{(1)}x \otimes y) = h\varphi(x \otimes y) = \varepsilon(h)\varphi(x \otimes y),$$

which is satisfied if

$$S(h_{(1)})h_{(2)} = \varepsilon(h) = h_{(1)}S(h_{(2)})$$

for all h in H.

If there is such a map S, then it is called an antipode, and H is a Hopf algebra. The desire for a monoidal category of modules with functorial tensor products and dual representations is therefore one motivation for the concept of a Hopf algebra.

Representations on an algebra

A Hopf algebra also has representations which carry additional structure, namely they are algebras. Let H be a Hopf algebra. If A is an algebra with the product operation ì : A —" A '! A, and ñ : H —" A '! A is a representation of H on A, then ñ is said to be a representation of H on an algebra if ì is H-equivariant. As special cases, Lie algebras, Lie superalgebras and groups can also have representations on an algebra.

10

Differential Calculus

In mathematics, differential calculus is a subfield of calculus concerned with the study of the rates at which quantities change. It is one of the two traditional divisions of calculus, the other being integral calculus, the study of the area beneath a curve.

The primary objects of study in differential calculus are the derivative of a function, related notions such as the differential, and their applications. The derivative of a function at a chosen input value describes the rate of change of the function near that input value. The process of finding a derivative is called differentiation. Geometrically, the derivative at a point is the slope of the tangent line to the graph of the function at that point, provided that the derivative exists and is defined at that point. For a real-valued function of a single real variable, the derivative of a function at a point generally determines the best linear approximation to the function at that point.

Differential calculus and integral calculus are connected by the fundamental theorem of calculus, which states that differentiation is the reverse process to integration.

Differentiation has applications to nearly all quantitative disciplines. For example, in physics, the derivative of the displacement of a moving body with respect to time is the velocity of the body, and the derivative of velocity with respect to time is acceleration. The derivative of the momentumof a body equals the force applied to the body; rearranging this derivative statement leads to the famous F = ma equation associated with Newton's second law of motion. The reaction rate of a chemical reaction is a derivative. In operations research, derivatives determine the most efficient ways to transport materials and design factories.

Derivatives are frequently used to find the maxima and minima of a function. Equations involving derivatives are called differential equations and are fundamental in describing natural phenomena. Derivatives and their generalizations appear in many fields of mathematics, such as complex analysis, functional analysis, differential geometry, measure theory, and abstract algebra.

HISTORY OF DIFFERENTIATION

The concept of a derivative in the sense of a tangent line is a very old one, familiar to Greek geometers such as Euclid (c. 300 BC), Archimedes (c. 287–212 BC) and Apollonius of Perga (c. 262–190 BC). Archimedes also introduced the use of infinitesimals, although these were primarily used to study areas and volumes rather than derivatives and tangents; see Archimedes' use of infinitesimals.

The use of infinitesimals to study rates of change can be found in Indian mathematics, perhaps as early as 500 AD, when the astronomer and mathematician Aryabhata (476–550) used infinitesimals to study the orbit of the Moon. The use of infinitesimals to compute rates of change was developed significantly by Bhâskara II (1114–1185); indeed, it has been argued that many of the key notions of differential calculus can be found in his work, such as "Rolle's theorem".

The Persian mathematician, Sharaf al-Dîn al-Tûsî (1135–1213), in his Treatise on Equations, established conditions for some cubic equations to have solutions, by finding the maxima of appropriate cubic polynomials. He proved, for example, that the maximum of the cubic $a x — x^3$ occurs when $x = 2a/3$, and concluded therefrom that the equation $a x^2 — x^3 = c$ has exactly one positive solution when $c = 4 a^3/27$, and two positive solutions whenever $0 < c < 4 a^3/27$. The historian of science, Roshdi Rashed,[8] has argued that al-Tûsî must have used the derivative of the cubic to obtain this result. Rashed's conclusion has been contested by other scholars, however, who argue that he could have obtained the result by other methods which do not require the derivative of the function to be known.

The modern development of calculus is usually credited to Isaac Newton (1643–1727) and Gottfried Wilhelm Leibniz (1646–1716), who provided independent and unified approaches to differentiation and derivatives. The key insight, however, that earned them this credit, was the fundamental theorem of calculus relating differentiation and integration: this rendered obsolete most previous methods for computing areas and volumes, which had not been significantly extended since the time of Ibn al-Haytham (Alhazen). For their ideas on derivatives, both Newton and Leibniz built on significant earlier work by mathematicians such as Pierre de Fermat (1607-1665), Isaac Barrow (1630–1677), René Descartes (1596–1650), Christiaan Huygens (1629–1695), Blaise Pascal (1623–1662) and John Wallis (1616–1703). Regarding Fermat's influence, Newton once wrote in a letter that "I had the hint of this method [of fluxions] from Fermat's way of drawing tangents, and by applying it to abstract equations, directly and invertedly, I made it general." Isaac Barrow is generally given credit for the early development of the derivative. Nevertheless, Newton and Leibniz remain key figures in the history of differentiation, not least because Newton was the first to apply differentiation to theoretical physics, while Leibniz systematically developed much of the notation still used today.

Since the 17th century many mathematicians have contributed to the theory of differentiation. In the 19th century, calculus was put on a much more rigorous footing by mathematicians such as Augustin Louis Cauchy (1789–1857), Bernhard Riemann (1826–1866), and Karl Weierstrass (1815–1897). It was also during this period that the differentiation was generalized to Euclidean space and the complex plane.

Applications of derivatives

Optimization

If f is a differentiable function on "! (or an open interval) and x is a local maximum or a local minimum of f, then the derivative of f at x is zero. Points where $f'(x) = 0$ are called critical points or stationary points (and the value of f at x is called a critical value). If f is not assumed to be everywhere differentiable, then points at which it fails to be differentiable are also designated critical points.

If f is twice differentiable, then conversely, a critical point x of f can be analysed by considering the second derivative of f at x :

- if it is positive, x is a local minimum;
- if it is negative, x is a local maximum;
- if it is zero, then x could be a local minimum, a local maximum, or neither. (For example, $f(x) = x^3$ has a critical point at $x = 0$, but it has neither a maximum nor a minimum there, whereas $f(x) = \pm x^4$ has a critical point at $x = 0$ and a minimum and a maximum, respectively, there.)

This is called the second derivative test. An alternative approach, called the first derivative test, involves considering the sign of the f' on each side of the critical point.

Taking derivatives and solving for critical points is therefore often a simple way to find local minima or maxima, which can be useful in optimization. By the extreme value theorem, a continuous function on a closed interval must attain its minimum and maximum values at least once. If the function is differentiable, the minima and maxima can only occur at critical points or endpoints.

This also has applications in graph sketching: once the local minima and maxima of a differentiable function have been found, a rough plot of the graph can be obtained from the observation that it will be either increasing or decreasing between critical points.

In higher dimensions, a critical point of a scalar valued function is a point at which the gradient is zero. The second derivative test can still be used to analyse critical points by considering the eigenvalues of the Hessian matrix of second partial derivatives of the function at the critical point. If all of the eigenvalues are positive, then the point is a local minimum; if all are negative, it is a local maximum. If there are some positive and some negative eigenvalues, then the critical point is called a "saddle point", and if none of these cases hold (i.e., some of the eigenvalues are zero) then the test is considered to be inconclusive.

Calculus of Variations

One example of an optimization problem is: Find the shortest curve between two points on a surface, assuming that the curve must also lie on the surface. If the surface is a plane, then the shortest curve is a line. But if the surface is, for example, egg-shaped, then the shortest path is not immediately clear. These paths are called geodesics, and one of the simplest problems in the calculus of variations is finding geodesics. Another example is: Find the smallest area surface filling in a closed curve in space. This surface is called a minimal surface and it, too, can be found using the calculus of variations.

Physics

Calculus is of vital importance in physics: many physical processes are described by equations involving derivatives, called differential equations. Physics is particularly concerned with the way quantities change and develop over time, and the concept of the "time derivative" — the rate of change over time — is essential for the precise definition of several important concepts. In particular, the time derivatives of an object's position are significant in Newtonian physics:

- velocity is the derivative (with respect to time) of an object's displacement (distance from the original position)
- acceleration is the derivative (with respect to time) of an object's velocity, that is, the second derivative (with respect to time) of an object's position.

Derivative

The derivative of a function of a real variable measures the sensitivity to change of the function value (output value) with respect to a change in its argument (input value). Derivatives are a fundamental tool of calculus. For example, the derivative of the position of a moving object with respect to time is the object's velocity: this measures how quickly the position of the object changes when time advances.

The derivative of a function of a single variable at a chosen input value, when it exists, is the slope of the tangent line to the graph of the function at that point. The tangent line is the best linear approximation of the function near that input value. For this reason, the derivative is often described as the "instantaneous rate of change", the ratio of the instantaneous change in the dependent variable to that of the independent variable.

Derivatives may be generalized to functions of several real variables. In this generalization, the derivative is reinterpreted as a linear transformation whose graph is (after an appropriate translation) the best linear approximation to the graph of the original function. The Jacobian matrix is the matrix that represents this linear transformation with respect to the basis given by the choice of independent and dependent variables. It can be calculated in terms of the partial derivatives with respect to the independent variables. For a real-valued function of several variables, the Jacobian matrix reduces to the gradient vector.

The process of finding a derivative is called differentiation. The reverse process is called antidifferentiation. The fundamental theorem of calculus states that antidifferentiation is the same as integration. Differentiation and integration constitute the two fundamental operations in single-variable calculus.[Note 1]

CONTINUITY AND DIFFERENTIABILITY

If f is differentiable at a, then f must also be continuous at a. As an example, choose a point a and let f be the step function that returns the value 1 for all x less than a, and returns a different value 10 for all x greater than or equal to a. f cannot have a derivative at a. If h is negative, then a + h is on the low part of the step, so the secant line from a to a + h is very steep, and as h tends to zero the slope tends to infinity. If h is positive, then a + h is on the high part of the step, so the secant line from a to a + h has slope zero. Consequently, the secant lines do not approach any single slope, so the limit of the difference quotient does not exist.[Note 2]

However, even if a function is continuous at a point, it may not be differentiable there. For example, the absolute value function given by $f(x) = |x|$ is continuous at $x = 0$, but it is not differentiable there. If h is positive, then the slope of the secant line from 0 to h is one, whereas if h is negative, then the slope of the secant line from 0 to h is negative one. This can be seen graphically as a "kink" or a "cusp" in the graph at $x = 0$. Even a function with a smooth graph is not differentiable at a point where its tangent is vertical: For instance, the function given by $f(x) = x^{1/3}$ is not differentiable at $x = 0$.

In summary: for a function f to have a derivative it is necessary for the function f to be continuous, but continuity alone is not sufficient.

Most functions that occur in practice have derivatives at all points or at almost every point. Early in the history of calculus, many mathematicians assumed that a continuous function was differentiable at most points. Under mild conditions, for example if the function is a monotone function or a Lipschitz function, this is true. However, in 1872 Weierstrass found the first example of a function that is continuous everywhere but differentiable nowhere. This example is now known as the Weierstrass function. In 1931, Stefan Banach proved that the set of functions that have a derivative at some point is a meager set in the space of all continuous functions. Informally, this means that hardly do any random continuous functions have a derivative at even one point.

The derivative as a function

Let f be a function that has a derivative at every point in its domain. We can then define a function that maps every point to the value of the derivative of. This function is written f2 and is called the derivative function or the derivative of f.

Sometimes f has a derivative at most, but not all, points of its domain. The function whose value at a equals f2 (a) whenever f2 (a) is defined and elsewhere is undefined is also called the derivative of f. It is still a function, but its domain is strictly smaller than the domain of f.

Using this idea, differentiation becomes a function of functions: The derivative is an operator whose domain is the set of all functions that have derivatives at every point of their domain and whose range is a set of functions. If we denote this operator by D, then D(f) is the function f2 . Since D(f) is a function, it can be evaluated at a point a. By the definition of the derivative function, D(f)(a) = f2 (a).

Higher derivatives

Let f be a differentiable function, and let f 2 be its derivative. The derivative of f 2 (if it has one) is written f 2 2 and is called the second derivative of f. Similarly, the derivative of the second derivative, if it exists, is written f 2 2 2 and is called the third derivative of f. Continuing this process, one can define, if it exists, the nth derivative as the derivative of the (n-1)th derivative. These repeated derivatives are called higher-order derivatives. The nth derivative is also called the derivative of order n.

If x(t) represents the position of an object at time t, then the higher-order derivatives of x have physical interpretations. The second derivative of x is the derivative of x2 , the velocity, and by definition this is the object's acceleration. The third derivative of x is defined to be the jerk, and the fourth derivative is defined to be the jounce.

Generalizations

The concept of a derivative can be extended to many other settings. The common thread is that the derivative of a function at a point serves as a linear approximation of the function at that point.

- An important generalization of the derivative concerns complex functions of complex variables, such as functions from (a domain in) the complex numbers C to C. The notion of the derivative of such a function is obtained by replacing real variables with complex variables in the definition. If C is identified with R^2 by writing a complex number z as x + iy, then a differentiable function from C to C is certainly differentiable as a function from R^2 to R^2 (in the sense that its partial derivatives all exist), but the converse is not true in general: the complex derivative only exists if the real derivative is complex linear and this imposes relations between the partial derivatives called the Cauchy–Riemann equations–see holomorphic functions.
- Another generalization concerns functions between differentiable or smooth manifolds. Intuitively speaking such a manifold M is a space that can be approximated near each point x by a vector space called its tangent space: the prototypical example is a smooth surface in R^3. The derivative (or differential) of a (differentiable) map f: M '! N between manifolds, at a point x in M, is then a linear map from the tangent space of M at x to the tangent space of N at f(x). The derivative function becomes a map between the tangent bundles of M and N. This definition is fundamental in differential geometry and has many uses – see pushforward (differential) and pullback (differential geometry).
- Differentiation can also be defined for maps between infinite dimensional vector spaces such as Banach spaces and Fréchet spaces. There is a generalization both of the directional derivative, called the Gâteaux derivative, and of the differential, called the Fréchet derivative.
- One deficiency of the classical derivative is that not very many functions are differentiable. Nevertheless, there is a way of extending the notion of the derivative so that all continuous functions and many other functions can be differentiated using a concept known as the weak derivative. The idea is to embed the continuous functions in a larger space called the space of distributions and only require that a function is differentiable "on average".
- The properties of the derivative have inspired the introduction and study of many similar objects in algebra and topology — see, for example, differential algebra.
- The discrete equivalent of differentiation is finite differences. The study of differential calculus is unified with the calculus of finite differences in time scale calculus.
- Also see arithmetic derivative.

HYPERBOLIC FUNCTIONS

If you work in engineering or physics, you have to work with a lot of models. Well, to solve modeling problems you'll need hyperbolic functions.

Certain even and odd combinations of the exponential functions e^x and e^{-x} arise so frequently in engineering, physics, and mathematics that they're given special names. Hyperbolic functions are analogous trigonometric functions in that they are named the same as trigonometric functions with the letter 'h' appended to each name.

These special functions are have the same relationship to the hyperbola that trigonometric functions have to the circle. For this reason, they are collectively known as hyperbolic functions and are individually called hyperbolic sine, hyperbolic cosine, and so on. In addition to modeling, they can be used as solutions to some types of partial differential equations.

Think of taking a rope, fixing the two ends, and then letting it hang under the force of gravity. The shape of the rope will naturally form a hyperbolic cosine curve.

Types & Graphs

Hyperbolic Sine: y = sinh(x)

This math statement is read as 'y equals hyperbolic sine x' or 'sinch x.'

The exponential formula is:

$$sinh(x) = \frac{e^x - e^{-x}}{2}$$

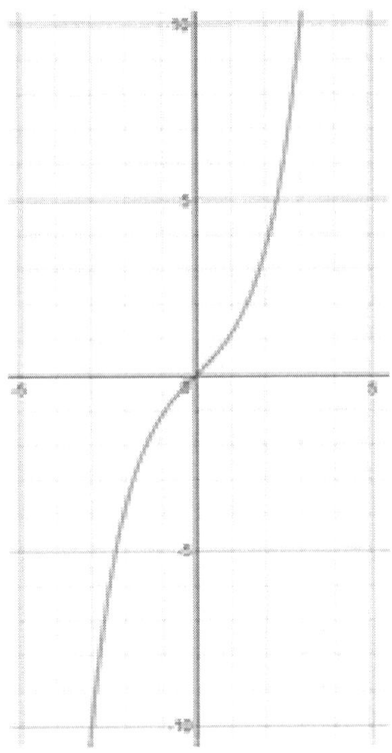

The graph of sinh

Note that the domain and range of this function are both the set of real numbers. This function is also one-to-one.

Hyperbolic Cosine: y = cosh(x)

This math statement is read as 'y equals hyperbolic cosine x' or 'cosh x.'

$$cosh(x) = \frac{e^x + e^{-x}}{2}$$

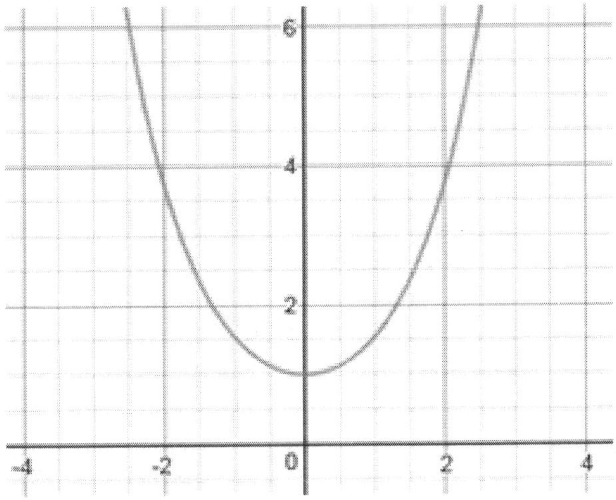

The graph of cosh(x)

Note that the graph is symmetric with respect to the y-axis and it has a minimum vertex at (0,1). The domain of this function is the set of real numbers and the range is any number equal to or greater than one.

Hyperbolic Tangent: y = tanh(x)

This math statement is read as 'y equals hyperbolic tangent x' or 'tanch x.'

$$tanh(x) = \frac{e^x - e^{-x}}{e^x + e^{-x}} = \frac{sinh(x)}{cosh(x)}$$

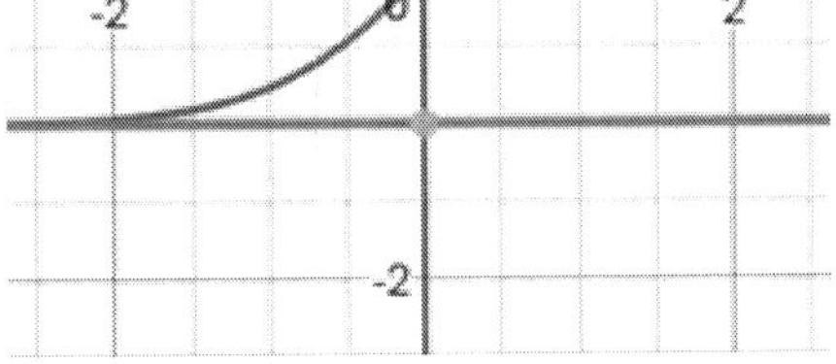

The graph of tanh(x)

Note that the graph is symmetric with respect to the origin. The lines y = 1 and y = -1 are both horizontal asymptotes. The domain of the function is the set of real numbers. The range is the open interval (-1,1).

Hyperbolic Cotangent: y = coth(x)

This math statement is read as 'y equals hyperbolic cotangent x' or 'coth x.'

$$coth(x) = \frac{e^x + e^{-x}}{e^x - e^{-x}} = \frac{cosh(x)}{sinh(x)}$$

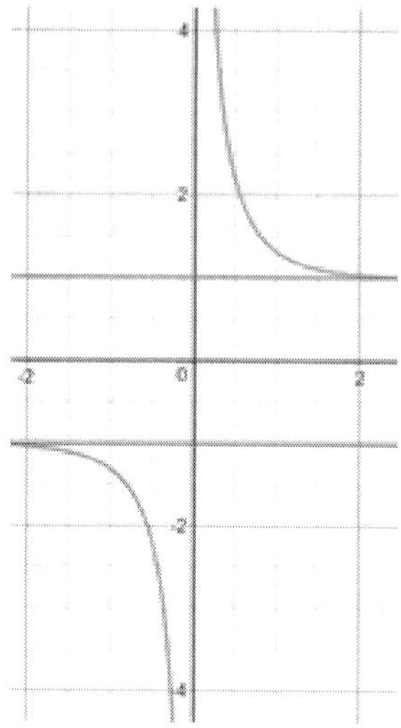

The graph of coth(x)

Note that the graph is symmetric with respect to the origin. The lines y = 1 and y = -1 are both horizontal asymptotes. The y-axis is a vertical asymptote. The domain of the function is the set of real numbers excluding zero. The range is (-infinity,-1) union (1,infinity).

Hyperbolic Secant: y = sech(x)

This math statement is read as 'y equals hyperbolic secant x.'

$$sech(x) = \frac{2}{e^x + e^{-x}} = \frac{1}{cosh(x)}$$

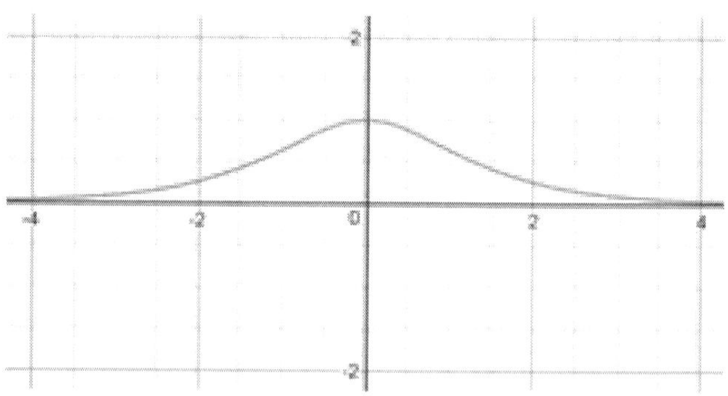

The graph of sech(x)

The x-axis is a horizontal asymptote. The graph is symmetric with respect to the y-axis. The domain of the function is the set of real numbers. The range of the function is the half-open interval (0,1].

Hyperbolic Cosecant: y = csch(x)

This math statement is read as 'y equals hyperbolic cosecant x.'

$$csch(x) = \frac{2}{e^x - e^{-x}} = \frac{1}{sinh(x)}$$

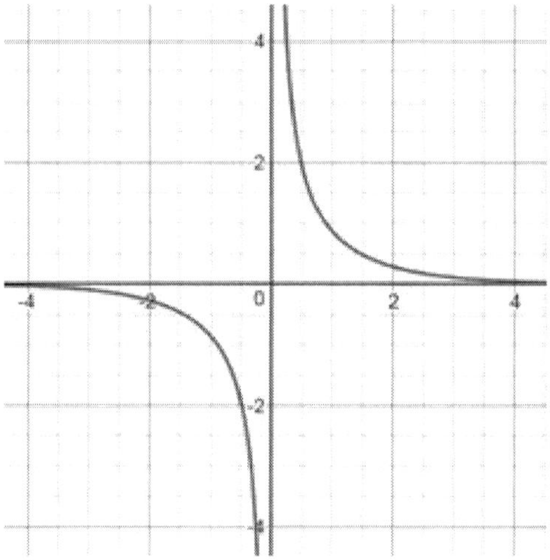

The graph of csch(x)

Note that both the x- and y-axes are asymptotes. The graph is symmetric with respect to the origin. Both the domain and range of the function are the set of real numbers excluding zero.

LOGARITHMIC DIFFERENTIATION

The technique is often performed in cases where it is easier to differentiate the logarithm of a function rather than the function itself. This usually occurs in cases where the function of interest is composed of a product of a number of parts, so that a logarithmic transformation will turn it into a sum of separate parts (which is much easier to differentiate). It can also be useful when applied to functions raised to the power of variables or functions. Logarithmic differentiation relies on the chain rule as well as properties of logarithms (in particular, the natural logarithm, or the logarithm to the base e) to transform products into sums and divisions into subtractions. The principle can be implemented, at least in part, in the differentiation of almost all differentiable functions, providing that these functions are non-zero.

Implicit Functions

We're used to seeing algebraic definitions of functions in which the output y is given explicitly in terms of some combination of the input x:

$$y = \sqrt{x^2 + 1}$$

$$y = 2\cos(\pi x) - \cos(2\pi x)$$

$$y = \frac{1}{1 + 2e^{-x}}$$

We think of these rules as accepting the input x, going through a series of steps that transform x, and then announcing the final result as y. The output y is not involved in the "business" of computing the function, it is only the end product.

What, then, do we make of equations like this:

$$x^2 + y^2 = 1$$

$$y^3 + x^3 = 3xy$$

$$2^{xy} = x + y$$

Here the xs and ys are all mixed-up together, on either side of the equation. There is no obvious way to proceed from an "input" x to an "output" y. It isn't even clear if each y is actually a function of the corresponding x.

We could, of course, attempt to solve each of these equations for y, writing y explicitly as a function of x, and so try to put ourselves back on familiar ground. In all three cases, however, we will fail.

Sometimes the reason we will fail is that y is not uniquely determined by the value of x: many ys will balance the equation when we plug in a particular x. (For example, solving the first equation for y leads to $y = \pm\sqrt{1 - x^2}$.) Sometimes the reason is a failure of algebra itself: we can not solve for y; it is just too mixed-up with the xs for us to be able to extract it. (Try, for example, to solve the last equation for y.)

The ys in such "mixed-up" equations are often called implicit functions of x. Using the word "function" here is tricky, because, as we have seen, there may in fact be many ys corresponding to a particular x. What is "implicit" (the dictionary defines this as "implied or understood although not directly expressed") is the understanding that many functions may be lurking in such equations, if only we could coax them out.

One way to get at the implicit functions embedded in a "mixed-up" equation is to graph all of the pairs (x, y) that balance the equation.

For example, we have noted that if

$$x^2 + y^2 = 1$$

then

$$y = f(x) = +\sqrt{1 - x^2} \quad \text{or} \quad y = g(x) = -\sqrt{1 - x^2}$$

Here the equation implicitly defines two perfectly normal functions, and it is a simple matter to plot them together:

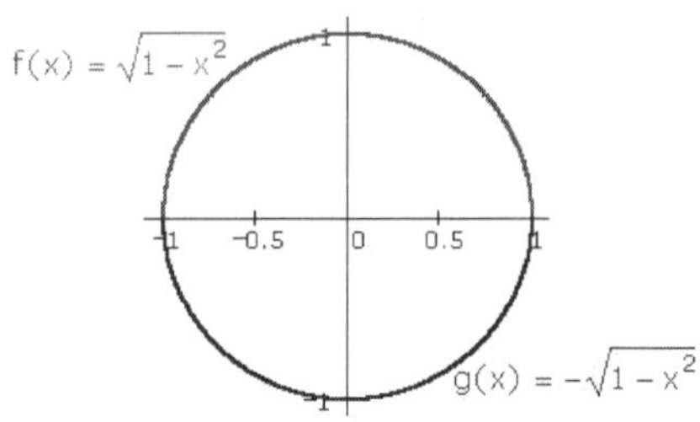

Other cases are harder. For example,

$$y^3 + x^3 = 3xy$$

This is the famous Folium of Descartes. While there is an algebraic method for solving this equation for the various functions that give y in terms of x, the results are extremely complicated.

To produce a graph of such an equation, we can divide a region of the xy-plane into a very fine grid. At each point (x, y) on the grid, we can plug the coordinate values into both the left and the right sides of the equation. If the two sides agree (within some pre-set tolerance), we add the point to the plot. If the two sides do not agree, we don't. This method may seem tedious and time consuming, but it is the kind of thing for which computers are tailor-made.

Here's a plot of the Folium of Descartes, produced in this manner:

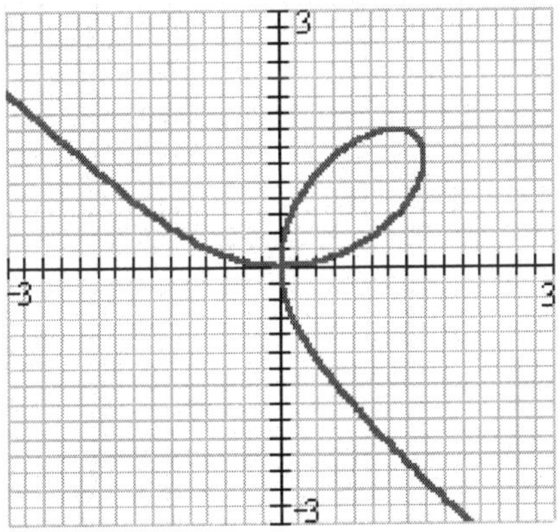

We see that the equation divides the xy-plane into three regions. Along the "tails" y can be given as an explicit function of x, but in the x neighborhood of the "loop" three separate functions of x are implicitly defined:

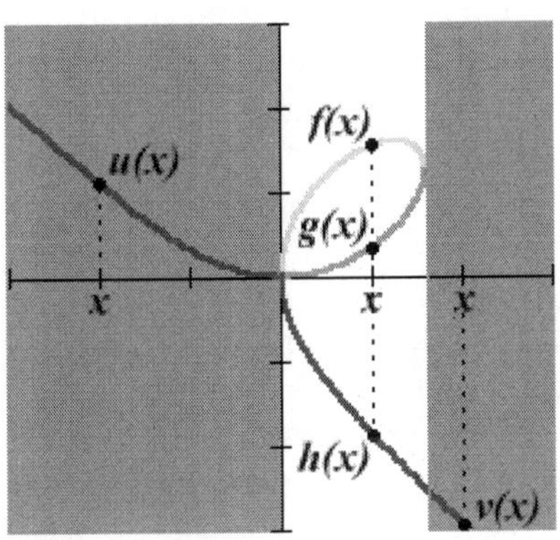

Parametric Equations And Curves

To this point (in both Calculus I and Calculus II) we've looked almost exclusively at functions in the form y=f(x)y=f(x) or x=h(y)x=h(y) and almost all of the formulas that we've developed require that functions be in one of these two forms. The problem is that not all curves or equations that we'd like to look at fall easily into this form.

Take, for example, a circle. It is easy enough to write down the equation of a circle centered at the origin with radius rr.

x2+y2=r2x2+y2=r2

However, we will never be able to write the equation of a circle down as a single equation in either of the forms above. Sure we can solve for xx or yy as the following two formulas show

y=±"r2"x2x=±"r2"y2y=±r2"x2x=±r2"y2

but there are in fact two functions in each of these. Each formula gives a portion of the circle.

y="r2"x2(top) x="r2"y2(right side)y=""r2"x2(bottom)x=""r2"y2(left side)y=r2"x2(top) x=r2"y2(right side)y="r2"x2(bottom)x="r2"y2(left side)

Unfortunately, we usually are working on the whole circle, or simply can't say that we're going to be working only on one portion of it. Even if we can narrow things down to only one of these portions the function is still often fairly unpleasant to work with.

There are also a great many curves out there that we can't even write down as a single equation in terms of only xx and yy. So, to deal with some of these problems we introduce parametric equations. Instead of defining yy in terms of xx (y=f(x)y=f(x)) or xx in terms of yy (x=h(y)x=h(y)) we define both xx and yy in terms of a third variable called a parameter as follows,

x=f(t)y=g(t)x=f(t)y=g(t)

This third variable is usually denoted by tt (as we did here) but doesn't have to be of course. Sometimes we will restrict the values of tt that we'll use and at other times we won't. This will often be dependent on the problem and just what we are attempting to do.

Each value of tt defines a point (x,y)=(f(t),g(t))(x,y)=(f(t),g(t)) that we can plot. The collection of points that we get by letting tt be all possible values is the graph of the parametric equations and is called the parametric curve.

To help visualize just what a parametric curve is pretend that we have a big tank of water that is in constant motion and we drop a ping pong ball into the tank. The point (x,y)=(f(t),g(t))(x,y)=(f(t),g(t)) will then represent the location of the ping pong ball in the tank at time tt and the parametric curve will be a trace of all the locations of the ping pong ball. Note that this is not always a correct analogy but it is useful initially to help visualize just what a parametric curve is.

Sketching a parametric curve is not always an easy thing to do. Let's take a look at an example to see one way of sketching a parametric curve. This example will also illustrate why this method is usually not the best.

We should give a small warning at this point. Because of the ideas involved in them we concentrated on parametric curves that retraced portions of the curve more than once. Do not, however, get too locked into the idea that this will always happen. Many, if not most parametric curves will only trace out once. The first one we looked at is a good example of this. That parametric curve will never repeat any portion of itself.

There is one final topic to be discussed in this section before moving on. So far we've started with parametric equations and eliminated the parameter to determine the parametric curve.

However, there are times in which we want to go the other way. Given a function or equation we might want to write down a set of parametric equations for it. In these cases we say that we parameterize the function.

If we take Examples 4 and 5 as examples we can do this for ellipses (and hence circles). Given the ellipse

x2a2+y2b2=1x2a2+y2b2=1

a set of parametric equations for it would be,

x=acosty=bsintx=acosa ty=bsina t

This set of parametric equations will trace out the ellipse starting at the point (a,0)(a,0) and will trace in a counter-clockwise direction and will trace out exactly once in the range 0d"td"2À0d"td"2À. This is a fairly important set of parametric equations as it used continually in some subjects with dealing with ellipses and/or circles.

Every curve can be parameterized in more than one way. Any of the following will also parameterize the same ellipse.

x=acos(Ét)y=bsin(Ét)x=asin(Ét)y=bcos(Ét)x=acos(Ét)y="bsin(Ét)x=acosa (Ét)y=bsina (Ét)x=asina (Ét)y=bcosa (Ét)x=acosa (Ét)y="bsina (Ét)

The presence of the ÉÉ will change the speed that the ellipse rotates as we saw in Example 5. Note as well that the last two will trace out ellipses with a clockwise direction of motion (you might want to verify this). Also note that they won't all start at the same place (if we think of t=0t=0 as the starting point that is).

There are many more parameterizations of an ellipse of course, but you get the idea. It is important to remember that each parameterization will trace out the curve once with a potentially different range of tt's. Each parameterization may rotate with different directions of motion and may start at different points.

You may find that you need a parameterization of an ellipse that starts at a particular place and has a particular direction of motion and so you now know that with some work you can write down a set of parametric equations that will give you the behavior that you're after.

Now, let's write down a couple of other important parameterizations and all the comments about direction of motion, starting point, and range of tt's for one trace (if applicable) are still true.

First, because a circle is nothing more than a special case of an ellipse we can use the parameterization of an ellipse to get the parametric equations for a circle centered at the origin of radius rr as well. One possible way to parameterize a circle is,

x=rcosty=rsintx=rcosa ty=rsina t

Finally, even though there may not seem to be any reason to, we can also parameterize functions in the form y=f(x)y=f(x) or x=h(y)x=h(y). In these cases we parameterize them in the following way,

x=tx=h(t)y=f(t)y=tx=tx=h(t)y=f(t)y=t

At this point it may not seem all that useful to do a parameterization of a function like this, but there are many instances where it will actually be easier, or it may even be required, to work with the parameterization instead of the function itself. Unfortunately, almost all of these instances occur in a Calculus III course.

11

Tangents and Normals

A tangent to a curve is a line that touches the curve at one point and has the same slope as the curve at that point.

A normal to a curve is a line perpendicular to a tangent to the curve.

A tangent at a point on the curve is a straight line that touches the curve at that point and whose slope is equal to the gradient/derivative of the curve at that point. From the definition, you can deduce how to find the equation of the tangent to the curve at any point. Given a function $y = f(x)$, the equation of the tangent to this curve at $x = x_0$ can be found in the following way:

- Find out the gradient/derivative of the curve at the point $x = x_0$: To do this one needs to calculate dydx#x=x0. Let us call this value m, in analogy to the slope of a straight line.
- Find the equation of the straight line passing through the point $(x_0, y(x_0))$ with slope m. This is quite straightforward and can be found out as
  y–y1x–x1=m
  You have found out the equation of the tangent to the curve at the given point!

NORMAL

A normal at a point on the curve is a straight line that intersects the curve at that point and is perpendicular to the tangent at that point. If its slope is given by n, and the slope of the tangent at that point or the value of the gradient/derivative at that point is given by m; then we have $m \times n = -1$. Steps for finding the normal to a given curve $y = f(x)$ at a point $x = x_0$:

- Find out the gradient/derivative of the curve at the point $x = x_0$: This first step is exactly the same as in the method of finding the equation of the tangent to the curve i.e. m = dydx#x=x0
- Find the slope 'n' of the normal: As the normal is perpendicular to the tangent, we have:
  n="1m
- Now, find the equation of the straight line passing through the point $(x_0, y(x_0))$ with slope n. The equation is given by:

y–y1x–x1=n

It might be quite noticeable that both the tangents and normals to a curve go hand in hand. Both are easily derivable from one another. Now take a look at the diagram below to visualize them better, and then proceed towards the solved example to clear your doubts.

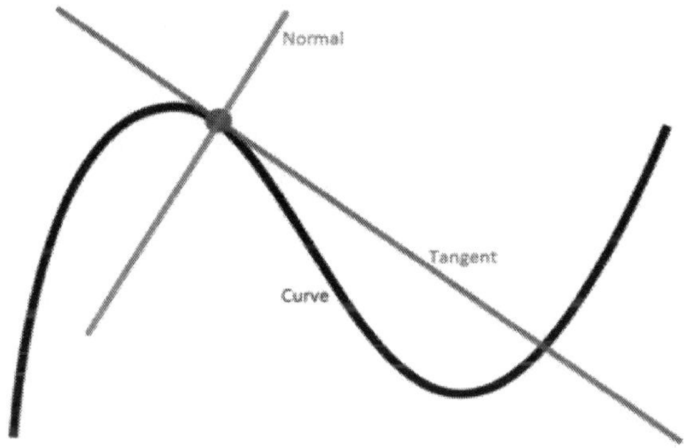

History

Euclid makes several references to the tangent (öáõôïìÝíç ephaptoménç) to a circle in book III of the Elements (c. 300 BC). In Apollonius work Conics (c. 225 BC) he defines a tangent as being a line such that no other straight line could fall between it and the curve.

Archimedes (c. 287 – c. 212 BC) found the tangent to an Archimedean spiral by considering the path of a point moving along the curve.

In the 1630s Fermat developed the technique of adequality to calculate tangents and other problems in analysis and used this to calculate tangents to the parabola. The technique of adeqality is similar to taking the difference between and dividing by a power of. Independently Descartes used his method of normals based on the observation that the radius of a circle is always normal to the circle itself.

These methods led to the development of differential calculus in the 17th century. Many people contributed. Roberval discovered a general method of drawing tangents, by considering a curve as described by a moving point whose motion is the resultant of several simpler motions. René-François de Sluse and Johannes Hudde found algebraic algorithms for finding tangents. Further developments included those of John Wallis and Isaac Barrow, leading to the theory of Isaac Newton and Gottfried Leibniz.

An 1828 definition of a tangent was "a right line which touches a curve, but which when produced, does not cut it". This old definition prevents inflection points from having any tangent. It has been dismissed and the modern definitions are equivalent to those of Leibniz who defined the tangent line as the line through a pair of infinitely close points on the curve.

Tangent line to a curve

The intuitive notion that a tangent line "touches" a curve can be made more explicit by considering the sequence of straight lines (secant lines) passing through two points, A and B, those that lie on the function curve. The tangent at A is the limit when point B approximates or tends to A. The existence and uniqueness of the tangent line depends on a certain type of mathematical smoothness, known as "differentiability." For example, if two circular arcs meet at a sharp point (a vertex) then there is no uniquely defined tangent at the vertex because the limit of the progression of secant lines depends on the direction in which "point B" approaches the vertex.

At most points, the tangent touches the curve without crossing it (though it may, when continued, cross the curve at other places away from the point of tangent). A point where the tangent (at this point) crosses the curve is called an inflection point. Circles, parabolas, hyperbolasand ellipses do not have any inflection point, but more complicated curves do have, like the graph of a cubic function, which has exactly one inflection point, or a sinusoid, which has two inflection points per each period of the sine.

Conversely, it may happen that the curve lies entirely on one side of a straight line passing through a point on it, and yet this straight line is not a tangent line. This is the case, for example, for a line passing through the vertex of a triangle and not intersecting it otherwise—where the tangent line does not exist for the reasons explained above. In convex geometry, such lines are called supporting lines.

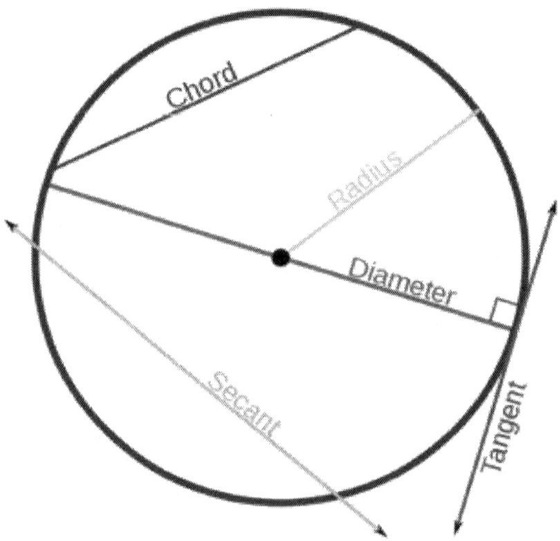

A tangent, a chord, and a secant to a circle

Analytical approach

The geometrical idea of the tangent line as the limit of secant lines serves as the motivation for analytical methods that are used to find tangent lines explicitly. The question of finding the tangent line to a graph, or the tangent line problem, was one of the central questions leading to the development of calculus in the 17th century. In the second book of his Geometry, René Descartes[8] said of the problem of constructing the tangent to a curve, "And I dare say that this is not only the most useful and most general problem in geometry that I know, but even that I have ever desired to know".

INTUITIVE DESCRIPTION

More rigorous description

To make the preceding reasoning rigorous, one has to explain what is meant by the difference quotient approaching a certain limiting value k. The precise mathematical formulation was given by Cauchy in the 19th century and is based on the notion of limit. Suppose that the graph does not have a break or a sharp edge at pand it is neither plumb nor too wiggly near p. Then there is a unique value of k such that, as h approaches 0, the difference quotient gets closer and closer to k, and the distance between them becomes negligible compared with the size of h, if h is small enough. This leads to the definition of the slope of the tangent line to the graph as the limit of the difference quotients for the function f. This limit is the derivative of the function f at x = a, denoted f 2 (a).

How the method can fail

Calculus also demonstrates that there are functions and points on their graphs for which the limit determining the slope of the tangent line does not exist. For these points the function f is non-differentiable. There are two possible reasons for the method of finding the tangents based on the limits and derivatives to fail: either the geometric tangent exists, but it is a vertical line, which cannot be given in the point-slope form since it does not have a slope, or the graph exhibits one of three behaviors that precludes a geometric tangent.

The graph $y = x^{1/3}$ illustrates the first possibility: here the difference quotient at a = 0 is equal to $h^{1/3}/h = h^{"2/3}$, which becomes very large as h approaches 0. This curve has a tangent line at the origin that is vertical.

The graph $y = x^{2/3}$ illustrates another possibility: this graph has a cusp at the origin. This means that, when h approaches 0, the difference quotient at a = 0 approaches plus or minus infinity depending on the sign of x. Thus both branches of the curve are near to the half vertical line for which y=0, but none is near to the negative part of this line. Basically, there is no tangent at the origin in this case, but in some context one may consider this line as a tangent, and even, in algebraic geometry, as a double tangent.

The graph $y = |x|$ of the absolute value function consists of two straight lines with different slopes joined at the origin. As a point q approaches the origin from the right, the secant line always has slope 1. As a point q approaches the origin from the left, the secant line always has slope "1. Therefore, there is no unique tangent to the graph at the origin. Having two different (but finite) slopes is called a corner.

Finally, since differentiability implies continuity, the contrapositive states discontinuity implies non-differentiability. Any such jump or point discontinuity will have no tangent line. This includes cases where one slope approaches positive infinity while the other approaches negative infinity, leading to an infinite jump discontinuity

Angle between curves

The angle between two curves at a point where they intersect is defined as the angle between their tangent lines at that point. More specifically, two curves are said to be tangent at a point if they have the same tangent at a point, and orthogonal if their tangent lines are orthogonal.

Multiple tangents at a point

The formulas above fail when the point is a singular point. In this case there may be two or more branches of the curve that pass through the point, each branch having its own tangent line. When the point is the origin, the equations of these lines can be found for algebraic curves by factoring the equation formed by eliminating all but the lowest degree terms from the original equation. Since any point can be made the origin by a change of variables, this gives a method for finding the tangent lines at any singular point.

When the curve is not self-crossing, the tangent at a reference point may still not be uniquely defined because the curve is not differentiable at that point although it is differentiable elsewhere. In this case the left and right derivatives are defined as the limits of the derivative as the point at which it is evaluated approaches the reference point from respectively the left (lower values) or the right (higher values). For example, the curve $y = |x|$ is not differentiable at $x = 0$: its left and right derivatives have respective slopes -1 and 1; the tangents at that point with those slopes are called the left and right tangents.

Sometimes the slopes of the left and right tangent lines are equal, so the tangent lines coincide. This is true, for example, for the curve $y = x^{2/3}$, for which both the left and right derivatives at $x = 0$ are infinite; both the left and right tangent lines have equation $x = 0$.

Tangent to a Circle

The line that joins two infinitely close points from a point on the circle is a Tangent. In other words, we can say that the lines that intersect the circles exactly in one single point are Tangents. Point of tangency is the point where the tangent touches the circle. At the point of tangency, a tangent is perpendicular to the radius. Several theorems are related to this because it plays a significant role in geometrical constructions and proofs. We will look at them one by one.

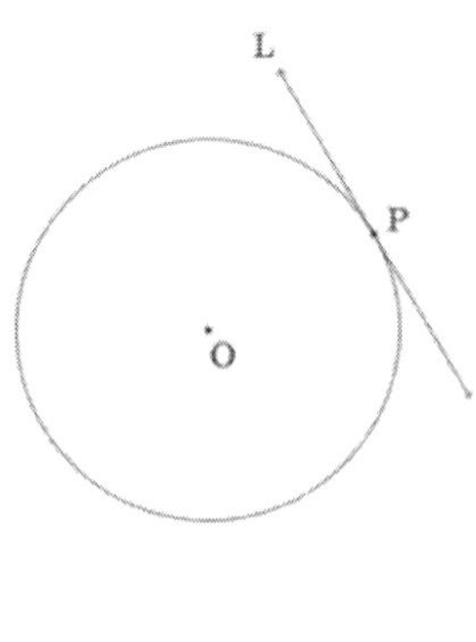

Tangents Formula

The formula for the tangent is given below. For the description of formula, please look at the following diagram.

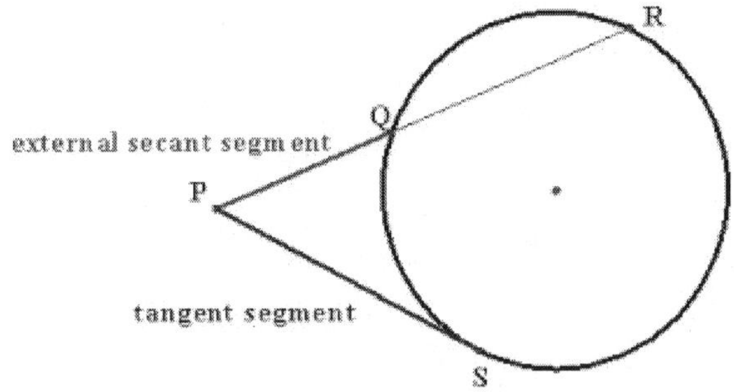

Here, we consider a circle where P is the exterior point. From that exterior point, the circle has the tangent at a points A and B. A straight line which cuts curve into two or more parts is known as a secant. So, here secant is PR is drawn and at Q, R intersects the circle as shown in the upper diagram. The formula for tangent-secant states that:

$PR/PS = PS/PQ$

$PS^2 = PQ.PR$

Properties of Tangents

Remember the following points about the properties of tangents-

- The tangent line never crosses the circle, it just touches the circle.
- At the point of tangency, it is perpendicular to the radius.
- A chord and tangent form an angle and this angle is same as that of tangent inscribed on the opposite side of the chord.
- From the same external point, the tangent segments to a circle are equal.

Theorems for Tangents to Circle

Theorem 1

A radius is obtained by joining the centre and the point of tangency. The tangent at a point on a circle is at right angles to this radius. Just follow this below diagram: Here AB¥"OP

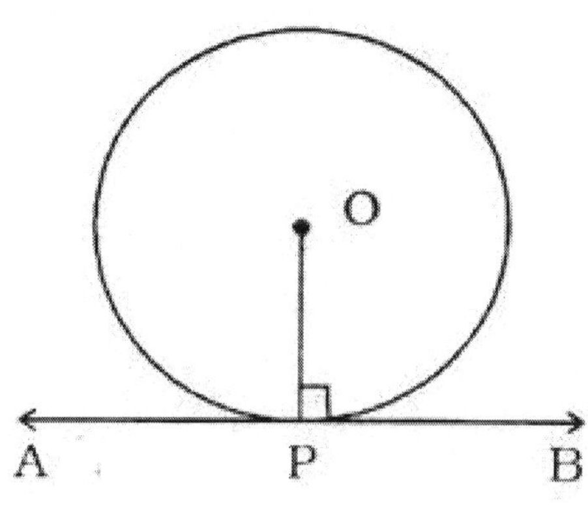

Theorem 2

This theorem states that if from one external point, two tangents are drawn to a circle then they have equal tangent segments. Tangent segment means line joining to the external point and the point of tangency. Consider the following diagram: He AC=BC.

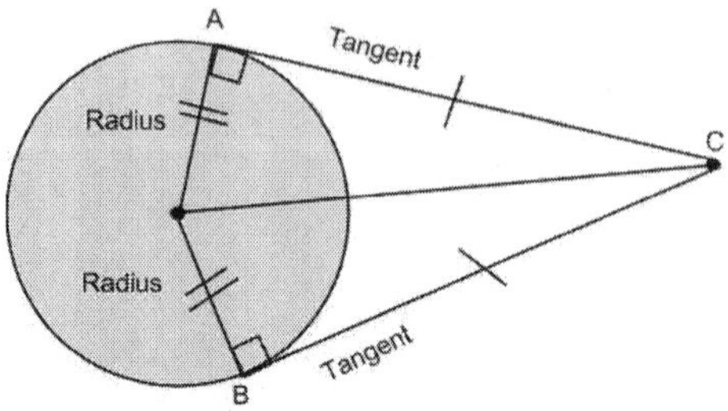

These are some of the basic theorems on tangents to a circle. In the next section, we will look at some of the basic properties of tangents.

Slope of a Straight Line

The tangent value of any trigonometric angle that a straight line makes with the positive direction of the x-axis in anticlockwise direction is called the slope or gradient of a straight line.

The angle of inclination of a line is the angle made by the line with the positive direction of the x-axis. It is usually measured from the positive x-axis in the anti-clockwise direction.

The slope of the line is generally denoted by 'm'. Thus, m = tan è. The gradient or slope of a line (not parallel to the axis of y) is the trigonometrical tangent of the angle which the line makes with the positive direction of the x-axis. Thus, if a line makes an angle è with the positive direction of the x-axis, then its slope will be tan è. The slop of a line is positive or negative according as è is acute or obtuse. Sine a line parallel to x-axis makes an angle of 0° with x-axis, therefore its slope is tan 0° = 0. A line parallel to y-axis that is i.e., perpendicular to x-axis makes an angle of 90° with x-axis, so its slope is tan À2δ2 = infinity. Also the slope of a line equally inclined with axes is 1 or -1 as it makes 45° or 135° angle with x-axis.

In short, the slope of a line is the trigonometrical tangent of its inclination.

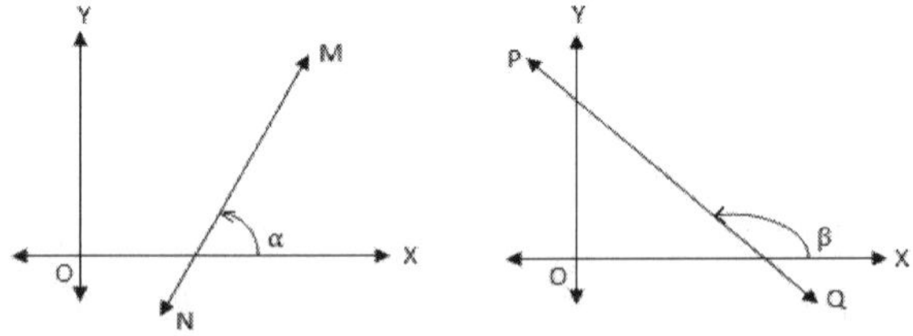

Slope of a Straight Line

In the above figure the inclination of the lines MN and PQ are á and â respectively.

Length of the Tangent to a Circle

Let the tangent drawn from the point $P(x_1, y_1)$ meet the circle at the point T as shown in the given diagram. The equation is given by

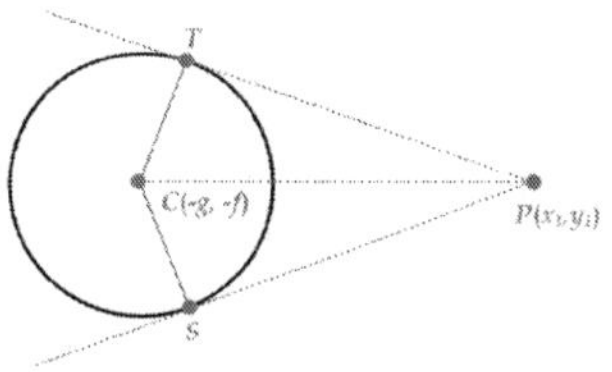

$$x^2+y^2+2gx+2fy+c=0 \quad ---(i)$$

Consider the triangle PTC formed in this way is a right triangle, so according to the given diagram we have

$$|PT|^2+|TC|^2=|PC|^2 \quad ---(ii)$$

It is observed that $|TC$ is the radius of the circle, so $|TC|^2=g^2+f^2-c.$

We also have

$$|PC|^2=(x_1-(-g))^2+(y_1-(-f))^2=(x_1+g)^2+(y_1+f)^2$$

Putting all these values in (ii), we get

This gives the length of the tangent from the point $P(x_1,y_1)$ to the circle

$$x^2+y^2+2gx+2fy+c=0.$$

Similarly, we can show that the PS is also of the same length.

Example: Find the length of the tangent from $(12,-9)$ to the circle

$$3x^2+3y^2-7x+22y+9=0$$

Dividing the equation of the circle by 3, we get the standard form

$$x^2+y^2-\frac{7}{3}x+\frac{22}{3}y+3=0$$

The required length of the tangent from is

$$\sqrt{(12)^2+(-9)^2-\frac{7}{3}(12)+\frac{22}{3}(-9)+3}=\sqrt{144+81-28-66+3}=\sqrt{134}$$

POLAR COORDINATE SYSTEM

In mathematics, the polar coordinate system is a two-dimensional coordinate system in which each point on a plane is determined by a distance from a reference point and an angle from a reference direction.

The reference point (analogous to the origin of a Cartesian coordinate system) is called the pole, and the ray from the pole in the reference direction is the polar axis. The distance from the pole is called the radial coordinate or radius, and the angle is called the angular coordinate, polar angle, or azimuth.

The concepts of angle and radius were already used by ancient peoples of the first millennium BC. The Greek astronomer and astrologer Hipparchus(190–120 BC) created a table of chord functions giving the length of the chord for each angle, and there are references to his using polar coordinates in establishing stellar positions. In On Spirals, Archimedes describes the Archimedean spiral, a function whose radius depends on the angle. The Greek work, however, did not extend to a full coordinate system.

From the 8th century AD onward, astronomers developed methods for approximating and calculating the direction to Mecca (qibla)—and its distance—from any location on the Earth. From the 9th century onward they were using spherical trigonometry and map projection methods to determine these quantities accurately. The calculation is essentially the conversion of the equatorial polar coordinates of Mecca (i.e. its longitude and latitude) to its polar coordinates (i.e. its qibla and distance) relative to a system whose reference meridian is the great circle through the given location and the Earth's poles, and whose polar axis is the line through the location and its antipodal point.

There are various accounts of the introduction of polar coordinates as part of a formal coordinate system. The full history of the subject is described in Harvard professor Julian Lowell Coolidge's Origin of Polar Coordinates. Grégoire de Saint-Vincent and Bonaventura Cavalieri independently introduced the concepts in the mid-seventeenth century. Saint-Vincent wrote about them privately in 1625 and published his work in 1647, while Cavalieri published his in 1635 with a corrected version appearing in 1653. Cavalieri first used polar coordinates to solve a problem relating to the area within an Archimedean spiral. Blaise Pascal subsequently used polar coordinates to calculate the length of parabolic arcs.

In Method of Fluxions (written 1671, published 1736), Sir Isaac Newton examined the transformations between polar coordinates, which he referred to as the "Seventh Manner; For Spirals", and nine other coordinate systems. In the journal Acta Eruditorum (1691), Jacob Bernoulli used a system with a point on a line, called the pole and polar axis respectively. Coordinates were specified by the distance from the pole and the angle from the polar axis. Bernoulli's work extended to finding the radius of curvature of curves expressed in these coordinates.

The actual term polar coordinates has been attributed to Gregorio Fontana and was used by 18th-century Italian writers. The term appeared in English in George Peacock's 1816 translation of Lacroix's Differential and Integral Calculus.[8] Alexis Clairaut was the first to think of polar coordinates in three dimensions, and Leonhard Euler was the first to actually develop them.

Conventions

The radial coordinate is often denoted by r or ñ, and the angular coordinate by ö, è, or t. The angular coordinate is specified as ö by ISO standard 31-11. However, in mathematical literature the angle is often denoted by è instead of ö.

Angles in polar notation are generally expressed in either degrees or radians (2ð rad being equal to 360°). Degrees are traditionally used in navigation, surveying, and many applied disciplines, while radians are more common in mathematics and mathematical physics.

The angle ö is defined to start at 0° from a reference direction, and to increase for rotations in either counterclockwise (ccw) or clockwise (cw) orientation. In mathematics, e.g., the reference direction is usually drawn as a ray from the pole horizontally to the right, and the polar angle increases to positive angles for ccw rotations, whereas in navigation (bearing, heading) the 0°-heading is drawn vertically upwards and the angle increases for cw rotations. The polar angles decrease towards negative values for rotations in the respectively opposite orientations.

Uniqueness of polar coordinates

Adding any number of full turns (360°) to the angular coordinate does not change the corresponding direction. Similarly, any polar coordinate is identical to the coordinate with the negative radial component and the opposite direction (adding 180° to the polar angle). Therefore, the same point (r, ö) can be expressed with an infinite number of different polar coordinates (r, ö + n × 360°) and ("r, ö + 180° + n × 360°) = ("r, ö + (2n + 1) × 180°), where n is an arbitrary integer. Moreover, the pole itself can be expressed as (0, ö) for any angle ö.

Where a unique representation is needed for any point besides the pole, it is usual to limit r to positive numbers (r > 0) and ö to the interval [0, 360°) or ("180°, 180°] (in radians, [0, 2δ) or ("δ, δ]). Another convention, in reverence to the usual codomain of the arctan-function, is to allow for arbitrary nonzero real values of the radial component and restrict the polar angle to ("90°, 90°]. In all cases a unique azimuth for the pole (r = 0) must be chosen, e.g., ö = 0.

Polar equation of a curve

The equation defining an algebraic curve expressed in polar coordinates is known as a polar equation. In many cases, such an equation can simply be specified by defining r as a function of ö. The resulting curve then consists of points of the form (r(ö), ö) and can be regarded as the graph of the polar function r. Note that, in contrast to Cartesian coordinates, the independent variable ö is the secondentry in the ordered pair.

Different forms of symmetry can be deduced from the equation of a polar function r. If r("ö) = r(ö) the curve will be symmetrical about the horizontal (0°/180°) ray, if r(δ " ö) = r(ö) it will be symmetric about the vertical (90°/270°) ray, and if r(ö " á) = r(ö) it will be rotationally symmetric by á clockwise and counterclockwise about the pole.

Because of the circular nature of the polar coordinate system, many curves can be described by a rather simple polar equation, whereas their Cartesian form is much more intricate. Among the best known of these curves are the polar rose, Archimedean spiral, lemniscate, limaçon, and cardioid.

For the circle, line, and polar rose below, it is understood that there are no restrictions on the domain and range of the curve.

Up to this point we've dealt exclusively with the Cartesian (or Rectangular, or x-y) coordinate system. However, as we will see, this is not always the easiest coordinate system to work in. So, in this section we will start looking at the polar coordinate system.

Coordinate systems are really nothing more than a way to define a point in space. For instance in the Cartesian coordinate system at point is given the coordinates (x,y)(x,y) and we use this to define the point by starting at the origin and then moving xx units horizontally followed by yy units vertically. This is shown in the sketch below.

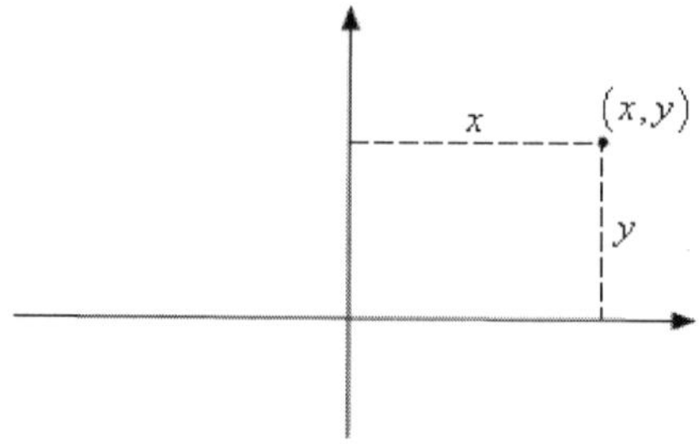

This is not, however, the only way to define a point in two dimensional space. Instead of moving vertically and horizontally from the origin to get to the point we could instead go straight out of the origin until we hit the point and then determine the angle this line makes with the positive xx-axis. We could then use the distance of the point from the origin and the amount we needed to rotate from the positive xx-axis as the coordinates of the point. This is shown in the sketch below.

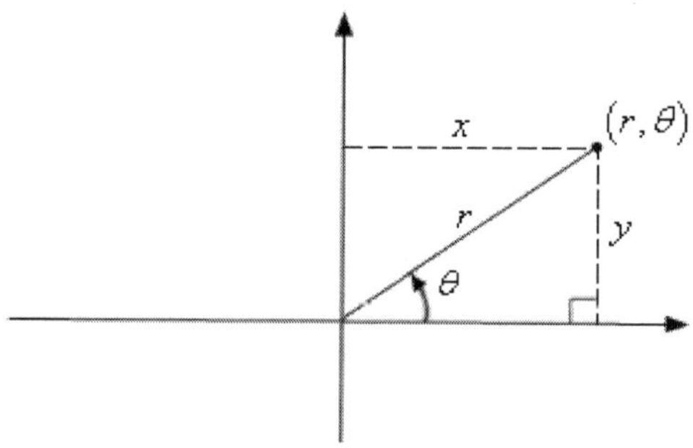

Coordinates in this form are called polar coordinates.

The above discussion may lead one to think that rr must be a positive number. However, we also allow rr to be negative. Below is a sketch of the two points (2,À6)(2,À6) and ("2,À6)("2,À6).

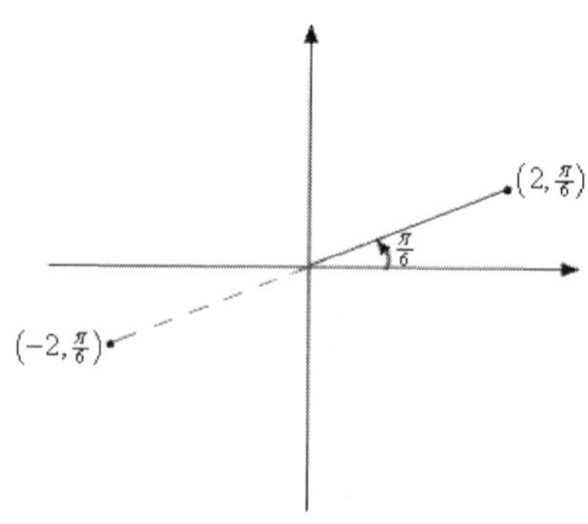

From this sketch we can see that if rr is positive the point will be in the same quadrant as ¸¸. On the other hand if rr is negative the point will end up in the quadrant exactly opposite ¸¸. Notice as well that the coordinates ("2,À6)("2,À6) describe the same point as the coordinates (2,7À6)(2,7À6) do. The coordinates

(2,7À6)(2,7À6) tells us to rotate an angle of 7À67À6 from the positive xx-axis, this would put us on the dashed line in the sketch above, and then move out a distance of 2.

This leads to an important difference between Cartesian coordinates and polar coordinates. In Cartesian coordinates there is exactly one set of coordinates for any given point. With polar coordinates this isn't true. In polar coordinates there is literally an infinite number of coordinates for a given point. For instance, the following four points are all coordinates for the same point.

(5,À3)=(5,″5À3)=(″5,4À3)=(″5,″2À3)(5,À3)=(5,″5À3)=(″5,4À3)=(″5,″2À3)

Here is a sketch of the angles used in these four sets of coordinates.

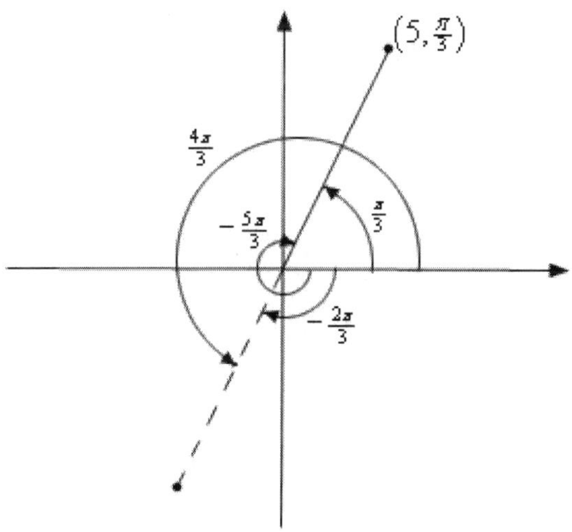

In the second coordinate pair we rotated in a clock-wise direction to get to the point. We shouldn't forget about rotating in the clock-wise direction. Sometimes it's what we have to do.

The last two coordinate pairs use the fact that if we end up in the opposite quadrant from the point we can use a negative rr to get back to the point and of course there is both a counter clock-wise and a clock-wise rotation to get to the angle.

These four points only represent the coordinates of the point without rotating around the system more than once. If we allow the angle to make as many complete rotations about the axis system as we want then there are an infinite number of coordinates for the same point. In fact, the point (r,¸)(r,¸) can be represented by any of the following coordinate pairs.

Next, we should talk about the origin of the coordinate system. In polar coordinates the origin is often called the pole. Because we aren't actually moving away from the origin/pole we know that r=0r=0. However, we can still rotate around the system by any angle we want and so the coordinates of the origin/pole are (0,¸)(0,¸).

Now that we've got a grasp on polar coordinates we need to think about converting between the two coordinate systems. Well start out with the following sketch reminding us how both coordinate systems work.

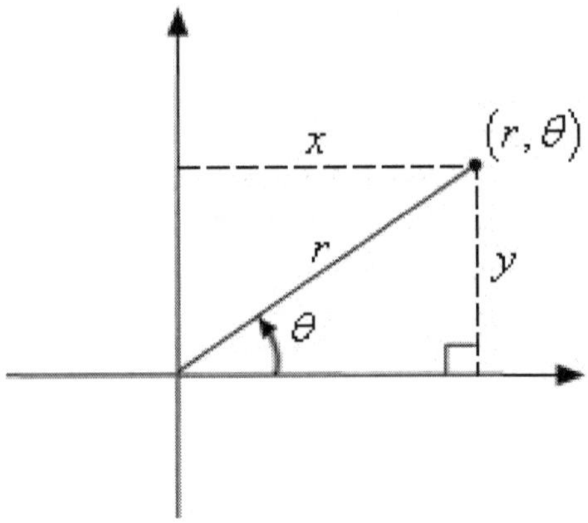

Note that we've got a right triangle above and with that we can get the following equations that will convert polar coordinates into Cartesian coordinates.

POLAR TO CARTESIAN CONVERSION FORMULAS

x=rcos¸y=rsin¸x=rcosa ¸y=rsina ¸

Converting from Cartesian is almost as easy. Let's first notice the following.

x2+y2=(rcos¸)2+(rsin¸)2=r2cos2¸+r2sin2¸=r2(cos2¸+sin2¸)=r2x2+y2=(rcosa ¸)2+(rsina ¸)2=r2cos2¸+r2sin2¸=r2(cos2¸+sin2¸)=r2

This is a very useful formula that we should remember, however we are after an equation for rr so let's take the square root of both sides. This gives,

r="x2+y2r=x2+y2

Note that technically we should have a plus or minus in front of the root since we know that rr can be either positive or negative. We will run with the convention of positive rr here.

Getting an equation for ¸¸ is almost as simple. We'll start with,

yx=rsin¸rcos¸=tan¸yx=rsina ¸rcosa ¸=tana ¸

Taking the inverse tangent of both sides gives,

¸=tan"1(yx)¸=tan"1(yx)

We will need to be careful with this because inverse tangents only return values in the range "À2<¸<À2"À2<¸<À2. Recall that there is a second possible angle and that the second angle is given by ¸+À¸+À.

Summarizing then gives the following formulas for converting from Cartesian coordinates to polar coordinates.

Any time we display a shape in ProcessingJS, we have to specify a pixel location, a set of x and y coordinates. These coordinates are known as Cartesian coordinates, named for René Descartes, the French mathematician who developed the ideas behind Cartesian space.

Another useful coordinate system known as polar coordinates describes a point in space as an angle of rotation around the origin and a radius from the origin. Thinking about this in terms of a vector:

- Cartesian coordinate—the x,y components of a vector
- Polar coordinate—the magnitude (length) and direction (angle) of a vector

However, the drawing functions in ProcessingJS don't understand polar coordinates. Whenever we want to display something in ProcessingJS, we have to specify locations as (x,y) Cartesian coordinates. However, sometimes it is a great deal more convenient for us to think in polar coordinates when designing. Happily for us, with trigonometry we can convert back and forth between polar and Cartesian, which allows us to design with whatever coordinate system we have in mind but always draw with Cartesian coordinates.

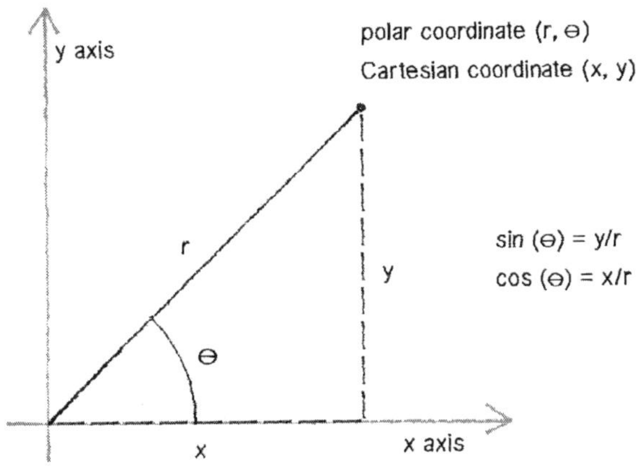

The Greek letter ¸ (theta) is often used to denote an angle, and a polar coordinate is conventionally referred to as (r, ¸) instead of (x, y). Thus, when dealing with polar coordinates, we'll now use "theta" as the preferred variable name for the angle.

sine(theta) = y/r '! y = r * sine(theta)

cosine(theta) = x/r '! x = r * cosine(theta)

For example, if r is 75 and theta is 45 degrees (or PI/4 radians), we can calculate x and y as below. The functions for sine and cosine in ProcessingJS are sin() and cos(), respectively. They each take one argument, an angle measured in degrees.

```
var r = 75;
var theta = 45;

// Convert polar to cartesian
var x = r * cos(theta);
var y = r * sin(theta);
```

This type of conversion can be useful in certain applications. For example, to move a shape along a circular path using Cartesian coordinates is not so easy. With polar coordinates, on the other hand, it's simple: increment the angle!

Conversion formulas

We can calculate the Cartesian coordinates of a point with polar coordinates (r,¸)(r,è) by forming the right triangle illustrated in the below figure. The hypotenuse is the line segment from the origin to the point, and its length is rr. The projection of this line segment on the xx-axis is the leg of the triangle adjacent to the angle ¸è, so x=rcos¸x=rcosa è. The yy-component is determined by the other leg, so y=rsin¸y=rsina è. Our conversion formula is

xy=rcos¸=rsin¸.(1)x=rcosa è(1)y=rsina è.

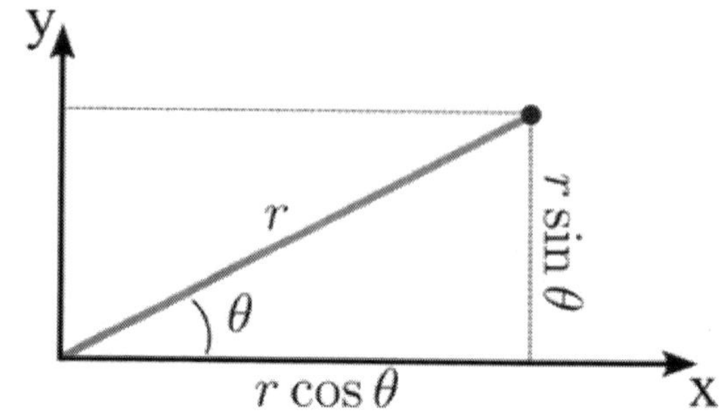

To go the other direction, one can use the same right triangle. Since rr is the distance from the origin to (x,y)(x,y), it is the magnitude r=x2+y2"″″″″″″″″r=x2+y2. Alternatively, from the equation (1)(1), one can calculate directly that

x2+y2=r2cos2¸+r2sin2¸=r2(cos2¸+sin2¸)=r2.x2+y2=r2cos2a è+r2sin2a è=r2(cos2a è+sin2a è)=r2.

Taking the ratio of yy and xx from equation (1)(1), one can obtain a formula for ¸è,

yx=rsin¸rcos¸=tan¸.yx=rsina èrcosa è=tana è.

One can also see this relationshp from the above right triangle. We can set ¸=arctanyxè=arctana yx, but a problem is that arctanarctan gives a value between ″À/2"ŏ/2 and À/2ŏ/2. One might neet to add Àŏ or 2À2ŏ to get the correct angle. With this caveat (and also mapping points where x=0x=0 to ¸=À/2è=ŏ/2 or ″À/ 2"ŏ/2), one obtains the following formula to convert from Cartesian to polar coordinates

r¸=x2+y2"″″″″″″″″=arctanyx.(2)r=x2+y2(2)è=arctana yx.

What about the point where (x,y)=(0,0)(x,y)=(0,0)? In this case, the angle ¸è isn't well defined. You could just take it to be ¸=0è=0 to be concrete.

Applications

Polar coordinates are two-dimensional and thus they can be used only where point positions lie on a single two-dimensional plane. They are most appropriate in any context where the phenomenon being considered is inherently tied to direction and length from a center point. For instance, the examples above show how elementary polar equations suffice to define curves—such as the Archimedean spiral—whose equation in the Cartesian coordinate system would be much more intricate. Moreover, many physical systems—such as those concerned with bodies moving around a central point or with phenomena originating from a central point—are simpler and more intuitive to model using polar coordinates. The initial motivation for the introduction of the polar system was the study of circular and orbital motion.

Position and navigation

Polar coordinates are used often in navigation as the destination or direction of travel can be given as an angle and distance from the object being considered. For instance, aircraftuse a slightly modified version of the polar coordinates for navigation. In this system, the one generally used for any sort of navigation, the 0°

ray is generally called heading 360, and the angles continue in a clockwise direction, rather than counterclockwise, as in the mathematical system. Heading 360 corresponds to magnetic north, while headings 90, 180, and 270 correspond to magnetic east, south, and west, respectively. Thus, an aircraft traveling 5 nautical miles due east will be traveling 5 units at heading 90 (read zero-niner-zero by air traffic control).

Modeling

Systems displaying radial symmetry provide natural settings for the polar coordinate system, with the central point acting as the pole. A prime example of this usage is the groundwater flow equation when applied to radially symmetric wells. Systems with a radial force are also good candidates for the use of the polar coordinate system. These systems include gravitational fields, which obey the inverse-square law, as well as systems with point sources, such as radio antennas.

Radially asymmetric systems may also be modeled with polar coordinates. For example, a microphone's pickup pattern illustrates its proportional response to an incoming sound from a given direction, and these patterns can be represented as polar curves. The curve for a standard cardioid microphone, the most common unidirectional microphone, can be represented as r = 0.5 + 0.5sin(Õ) at its target design frequency. The pattern shifts toward omnidirectionality at lower frequencies.

Angle between two curves

In order to measure the angle between two curves, we measure the angle between the tangents to the curves at that point. And that is obtained by the formula below:

$$\tan è = \left| \frac{m_1 - m_2}{1 + m_1 m_2} \right|$$

where è is the angle between the 2 curves, and m_1 and m_2 are slopes or gradients of the tangents to the curve at the point of intersection.

Let's look at this through an example below:

Find the acute angle between the curves $y = x^2$ and $y = (x - 3)^2$

The curve intersects at the point (where the two equations are equal). So

$x^2 = (x - 3)^2$

$x^2 = x^2 - 6x + 9$

$6x - 9 = 0$

$$x = \frac{9}{6} = \frac{3}{2}$$

$$y = x^2 = \left(\frac{3}{2}\right)^2 = \frac{9}{4}$$

The intersection point is $\left(\frac{3}{2}, \frac{9}{4}\right)$

The tangent to $y = x^2$ at has a gradient of
$= 2x = \quad = 3 = m_1$
The tangent to $y = (x - 3)^2$ at has a gradient of
$= 2(x - 3)^1 = \quad = \quad = -3 = m_2$
= 36° 522 (to the nearest minute)
Differential coefficient

In mathematics, differential coefficient is a term out of use now. The former differential coefficient of a function f(x) is now called its derivative df(x)/dx. It is the (not necessarily constant) multiplicative factor or coefficient of the differential dx in the differential df(x).

The differential coefficient of f is a constant function only if f is a linear function.

Early editions of Silvanus P. Thompson's Calculus Made Easy use the older term. In his 1998 update of this text, Martin Gardner puts "derivative" instead of "differential coefficient". He speaks of it only in context with Thompson's criticism of the term.

12

Curvature

In mathematics, curvature is any of a number of loosely related concepts in different areas of geometry. Intuitively, curvature is the amount by which a geometric object such as a surface deviates from being a flat plane, or a curve from being straight as in the case of a line, but this is defined in different ways depending on the context. There is a key distinction between extrinsic curvature, which is defined for objects embedded in another space (usually a Euclidean space) – in a way that relates to the radius of curvature of circles that touch the object – and intrinsic curvature, which is defined in terms of the lengths of curves within a Riemannian manifold. This article deals primarily with extrinsic curvature. Its canonical example is that of a circle, which has a curvature equal to the reciprocal of its radius everywhere. Smaller circles bend more sharply, and hence have higher curvature. The curvature of a smooth curve is defined as the curvature of its osculating circle at each point.

Curvature is normally a scalar quantity, but one may also define a curvature vector that takes into account the direction of the bend in addition to its magnitude. The curvature of more complex objects (such as surfaces or even curved n-dimensional spaces) is described by more complex objects from linear algebra, such as the general Riemann curvature tensor.

This article sketches the mathematical framework which describes the curvature of a curve embedded in a plane and the curvature of a surface in Euclidean space.

CURVATURE OF PLANE CURVES

Augustin-Louis Cauchy defined the center of curvature C as the intersection point of two infinitely close normals to the curve, the radius of curvature as the distance from the point to C, and the curvature itself as the inverse of the radius of curvature.

Let C be a plane curve (the precise technical assumptions are given below). The curvature of C at a point is a measure of how sensitive its tangent line is to moving the point to other nearby points. There are a number of equivalent ways that this idea can be made precise.

One way is geometrical. It is natural to define the curvature of a straight line to be constantly zero. The curvature of a circle of radius Rshould be large if R is small and small if R is large.

Another way to understand the curvature is physical. Suppose that a particle moves along the curve with unit speed. Taking the time s as the parameter for C, this provides a natural parametrization for the curve. The unit tangent vector T (which is also the velocity vector, since the particle is moving with unit speed) also depends on time. The curvature is then the magnitude of the rate of change of T. Symbolically,

This is the magnitude of the acceleration of the particle and the vector dT/ds is the acceleration vector. Geometrically, the curvature ê measures how fast the unit tangent vector to the curve rotates. If a curve keeps close to the same direction, the unit tangent vector changes very little and the curvature is small; where the curve undergoes a tight turn, the curvature is large.

These two approaches to the curvature are related geometrically by the following observation. In the first definition, the curvature of a circle is equal to the ratio of the angle of an arc to its length. Likewise, the curvature of a plane curve at any point is the limiting ratio of dè, an infinitesimal angle (in radians) between tangents to that curve at the ends of an infinitesimal segment of the curve, to the length of that segment ds, i.e., dè/ds. If the tangents at the ends of the segment are represented by unit vectors, it is easy to show that in this limit, the magnitude of the difference vector is equal to dè, which leads to the given expression in the second definition of curvature.

Precise definition

Signed curvature

The sign of the signed curvature k indicates the direction in which the unit tangent vector rotates as a function of the parameter along the curve. If the unit tangent rotates counterclockwise, then $k > 0$. If it rotates clockwise, then $k < 0$. So, for example, the sign of the curvature of the graph of a function is the same as the sign of the second derivative.

The signed curvature depends on the particular parametrization chosen for a curve. For example, the unit circle can be parametrised by (cos è, sin è) (counterclockwise, with $k > 0$), or by (cos("è), sin("è)) (clockwise, with $k < 0$). More precisely, the signed curvature depends only on the choice of orientation of an immersed curve. Every immersed curve in the plane admits two possible orientations.

Shape operator

Further information: Shape operator

A related notion of curvature is the shape operator, which is a linear operator from the tangent plane to itself. When applied to a tangent vector X to the surface, the shape operator is the tangential component of the rate of change of the normal vector when moved along a curve on the surface tangent to X. The principal curvatures are the eigenvalues of the shape operator, and in fact the shape operator and second fundamental form have the same matrix representation with respect to a pair of orthonormal vectors of the tangent plane. The Gauss curvature is thus the determinant of the shape tensor and the mean curvature is half its trace.

Curvature of space

By extension of the former argument, a space of three or more dimensions can be intrinsically curved. The curvature is intrinsic in the sense that it is a property defined at every point in the space, rather than a property defined with respect to a larger space that contains it. In general, a curved space may or may not be conceived as being embedded in a higher-dimensional ambient space; if not then its curvature can only be defined intrinsically.

After the discovery of the intrinsic definition of curvature, which is closely connected with non-Euclidean geometry, many mathematicians and scientists questioned whether ordinary physical space might be curved, although the success of Euclidean geometry up to that time meant that the radius of curvature must be astronomically large. In the theory of general relativity, which describes gravity and cosmology, the idea is slightly generalised to the "curvature of spacetime"; in relativity theory spacetime is a pseudo-Riemannian manifold. Once a time coordinate is defined, the three-dimensional space corresponding to a particular time is generally a curved Riemannian manifold; but since the time coordinate choice is largely arbitrary, it is the underlying spacetime curvature that is physically significant.

Although an arbitrarily curved space is very complex to describe, the curvature of a space which is locally isotropic and homogeneous is described by a single Gaussian curvature, as for a surface; mathematically these are strong conditions, but they correspond to reasonable physical assumptions (all points and all directions are indistinguishable). A positive curvature corresponds to the inverse square radius of curvature; an example is a sphere or hypersphere. An example of negatively curved space is hyperbolic geometry. A space or space-time with zero curvature is called flat. For example, Euclidean space is an example of a flat space, and Minkowski space is an example of a flat spacetime. There are other examples of flat geometries in both settings, though. A torus or a cylinder can both be given flat metrics, but differ in their topology. Other topologies are also possible for curved space. See also shape of the universe.

GENERALIZATIONS

The mathematical notion of curvature is also defined in much more general contexts. Many of these generalizations emphasize different aspects of the curvature as it is understood in lower dimensions.

One such generalization is kinematic. The curvature of a curve can naturally be considered as a kinematic quantity, representing the force felt by a certain observer moving along the curve; analogously, curvature in higher dimensions can be regarded as a kind of tidal force (this is one way of thinking of the sectional curvature). This generalization of curvature depends on how nearby test particles diverge or converge when they are allowed to move freely in the space; see Jacobi field.

Another broad generalization of curvature comes from the study of parallel transport on a surface. For instance, if a vector is moved around a loop on the surface of a sphere keeping parallel throughout the motion, then the final position of the vector may not be the same as the initial position of the vector. This phenomenon is known as holonomy.[8] Various generalizations capture in an abstract form this idea of curvature as a measure of holonomy; see curvature form. A closely related notion of curvature comes from gauge theory in physics, where the curvature represents a field and a vector potential for the field is a quantity that is in general path-dependent: it may change if an observer moves around a loop.

Two more generalizations of curvature are the scalar curvature and Ricci curvature. In a curved surface such as the sphere, the area of a disc on the surface differs from the area of a disc of the same radius in flat space. This difference (in a suitable limit) is measured by the scalar curvature. The difference in area of a sector of the disc is measured by the Ricci curvature. Each of the scalar curvature and Ricci curvature are defined in analogous ways in three and higher dimensions. They are particularly important in relativity theory, where they both appear on the side of Einstein's field equations that represents the geometry of spacetime (the other side of which represents the presence of matter and energy). These generalizations of curvature underlie, for instance, the notion that curvature can be a property of a measure; see curvature of a measure.

Another generalization of curvature relies on the ability to compare a curved space with another space that has constant curvature. Often this is done with triangles in the spaces. The notion of a triangle makes senses in metric spaces, and this gives rise to CAT(k) spaces.

Curvature, Intrinsic and Extrinsic

The extrinsic curvature ê of a plane curve at a given point on the curve is defined as the derivative of the curve's tangent angle with respect to position on the curve at that point. In other words, if è(s) denotes the angle which the curve makes with some fixed reference axis as a function of the path length s along the curve, then ê = dè/ds. In terms of orthogonal and naturally scaled coordinates X,Y we have tan(è) = dX/dY. If the X axis is tangent to the curve at the point in question, then tan(è) approaches è and dX approaches ds, so in terms of such tangent normal coordinates the curvature can equivalently be defined as simply the second derivative, ê = d^2Y/dX^2.

One way of specifying a plane curve is by giving a function Y = f(X) where X and Y are naturally scaled orthogonal coordinates. Natural scaling means $(ds)^2 = (dX)^2 + (dY)^2$, so we have

$$\frac{ds}{dX} = \sqrt{1+\left(\frac{dY}{dX}\right)^2}$$

The curvature can easily be determined by directly evaluating the derivative dè/ds as follows

$$\frac{d\theta}{ds} = \frac{dX}{ds}\frac{d\left(\operatorname{invtan}\left(\frac{dY}{dX}\right)\right)}{dX} = \frac{dX}{ds}\left(\frac{\frac{d^2Y}{dX^2}}{1+\left(\frac{dY}{dX}\right)^2}\right) = \frac{\frac{d^2Y}{dX^2}}{\left(1+\left(\frac{dY}{dX}\right)^2\right)^{3/2}}$$

Likewise if the curve is specified parametrically by the functions X(t) and Y(t) for some arbitrary path parameter t, we have ds/dt = $(X_t^2 + Y_t^2)^{1/2}$ where subscripts denote derivatives, and the curvature is

$$\frac{d\theta}{ds} = \frac{dt}{ds}\frac{d\left(\text{inv}\tan\left(\frac{Y_t}{X_t}\right)\right)}{dt} = \frac{dt}{ds}\left(\frac{\frac{X_tY_{tt} - Y_tX_{tt}}{X_t^2}}{1+\left(\frac{Y_t}{X_t}\right)^2}\right) = \frac{X_tY_{tt} - Y_tX_{tt}}{\left(X_t^2 + Y_t^2\right)^{3/2}}$$

Although these derivations are quite simple and satisfactory for the case of plane curves, it's worthwhile to examine both of them more closely to clarify the application to higher dimensional cases, where it is more convenient to use the definition of curvature based on the second derivative with respect to tangent coordinates. First, let's return to the case where the plane curve was specified by an explicit function Y = f(X) for naturally scaled orthogonal coordinates X,Y. Expanding this function into a power series (up to second order) about the point of interest, we have constants A,B,C such that $Y = AX^2 + BX + C$. The constant C is just a simple displacement, so it's irrelevant to the shape of the curve. Thus we need only consider the curve $Y = AX^2 + BX$. If B is non-zero this curve is not tangent to the X axis at the origin. To remedy this we can consider the curve with respect to a rotated system of coordinates x,y, related to the original coordinates by the transformation equations

$$X = \cos(\alpha)x + \sin(\alpha)y \qquad Y = -\sin(\alpha)x + \cos(\alpha)y$$

Substituting these expressions for X and Y into the equation $Y = AX^2 + BX$ and re-arranging terms gives

$$y = \frac{A\cos(\alpha)^2 x^2 + (B\cos(\alpha) + \sin(\alpha))x}{(\cos(\alpha) - B\sin(\alpha)) - A\sin(\alpha)(2\cos(\alpha)x + \sin(\alpha)y)}$$

If we select an angle á such that the coefficient of the linear term in the numerator vanishes, i.e., if we set Bcos(á) + sin(á) = 0 by putting á = invtan("B), then the numerator is purely second order. If we then expand the denominator into a power series in x and y, the product of this series with the numerator yields just a constant times the numerator plus terms of third and higher order in x and y. Hence the non-constant terms in the denominator are insignificant up to second order, so the denominator is effectively just equal to the constant term. Inserting the value of á into the above equation gives

$$y = \frac{A}{\left(1+B^2\right)^{3/2}}x^2$$

The curvature ê at the origin is just the second derivative, so we have

$$\kappa = \frac{2A}{\left(1+B^2\right)^{3/2}} = \frac{f_{XX}}{\left(1+f_X^2\right)^{3/2}}$$

where subscripts denote derivatives, and we have used the facts that, for the original function f(X) at the origin we have $f_X = B$ and $f_{XX} = 2A$. This shows how we can arrive (somewhat laboriously) at our previous result by using the "second derivative" definition of curvature and an explicitly defined curve Y = f(X).

A plane curve can be expressed parametrically as a function of the path length s by the functions x(s), y(s). Since $(ds)^2 = (dx)^2 + (dy)^2$, it follows that $x_s^2 + y_s^2 = 1$ (where again subscripts denote derivatives). The vector (x_s,y_s) is tangent to the curve, so the perpendicular vector ("y_s,x_s) is normal to the curve. The vector (x_{ss},y_{ss}) represents the rate of change of the tangent direction of the curve with respect to s. Recall that the curvature ê of a line in the plane is defined as the rate of change of the angle of the curve as a function of distance along the curve, but since tan(è) approaches è to the second order as è goes to zero, we can just as well define curvature as the rate of change of the tangent. Noting that y_s = (1"$x_s^2)^{1/2}$ we have y_{ss} = "$x_s x_{ss}$/(1"$x_s^2)^{1/2}$ and hence $y_s y_{ss}$ = "$x_s x_{ss}$. Thus we have y_{ss}/x_{ss} = "y_s/x_s, which means the vector (x_{ss},y_{ss}) is perpendicular to the curve. The magnitude of this vector is |ê| = $(x_{ss}^2 + y_{ss}^2)^{1/2}$, and we can define the signed magnitude as the dot product of (x_{ss},y_{ss}) with the vector ("y_s,x_s), and normalize this to the length of this vector, which happens to be $(x_s^2 + y_s^2)^{1/2} = 1$. This gives the signed curvature

$$\kappa = \{x_{ss}, y_{ss}\} \cdot \{-y_s, x_s\} = x_s y_{ss} - y_s x_{ss}$$

The center of curvature of the curve at the point (x,y) is at the point (x " y_s/|ê|, y + x_s/|ê|). To illustrate, a circle of radius R centered at the origin can be expressed by the parametric equations x(s) = Rcos(s/R) and y(s) = Rsin(s/R), and the first derivatives are x_s = "sin(s/R) and y_s = cos(s/R). The second derivatives are x_{ss} = "(1/R)cos(s/R) and y_{ss} = "(1/R)sin(s/R). From this we have the magnitude of the curvature |ê| = 1/R and the signed curvature +1/R. The sign is based on the path direction being positive in the counter-clockwise direction. The center of curvature for every point on this curve is the origin (0,0).

The preceding parametric derivation was based on the path length parameter s, but we can also define a curve in terms of an arbitrary parameter t, not necessarily the path length. In this case we have the functions x(t), y(t), and s(t). Again we have $(ds)^2 = (dx)^2 + (dy)^2$, so the derivatives of these three functions are related by $x_t^2 + y_t^2 = s_t^2$. We also have $x_s = x_t/s_t$ and $y_s = y_t/s_t$, and the second derivatives are

$$x_{ss} = \frac{1}{s_t}\frac{dx_s}{dt} = \frac{1}{s_t}\frac{d(x_t/s_t)}{dt} = \frac{s_t x_{tt} - x_t s_{tt}}{s_t^3}$$

and the similar expression for y_{ss}. Substituting into the previous formula for the signed curvature we get

$$\begin{aligned}\kappa &= \{x_{ss}, y_{ss}\} \cdot \{y_s, -x_s\} \\ &= \left\{\frac{s_t x_{tt} - x_t s_{tt}}{s_t^3}, \frac{s_t y_{tt} - y_t s_{tt}}{s_t^3}\right\} \cdot \left\{\frac{-y_t}{s_t}, \frac{x_t}{s_t}\right\} \\ &= \frac{x_t y_{tt} - y_t x_{tt}}{\left(x_t^2 + y_t^2\right)^{3/2}}\end{aligned}$$

The techniques described above for determining the curvature of plane curves can be used to determine the sectional curvatures of a two-dimensional surface embedded in three-dimensional space. The general power series expansion of a curve defined by the function f(x) is $f(x) = c_0 + c_1 x + c_2 x^2 + c_3 x^3 + \ldots$, but by choosing coordinates so that the curve passes through the origin tangent to the x axis at the point in question we can arrange to make $c_0 = c_1 = 0$, so the expansion of the curve about this point can be put into the form $f(x) = c_2 x^2 + c_3 x^3 + \ldots$ Also, since the 2nd derivative is $f''(x) = 2c_2 + 6c_3 x \ldots$, evaluating this at x = 0 gives simply $f''(0) = 2c_2$, so it's clear that only the 2nd-order term is significant in determining the curvature with respect to tangent normal coordinates, i.e., it is sufficient to represent the curve as $f(x) = ax^2$. Similarly if we consider the extrinsic curvature of a cross-section of a two-dimensional surface in space, we see that at any given point on the surface we can construct an orthogonal "xyz" coordinate system such that the xy plane is tangent to the surface and the z axis is perpendicular to the surface at that point. In general the equation of our surface can be expanded about this point into a polynomial giving the "height" z as a function of x and y. As in the one-dimensional case, the constant and 1st-order terms of this polynomial will be zero (because we defined our coordinates tangent to the surface with the origin at the point in question), and the 3rd and higher order terms don't affect the second derivative at the origin, so we can represent our surface by just the 2nd-order terms of the expansion, i.e.,

$$f(x,y) = ax^2 + bxy + cy^2$$

The second (partial) derivatives of this function with respect to x and y are 2a and 2c respectively, so these numbers give us some information about the curvature of the surface. However, we'd really like to know the curvature of the surface evaluated in any direction, not just in the x and y directions. (Note that the tangency condition uniquely determines the direction of the z axis, but the x and y axes can be rotated anywhere in the tangent plane.)

In general we can evaluate the curvature of the surface in the direction of the line y = qx for any constant q. The equation of the surface in this direction is simply

$$f(x,qx) = \left(a + bq + cq^2\right)x^2$$

but we want to evaluate the derivatives with respect to changes along this direction, rather than changes in the pure x direction. Parametrically the distance along the tangent plane in the y = qx direction is $s^2 = x^2 + y^2 = (1 + q^2)\,x^2$, so we can substitute for x^2 in the preceding equation to give the value of f as a function of the distance s

$$f(s) = \frac{a + bq + cq^2}{1 + q^2}\,s^2$$

The second derivative of this function gives the extrinsic curvature ê(q) of the surface in the "q" direction:

$$\kappa(q) = 2\frac{a+bq+cq^2}{1+q^2}$$

The directions of extreme (minimum and maximum) curvature are found by setting the derivative of ê(q) to zero, which gives

$$q^2 - 2mq - 1 = 0$$

where m = (c - a)/b. The two roots are $q_1 = m + (m^2 + 1)^{1/2}$ and $q_2 = m - (m^2 + 1)^{1/2}$. The product of these two roots is “1, so they are negative reciprocals of each other, and hence the lines of minimum and maximum curvature are of the form $y = q_1x$ and $y = (“1/q_1)x$, which shows that the two directions are perpendicular.

Substituting q_1 and q_2 into the equation for ê(q) gives the two “principal curvatures” of the surface

$$\kappa_1 = (a+c) + \sqrt{(c-a)^2 + b^2} \qquad \kappa_2 = (a+c) - \sqrt{(c-a)^2 + b^2}$$

The product of these two is called the “Gaussian curvature” of the surface at that point, and is given by

$$K = \kappa_1\kappa_2 = 4ac - b^2$$

which is just the (negative) discriminant of the quadratic form $ax^2 + bxy + cy^2$. For example, the surface of a sphere of radius R satisfies $x^2 + y^2 + z^2 = R^2$, so we have $z = (R^2 - x^2 - y^2)^{1/2}$, and we can factor out R^2 from the square root and expand to a series to give the elevation $h = R - z = x^2/(2R) + y^2/(2R)$ of the surface in the infinitesimal region around the “south pole” relative to the tangent plane at that point. Thus we have a = c = 1/(2R) and b = 0, so the curvature is $4ac - b^2 = 1/R^2$.

Another measure of the curvature of a surface is called the “mean curvature”, which, as the name suggests, is the mean value of the curvature over all possible directions. Since we want to give all the directions equal weight, we insert tan(è) for q in the equation for ê(q) and then integrate over è, giving the mean curvature

$$H = \frac{\kappa_1 + \kappa_2}{2} = a + c$$

(We could also infer this mean value directly as the average of $ê_1$ and $ê_2$ since ê is symmetrically distributed.) Notice that the mean curvature occurs along two perpendicular directions, and these are oriented at 45 degrees relative to the “principal” directions. This can be verified by setting the derivative of the product ê(q) ê(“1/q) to zero and noting that the resulting quartic in q factors into two quadratics, one giving the two principal directions, and the other giving the directions of the mean curvature. (The product ê(q) ê(“1/q) is obviously a maximum when both terms have the mean value, and a minimum when the terms have their extreme values.)

Examples of surfaces with constant Gaussian curvature are the sphere, the plane, and the pseudosphere, which have positive, zero, and negative curvature respectively. (Negative Gaussian curvature signifies that the two principal curvatures have opposite signs, meaning the surface has a "saddle" shape.) Surfaces with vanishing mean curvature are called "minimal surfaces", and represent the kinds of surfaces that are formed by a "soap films". For many years the only complete and non-self-intersecting minimal surfaces known were the plane, the catenoid, and the helicoid, but recently an infinite family of such minimal surfaces was discovered.

The above discussion was based on extrinsic properties of surfaces, i.e., measuring the rate of deviation between one surface and another. However, we can also look at curvature from an intrinsic standpoint, in terms of the relations between points within the surface itself. For example, if we were confined to the surface of a sphere of radius R, we would find that the ratio Q of the circumference to the "radius" of a circle as measured on the surface of the sphere would not be constant but would depend on the circle's radius r according to the relation Q = ő(R/r) sin(r/R). Evaluating the second derivative of Q with respect to r in the limit as r goes to zero we have

$$\frac{d^2Q}{dr^2} = -\frac{\pi}{3R^2}$$

Thus we can infer the radius of our sphere entirely from local measurements over a small region of the surface. The results of such local measurements of intrinsic distances on a surface can be encapsulated in the form of a "metric tensor" relative to any chosen system of coordinates on the surface.

In general, any two-dimensional surface embedded in three-dimensional space can be represented over a sufficiently small region by an expression of the form Z = f(X,Y) where X,Y,Z are orthogonal coordinates. The expansion of this function up to second order is

$$Z = AX^2 + BXY + CY^2 + DX + EY \tag{1}$$

where A,B..,E are constants. If the coefficients D and E are zero, the surface is tangent to the XY plane, and we can immediately compute the Gaussian curvature and the metric tensor as discussed previously. However, if D and E are not zero, we need to rotate our coordinates so that the XY plane is tangent to the surface. To accomplish this we can apply the usual Euler rotation matrix for a rotation through an angle á about the z axis followed by a rotation through an angle â about the (new) X axis. Thus we have a new system of orthogonal coordinates x,y,z related to the original coordinates by

$$X = \cos(\alpha)x + \sin(\alpha)\cos(\beta)y + \sin(\alpha)\sin(\beta)z$$

$$Y = -\sin(\alpha)x + \cos(\alpha)\cos(\beta)y + \cos(\alpha)\sin(\beta)z$$

$$Z = 0x - \sin(\beta)y + \cos(\beta)z$$

Making these substitutions for X,Y, and Z in (1) gives the equation of the surface in terms of the rotated coordinates. The coefficients of the linear terms in x and y in this transformed equation are

$$D\cos(\alpha) - E\sin(\alpha) \qquad\qquad D\sin(\alpha)\cos(\beta) + E\cos(\alpha)\cos(\beta) + \sin(\beta)$$

respectively. To make these coefficients vanish we must set

$$\alpha = \text{invtan}\left(\frac{D}{E}\right) \qquad\qquad \beta = -\text{invtan}\left(\sqrt{D^2 + E^2}\right)$$

Substituting these angles into the full expression gives

$$\sqrt{1+D^2+E^2}\, z = \frac{AE^2 - BED + CD^2}{D^2+E^2} x^2 + \frac{B\left(E^2 - D^2\right) + 2DE(A-C)}{\left(D^2+E^2\right)\sqrt{1+D^2+E^2}} xy + \frac{AD^2 + BDE + CE^2}{\left(D^2+E^2\right)\left(1+D^2+E^2\right)} y^2$$

The cross-product terms involving xz, yz, and z^2 have been omitted, because if we bring these over to the left side and factor out z, we can then divide both sides by the factor $(k_1 + k_2x + k_3y + k_4z)$, and the power series expansion of this, multiplied by the second-order terms in x and y, gives just a constant times those terms, plus terms of third and higher order in x,y, and z, which do not affect the curvature at the origin. Therefore, the second-order terms involving z drop out, and we're left with the above quadratic for z. This describes a surface tangent to the xy plane at the origin, i.e., a surface of the form $z = ax^2 + bxy + cy^2$, and the curvature of such a surface equals 4ac"b^2 at the origin, so the curvature of the above surface at the origin is

$$K = \frac{4AC - B^2}{\left(1 + D^2 + E^2\right)^2}$$

Remember that we began with a surface defined by the function Z = f(X,Y), and from equation (1) we see that the partial derivatives of the function f with respect to X and Y at the origin are

$$f_X = D \qquad f_Y = E \qquad f_{XX} = 2A \qquad f_{XY} = B \qquad f_{YY} = 2C$$

Consequently, the equation for the curvature of the surface can be written as

$$K = \frac{f_{XX}f_{YY} - f_{XY}^{\;2}}{\left(1 + f_X^{\;2} + f_Y^{\;2}\right)^2}$$

In addition, if we take the differentials of both sides of (1) we have

$$dZ = 2AXdX + BXdY + BYdX + 2CYdY + DdX + EdY$$

Inserting this for dZ into the metrical expression $(ds)^2 = (dX)^2 + (dY)^2 + (dZ)^2$ gives the metric at the origin on the surface with respect to the XY coordinates projected onto the surface:

$$(ds)^2 = g_{XX}(dX)^2 + g_{XY}(dX)(dY) + g_{YX}(dY)(dX) + g_{YY}(dY)^2$$

where

$$g_{XX} = 1 + D^2 \qquad g_{XY} = g_{YX} = DE \qquad g_{YY} = 1 + E^2$$

Thus the curvature of the surface can also be written in the form

$$K = \frac{f_{XX}f_{YY} - f_{XY}^2}{\left(g_{XX}g_{YY} - g_{XY}^2\right)^2} = \frac{\left(\frac{f_{XX}}{\sqrt{g}}\right)\left(\frac{f_{YY}}{\sqrt{g}}\right) - \left(\frac{f_{XY}}{\sqrt{g}}\right)^2}{g_{XX}g_{YY} - g_{XY}^2}$$

where $g = g_{XX}g_{YY}$ " g_{XY}^2. The quantities in the numerator of the right hand expression are the coefficients of the "second groundform" of the surface, and the metric line element is called the first groundform. Hence the curvature is simply the ratio of the determinants of the two groundforms.

The preceding was based on treating the 2D surface embedded in 3D space defined by giving Z explicitly as a function of X and Y. This is analogous to our treatment of curves in the plane based on giving Y as an explicit function of X. However, we found that a more general and symmetrical expression for the curvature of a plane curve was found by considering the curve defined parametrically, i.e., giving x(u) and y(u) as functions of an arbitrary path parameter u. Similarly we can define a 2D surface in 3D space by giving x(u,v), y(u,v) and z(u,v) as functions of two arbitrary coordinates on the surface. From the Euclidean metric of the embedding 3D space we have

$$s_u^2 = x_u^2 + y_u^2 + z_u^2 \qquad s_v^2 = x_v^2 + y_v^2 + z_v^2$$

where subscripts denote partial derivatives. We also have the total differentials

$$dx = x_u du + x_v dv \qquad dy = y_u du + y_v dv \qquad dz = z_u du + z_v dv$$

which can be substituted into the basic 3D Euclidean metric $(ds)^2 = (dx)^2 + (dy)^2 + (dz)^2$ to give the 2D metric of the surface with respect to the arbitrary surface coordinates u,v

$$(ds)^2 = g_{uu}(du)^2 + g_{uv}(du)(dv) + g_{vu}(dv)(du) + g_{vv}(dv)^2$$

where

$$g_{uu} = x_u^2 + y_u^2 + z_u^2 \qquad g_{uv} = x_u x_v + y_u y_v + z_u z_v \qquad g_{vv} = x_v^2 + y_v^2 + z_v^2$$

and $g_{vu} = g_{uv}$. The space-vectors $[x_u, y_u, z_u]$ and $[x_v, y_v, z_v]$ are tangent to the surface and point along the u and v directions, respectively, so the cross-product of these two vectors is a vector normal to the surface

$$[x_u, y_u, z_u] \times [x_v, y_v, z_v] = [(y_u z_v - z_u y_v), (z_u x_v - x_u z_v), (x_u y_v - y_u x_v)]$$

whose magnitude is

$$\sqrt{(y_u z_v - z_u y_v)^2 + (z_u x_v - x_u z_v)^2 + (x_u y_v - y_u x_v)^2} = \sqrt{g_{uu} g_{vv} - g_{uv}^2}$$

The space-vectors $[x_{uu}, y_{uu}, z_{uu}]$ and $[x_{vv}, y_{vv}, z_{vv}]$ represent the rates of change of the tangent vectors to the surface along the u and v directions, and the vector $[x_{uv}, y_{uv}, z_{uv}]$ represents the rate of change of the u tangent with respect to v, and vice versa. Thus if we take the dot products of each of these vectors with the unit vector normal to the surface, we will get the signed coefficients of an expression for the surface of the pure quadratic form $h(u,v) = au^2 + buv + cv^2$, where "h" can be regarded as the height above the tangent plane at the origin, and the three scaled triple products correspond to $h_{uu} = 2a$, $h_{uv} = b$, and $h_{vv} = 2c$.

If u and v were projections of orthogonal coordinates (as were x and y in our prior discussion), the determinant of the surface metric at the origin would be 1, and the curvature would simply be 4ac"b². However, in general we allow u and v to be any surface coordinates, not necessarily orthogonal, and not necessarily scaled to equal the path length along constant coordinate lines. Given orthogonal metrically scaled tangent coordinates X,Y, there exist coefficients A,B,C such that the height h above the tangent plane is $h(X,Y) = AX^2 + BXY + CY^2$, and the curvature K at the origin is simply 4AC"B². Also, for points sufficiently near the origin we have

$$X = X_u u + X_v v \qquad\qquad Y = Y_u u + Y_v v$$

Substituting these expressions into h(X,Y) gives $h(u,v) = au^2 + buv + cv^2$ where

$$\begin{aligned} a &= AX_u^2 + BX_u X_v + CX_v^2 \\ b &= 2AX_u Y_u + B(X_u Y_v + X_v Y_u) + 2CX_v Y_v \\ c &= AY_u^2 + BY_u Y_v + CY_v^2 \end{aligned}$$

With these coefficients we find

$$4ac - b^2 = \left(4AC - B^2\right)(X_u Y_v - X_u Y_v)^2$$

In addition, we know that the surface is asymptotic to the tangent plane at the origin, so the metric in terms of X,Y is simply $(ds)^2 = (dX)^2 + (dY)^2$. Substituting the expressions for dX and dY in terms of du and dv, the metric at the origin in terms of the u,v coordinates is

$$(ds)^2 = \left[X_u^2 + Y_u^2\right](du)^2 + 2[X_u X_v + Y_u Y_v](du)(dv) + \left[X_v^2 + Y_v^2\right](dv)^2$$

From this we have the determinant of the metric

$$g = \left[X_u^2 + Y_u^2\right]\left[X_v^2 + Y_v^2\right] - [X_u X_v + Y_u Y_v]^2 = (X_u Y_v - X_u Y_v)^2$$

This shows that the intrinsic curvature K is related to the quantity 4ac"b^2 by the equation

$$K = \left(4AC - B^2\right) = \frac{4ac - b^2}{g}$$

We saw previously that the coefficients 2a,b,2c are given by triple vector products divided by the normalizing factor ™g. Writing out the triple products in determinant form, we have

$$2a = \frac{1}{\sqrt{g}} \begin{vmatrix} x_{uu} & y_{uu} & z_{uu} \\ x_u & y_u & z_u \\ x_v & y_v & z_v \end{vmatrix}$$

$$b = \frac{1}{\sqrt{g}} \begin{vmatrix} x_{uv} & y_{uv} & z_{uv} \\ x_u & y_u & z_u \\ x_v & y_v & z_v \end{vmatrix}$$

$$2c = \frac{1}{\sqrt{g}} \begin{vmatrix} x_{vv} & y_{vv} & z_{vv} \\ x_u & y_u & z_u \\ x_v & y_v & z_v \end{vmatrix}$$

Therefore the Gaussian curvature is given by

$$K = \frac{1}{g^2}\left[\begin{vmatrix} x_{uu} & y_{uu} & z_{uu} \\ x_u & y_u & z_u \\ x_v & y_v & z_v \end{vmatrix} \begin{vmatrix} x_{vv} & y_{vv} & z_{vv} \\ x_u & y_u & z_u \\ x_v & y_v & z_v \end{vmatrix} - \begin{vmatrix} x_{uv} & y_{uv} & z_{uv} \\ x_u & y_u & z_u \\ x_v & y_v & z_v \end{vmatrix}^2 \right]$$

Recalling that the determinant of the transpose of a matrix is the same as of the matrix itself, we can transpose the second factor in each determinant product to give the equivalent expression

$$K = \frac{1}{g^2}\left[\begin{vmatrix} x_{uu} & y_{uu} & z_{uu} \\ x_u & y_u & z_u \\ x_v & y_v & z_v \end{vmatrix} \begin{vmatrix} x_{vv} & x_u & x_v \\ y_{vv} & y_u & y_v \\ z_{vv} & z_u & z_v \end{vmatrix} - \begin{vmatrix} x_{uv} & y_{uv} & z_{uv} \\ x_u & y_u & z_u \\ x_v & y_v & z_v \end{vmatrix} \begin{vmatrix} x_{uv} & x_u & x_v \\ y_{uv} & y_u & y_v \\ z_{uv} & z_u & z_v \end{vmatrix} \right]$$

The determinant of a product of matrices is the same as the product of the determinants of those matrices, so we can carry out the matrix multiplications inside the determinant symbols. The first product of determinants can therefore be written as the single determinant

$$\begin{vmatrix} x_{uu}x_{vv}+y_{uu}y_{vv}+z_{uu}z_{vv} & x_u x_{uu}+y_u y_{uu}+z_u z_{uu} & x_v x_{uu}+y_v y_{uu}+z_v z_{uu} \\ x_u x_{vv}+y_u y_{vv}+z_u z_{vv} & x_u^2+y_u^2+z_u^2 & x_u x_v+y_u y_v+z_u z_v \\ x_v x_{vv}+y_v y_{vv}+z_v z_{vv} & x_u x_v+y_u y_v+z_u z_v & x_v^2+y_v^2+z_v^2 \end{vmatrix}$$

Notice that several of the entries in this matrix can be expressed purely in terms of the components g_{uu}, g_{uv}, and g_{vv} of the metric tensor and their partial derivatives, so we can write this determinant as

$$\begin{vmatrix} x_{uu}x_{vv}+y_{uu}y_{vv}+z_{uu}z_{vv} & \frac{1}{2}g_{uu,u} & g_{uv,u}-\frac{1}{2}g_{uu,v} \\ g_{uv,v}-\frac{1}{2}g_{vv,u} & g_{uu} & g_{uv} \\ \frac{1}{2}g_{vv,v} & g_{vu} & g_{vv} \end{vmatrix}$$

In a similar manner we can expand the second product of determinants into a single determinant and express most of the resulting components in terms of the metric to give

$$\begin{vmatrix} x_{uv}^2+y_{uv}^2+z_{uv}^2 & \frac{1}{2}g_{uu,v} & \frac{1}{2}g_{vv,u} \\ \frac{1}{2}g_{uu,v} & g_{uu} & g_{uv} \\ \frac{1}{2}g_{vv,u} & g_{vu} & g_{vv} \end{vmatrix}$$

The curvature is just $1/g^2$ times the difference between these two determinants. In both cases we have been able to express all the matrix components in terms of the metric, with the exception of the upper-left entries. However, notice that the cofactors of these two entries in their respective matrices are identical (namely g), so when we take the difference of these determinants the upper-left entries both appear simply as multiples of g. Thus we need only consider the difference of these two entries, which can indeed be written purely in terms of the metric coefficients and their derivatives as follows

$$(x_{uu}x_{vv}+y_{uu}y_{vv}+z_{uu}z_{vv})-(x_{uv}^2+y_{uv}^2+z_{uv}^2)=g_{uv,uv}-\frac{1}{2}g_{uu,vv}-\frac{1}{2}g_{vv,uu}$$

Consequently, we can express the Gaussian curvature K entirely in terms of the intrinsic metric with respect to arbitrary two-dimensional coordinates on the surface, as follows

$$
\begin{aligned}
K = {} & \frac{1}{g}\left(g_{uv,uv} - \frac{1}{2}g_{uu,vv} - \frac{1}{2}g_{vv,uu}\right) \\
& + \frac{g_{uu}}{g^2}\left(\frac{g_{vv,u}^{\ 2}}{4} + \frac{g_{vv,v}g_{uu,v}}{4} - \frac{g_{uv,u}g_{vv,v}}{2}\right) \\
& + \frac{g_{vv}}{g^2}\left(\frac{g_{uu,v}^{\ 2}}{4} + \frac{g_{uu,u}g_{vv,u}}{4} - \frac{g_{uv,v}g_{uu,u}}{2}\right) \\
& + \frac{g_{uv}}{g^2}\left(\frac{g_{uu,u}g_{vv,v}}{4} - \frac{g_{uu,v}g_{vv,u}}{4} - \frac{g_{uv,v}g_{uu,v}}{2} - \frac{g_{vv,u}g_{uv,u}}{2} + g_{uv,v}g_{uv,u}\right)
\end{aligned}
$$

This formula was first presented by Gauss in his famous paper "Disquisitiones generales circa superficies curvas" (General Investigations of Curved Surfaces), published in 1827. Gauss regarded this result as quite remarkable (egregium in Latin), so it is commonly known as the Theorema Egregium. The reason for Gauss' enthusiasm is that this formula proves the Gaussian curvature of a surface is indeed intrinsic, i.e., it is not dependent on the embedding of the surface in higher dimensional space. Operating entirely within the surface we can lay out arbitrary curvilinear coordinates u,v, and then determine the metric coefficients (and their derivatives) with respect to those coordinates, and from this information alone we can compute the intrinsic curvature of the surface. The Gaussian curvature K is defined as the product of the two principle extrinsic sectional curvatures $\hat{e}_1$ and $\hat{e}_2$, neither of which is an intrinsic metrical property of the surface, but the product of these two numbers is an intrinsic metrical property.

In Section 5.7 the full Riemann curvature tensor R_{abcd} for manifolds of any number of dimensions is defined, and we show that Gauss' surface curvature K is equal to R_{uvuv}/g, which completely characterizes the curvature of a two-dimensional surface. To highlight the correspondence between Gauss' formula and the full curvature tensor, we can re-write the above formula as

$$
\begin{aligned}
gK = {} & \frac{1}{2}\left(g_{uv,uv} + g_{uv,uv} - g_{uu,vv} - g_{vv,uu}\right) \\
& + g^{uu}\left[\left(\frac{g_{uu,v}}{2}\right)^2 - \left(\frac{g_{uu,u}}{2}\right)\left(\frac{2g_{uv,v} - g_{vv,u}}{2}\right)\right] \\
& + g^{uv}\left[\left(\frac{g_{uu,v}}{2}\right)\left(\frac{g_{vv,u}}{2}\right) - \left(\frac{g_{uu,u}}{2}\right)\left(\frac{g_{vv,v}}{2}\right)\right] \\
& + g^{vu}\left[\left(\frac{g_{vv,u}}{2}\right)\left(\frac{g_{uu,v}}{2}\right) - \left(\frac{2g_{vu,u} - g_{uu,v}}{2}\right)\left(\frac{2g_{uv,v} - g_{vv,u}}{2}\right)\right] \\
& + g^{vv}\left[\left(\frac{g_{vv,u}}{2}\right)^2 - \left(\frac{g_{vv,v}}{2}\right)\left(\frac{2g_{uv,u} - g_{uu,v}}{2}\right)\right]
\end{aligned}
$$

where we have used the facts that $g_{uu}/g = g^{vv}$, $g_{vv}/g = g^{uu}$, and $g_{uv}/g =$ "g^{uv}. Notice that if we define the symbol

$$\Gamma_{abc} = \frac{1}{2}\left(g_{ab,c} + g_{ac,b} - g_{bc,a}\right)$$

for any three indices a,b,c, then Gauss' formula for the curvature of a surface can be written more succinctly as

$$\begin{aligned} gK &= \frac{1}{2}\left(g_{uv,uv} + g_{uv,uv} - g_{uu,vv} - g_{vv,uu}\right) \\ &+ g^{uu}\left[\Gamma_{uuv}\Gamma_{uvu} - \Gamma_{uuu}\Gamma_{uvv}\right] + g^{uv}\left[\Gamma_{uuv}\Gamma_{vvu} - \Gamma_{uuu}\Gamma_{vvv}\right] \\ &+ g^{vu}\left[\Gamma_{vuv}\Gamma_{uvu} - \Gamma_{vuu}\Gamma_{uvv}\right] + g^{vv}\left[\Gamma_{vuv}\Gamma_{vvu} - \Gamma_{vuu}\Gamma_{vvv}\right] \end{aligned}$$

No summations are implied here, but to abbreviate the notation even further, we could designate the symbols á and â as "wild card" indices, with implied summation of every term in which they appear over all possible indices (i.e., over u and v). On this basis the formula is

$$gK = \frac{1}{2}\left(g_{uv,uv} + g_{uv,uv} - g_{uu,vv} - g_{vv,uu}\right) + g^{\alpha\beta}\left[\Gamma_{\alpha uv}\Gamma_{\beta vu} - \Gamma_{\alpha uu}\Gamma_{\beta vv}\right]$$

As discussed in Section 5.7, this is precisely the formula for the component R_{uvuv} of the full Riemann curvature tensor in n dimensions, which makes it clear how directly Gauss' result for two-dimensional surfaces generalizes to n dimensions. Naturally this formula for K reduces to $\hat{e}_1\hat{e}_2 = 4ac - b^2$ where $\hat{e}_1$ and $\hat{e}_2$ are the two principal extrinsic curvatures relative to a flat plane tangent to the surface at the point of interest. The reason this formula is so complicated is that it applies to any system of coordinates (rather than just projected tangent normal coordinates), and is based entirely on the intrinsic properties of the surface.

To illustrate this approach, consider the two-dimensional surface defined as the locus of points at a height h above the xy plane, where h is given by the equation

$$h = ax^2 + bxy + cy^2$$

with arbitrary constants a, b, and c. For example, with a = c = 0 and b = 1 this gives the simple surface h = xy shown below:

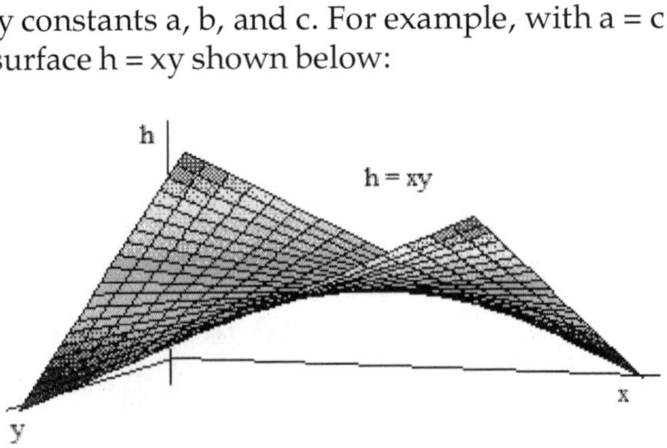

For other values of a,b,c this surface can have various shapes, such as paraboloids. The function h(x,y) is single-valued over the entire xy plane, so it's convenient to simply project the xy grid onto the surface and use this as our coordinates on the surface. (Any other system of curvilinear coordinates would serve just as well.)

Over a sufficiently small interval on this surface the distance ds along a path is related to the incremental changes dx, dy, and dh according to the usual Pythagorean relation

$$(ds)^2 = (dx)^2 + (dy)^2 + (dh)^2$$

Also the equation of the surface allows us to express the increment dh in terms of dx and dy as follows

$$dh = (2ax + by)dx + (bx + 2cy)dy$$

Therefore we have

$$\begin{aligned}(dh)^2 = {} & (2ax + by)^2 (dx)^2 \\ & + 2(2ax + by)(bx + 2cy)(dx)(dy) \\ & + (bx + 2cy)^2 (dy)^2\end{aligned}$$

Substituting this into the equation for the line element $(ds)^2$ gives the basic metrical equation of the surface

$$(ds)^2 = g_{xx}(dx)^2 + 2g_{xy}(dx)(dy) + g_{yy}(dy)^2$$

where the components of the "metric tensor" are

$$\begin{aligned} g_{xx} &= 1 + (2ax + by)^2 \\ g_{xy} &= g_{yx} = (2ax + by)(bx + 2cy) \\ g_{yy} &= 1 + (bx + 2cy)^2 \end{aligned}$$

We can, in principle, directly measure the incremental distance ds for any given increments dx and dy without ever leaving the surface, so the metric components are purely intrinsic properties of the surface. In general the metric tensor is a symmetric covariant tensor of second order, and is usually written in the form of a matrix. Thus, for our simple example we can write the metric as

$$g_{uv} = \begin{bmatrix} 1 + (2ax + by)^2 & (2ax + by)(bx + 2cy) \\ (2ax + by)(bx + 2cy) & 1 + (bx + 2cy)^2 \end{bmatrix}$$

The determinant of this matrix at the point (x,y) is

$$g = 1 + (2ax + by)^2 + (bx + 2cy)^2$$

The inverse of the metric tensor is denoted by g^{uv}, where the superscripts are still indices, not exponents. In our example the inverse metric tensor is

$$g^{uv} = \frac{1}{g}\begin{bmatrix} 1+(bx+2cy)^2 & -(2ax+by)(bx+2cy) \\ -(2ax+by)(bx+2cy) & 1+(2ax+by)^2 \end{bmatrix}$$

Substituting these metric components into the general formula for the Gaussian curvature K gives

$$K(x, y) = \frac{4ac - b^2}{\left(1 + (2ax + by)^2 + (bx + 2cy)^2\right)^2}$$

in agreement with our earlier result for surfaces specified explicitly in the form $z = f(x,y)$. At the origin, where $x = y = 0$, this gives $K = 4ac - b^2$, i.e., the product of the two principal extrinsic curvatures. In addition, the formula gives the Gaussian curvature for any point on the surface, so we don't have to go to the trouble of laboriously constructing a tangent plane at each point and finding the quadratic expansion of the surface about that point.

We can see from this formula that the curvature at every point of this simple two-dimensional surface always has the same sign as the discriminant $4ac - b^2$. Also, the shape of the constant-curvature lines on this surface can be determined by re-arranging the terms of the above equation, from which we find that the curvature equals K on the locus of points satisfying the equation

$$\left(4a^2 + b^2\right)x^2 + 4b(a+c)xy + \left(b^2 + 4c^2\right)y^2 = \sqrt{\frac{4ac - b^2}{K}} - 1$$

This is the equation of a conic with discriminant $+4(4ac - b^2)^2$. The case of zero curvature occurs only when the discriminant vanishes, which implies that $b = 2\sqrt{ac}$, and so the equation of the surface factors as

$$h = \left(\sqrt{a}\,x + \sqrt{b}\,y\right)^2$$

The quantity inside the parentheses is a planar function, so the surface is a parabolic "valley", which has no intrinsic curvature (like the walls of a cylinder).

It follows from the preceding conic equation that the lines of constant curvature (if there is any curvature) must be ellipses centered on the origin. However, this is not the most general form of curvature possible on a two-dimensional surface, it's just the most general form for a surface embedded in three-dimensional Euclidean space. Suppose we embed our two-dimensional surface in four dimensional Euclidean space. We can still, at any given point on the surface, construct a two-dimensional tangent plane with orthogonal xy coordinates, and expand the equation of the surface up to second degree about that point, but now instead of just a single perpendicular height h(x,y) we allow two mutually perpendicular heights, which we may call $h_1(x,y)$ and $h_2(x,y)$. Our surface can now be defined (in the neighborhood of the origin at the point of tangency) by the equations

$$h_1 = Ax^2 + Bxy + Cy^2 \qquad h_2 = ax^2 + bxy + cy^2$$

Following the same procedure as before, determining the components of the metric tensor for this surface and inserting them into Gauss's formula, we find that the intrinsic curvature of this surface is

$$K = \frac{(4AC - B^2)(1 + m^2 + n^2) + (4ac - b^2)(1 + M^2 + N^2) - 2(Mm + Nn)(2Ac + 2aC - Bb)}{\left[1 + M^2 + N^2 + m^2 + n^2 + (Mn - mN)^2\right]^2}$$

where

$$M = 2Ax + By \qquad N = Bx + 2Cy \qquad m = 2ax + by \qquad n = bx + 2cy$$

The lines of constant curvature on this surface can be much more diverse than for a surface embedded in just three dimensional space. As an example, if we define the surface with the equations

$$h_1 = x^2 + y^2 \qquad h_2 = xy$$

then the lines of constant curvature are as indicated in the figure below.

We have focused on two-dimensional surfaces in this section, but the basic idea of intrinsic curvature remains essentially the same in any number of dimensions. We'll see in subsequent sections that Riemann generalized Gauss's notion of intrinsic curvature by noting that any two (distinct) directional rays emanating from a given point P, if continued geodesically and with parallel transport (both of which we will discuss in detail), single out a two-dimensional surface within the manifold, and we can determine the "sectional" curvature of that surface in the same way as described in this section. Of course, in a manifold of three or more dimensions there are infinitely many two-dimensional surfaces passing through any given point, but Riemann showed how to encode enough information about the manifold at each point so that we can compute the sectional curvature on any surface.

For spaces of n>2 dimensions, we can proceed in essentially the same way, by imagining a flat n-dimensional Euclidean space tangent to the space at the point of interest, with a Cartesian coordinate system, and then evaluating how the curved space deviates from the flat space into another set of n(n"1)/2 orthogonal dimensions, one for each pair of dimensions in the flat tangent space. This is obviously just a generalization of our approach for n = 2 dimensions, when we considered a flat 2D space with Cartesian coordinates x,y tangent to the surface, and described the curved surface in the region around the tangent point in terms of the "height" h(x,y) perpendicular to the surface. Since we have chosen a flat baseline space tangent to the curved surface, it follows that the constant and first-order terms of h(x,y) are zero. Also, since we are not interested in any derivatives higher than the second, we can neglect all terms of h(x,y) above second order. Consequently we can write h(x,y) as a homogeneous second-order expression, i.e.,

$$h(x, y) = Ax^2 + Bxy + Cy^2$$

We saw that embedding a curved 2D surface in four dimensions allows even more freedom for the shape of the surface, but in the limit as the region becomes smaller and smaller, the surface approaches a single height. Similarly for a space of three dimensions we can imagine a flat three-dimensional space with x,y,z Cartesian coordinates tangent to the curved surface, and consider three perpendicular "heights" $h_1(x,y)$, $h_2(x,z)$, and $h_3(y,z)$.

There are obvious similarities between intrinsic curvature and ordinary spatial rotations, neither of which are possible in a space of just one dimension, and both of which are " in a sense " inherently two-dimensional phenomena, even when they exist in a space of more than two dimensions. Another similarity is the non-commutativity exhibited by rotations as well as by translations on a curved surface. In fact, we could define curvature as the degree to which translations along two given directions do not commute. The reason for this behavior is closely connected to the fact that rotations in space are non-commutative, as can be seen most clearly by imagining a curved surface embedded in a higher dimensional space, and noting that the translations on the surface actually involve rotations, i.e., angular displacements in the embedding space. Hence it's inevitable that such displacements don't commute.

CURVATURE AND RADIUS OF CURVATURE

The next important feature of interest is how much the curve differs from being a straight line at position s.

We measure this by the curvature $\kappa(s)$, which is defined by

$$\kappa = \left|\frac{dT(t(s))}{ds}\right|$$

which is, the magnitude of the change in unit tangent vector per unit change in distance along the curve.

The vector T being a unit vector has no dimension; that is, it is unaffected by a uniform change in scale of all coordinates. s on the other hand, is a length; therefore

$\frac{dT}{ds}$ has the dimension of the reciprocal of a length that is, of the reciprocal of a distance.

There is a second way to describe the information represented by the curvature. The center of curvature of the curve at parameter t is the point q(t) such that a circle centered at q which meets our curve at r(t), will have the same slope and curvature as our curve has there.

The radius of that circle is called the radius of curvature of our curve at argument t.

We will see that the radius of curvature, which is a length is exactly $\frac{1}{\kappa(t)}$, the reciprocal of the curvature.

In the applet in the next section you can enter your favorite parametrized curve and see the circle of curvature.

The center of curvature and the tangent vector to the curve, T(t), determine a plane called the plane of curvature.

Since the radius of a circle is always normal to a vector tangent to it, a line from r(t) toward the center of curvature will be normal to T. The vector N(t), called the normal vector to the curve, is a unit vector pointing from r(t) toward the center of curvature.

B(t), the "binormal vector" is a unit vector normal to both T and N, that is to the plane of curvature. By convention its direction is that of $T \times N$.

We define a(t) (the acceleration) to be the derivative with respect to t of v(t), (the velocity). In these terms, T is a unit vector in the direction of v, N is a unit vector in the direction of the projection of a normal to v, while B is a unit vector in the direction of va.

The "Frenet Frame" defined by a curve at "time" t is the set of unit vectors, T(t), N(t) and B(t).

PARAMETRIC CURVES

Curves and surfaces can have explicit, implicit, and parametric representations. Parametric representations are the most common in computer graphics.

Reparameterization

Parameterizations are in general not unique. Consider the following parameterizations for a line:

L(P0,P1) = P0 + u(P1-P0), u = [0...1]

L(P0,P1) = v(P1-P0)/2 + (P1+P0)/2, v = [-1...1]

Parameterizations can be changed to lie between desired bounds. To reparameterize from u = [a...b] to w = [0...1], we can use w = (u-a)/(b-a), which gives u = w(b-a) + a. Thus, we have:

P(u), u = [a...b] = P(w(b-a)+a), w = [0...1]

Parametric Cubic Curves

Cubic curves are commonly used in graphics because curves of lower order commonly have too little flexibility, while curves of higher order are usually considered unnecessarily complex and make it easy to introduce undesired wiggles.

A parametric cubic curve in 3D is defined by:

$$x(t) = a_3t^3 + a_2t^2 + a_1t + a_0$$
$$y(t) = b_3t^3 + b_2t^2 + b_1t + b_0$$
$$z(t) = c_3t^3 + c_2t^2 + c_1t + c_0$$

Usually, we consider t = [0...1].

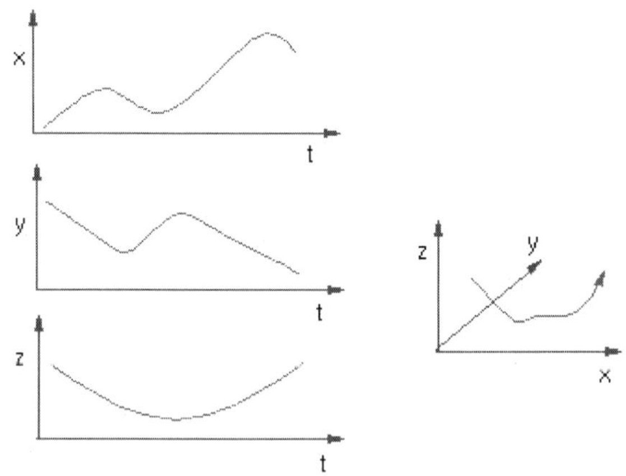

A compact version of the parametric equations can be written as follows:

$$x(t) = \begin{bmatrix} t^3 & t^2 & t & 1 \end{bmatrix} \begin{bmatrix} a_3 \\ a_2 \\ a_1 \\ a_0 \end{bmatrix}$$

$$x(t) = T \cdot A$$

Similarly, we can write

y(t) = T B

z(t) = T C

Each dimension is treated independently, so we can deal with curves in any number of dimensions.

The derivatives of the curve with respect to t can be expressed as follows:

x'(t) = [3t^2 2t 1 0] A

It is often convenient to think of the parameter t as being time in order to visualize some of the properties of the curve. The derivatives also have an intuitive interpretation in the cartesian space of the curve:

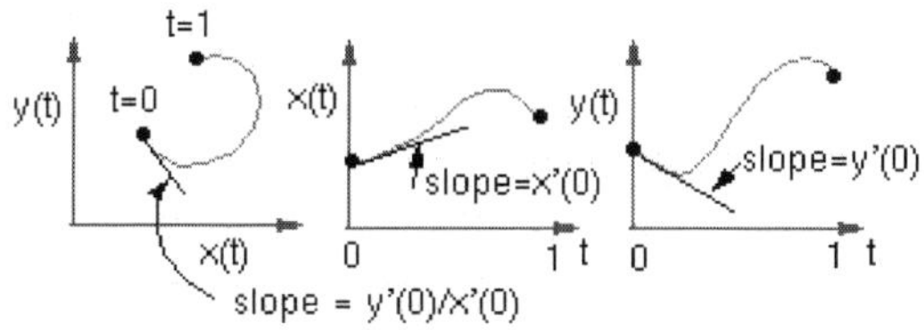

A First Example

Suppose we wish to define a cubic curve such that the user specifies the position of two endpoints and a midpoint, as well as the tangent at the midpoint. The following figure illustrates some sample curves.

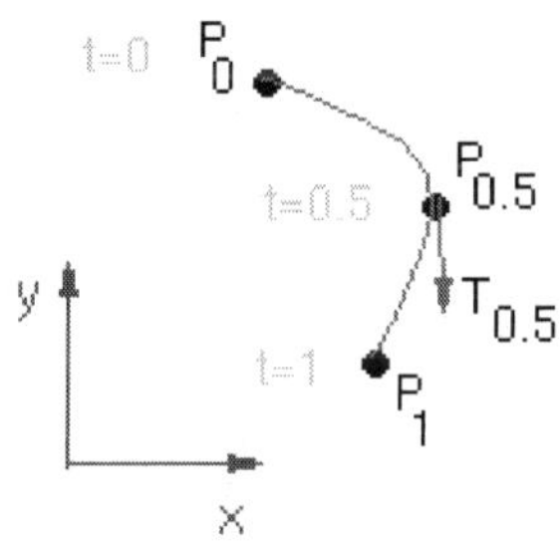

We'll first construct the cubic curve for x(t). There are four constraints that are to be met and four unknowns:

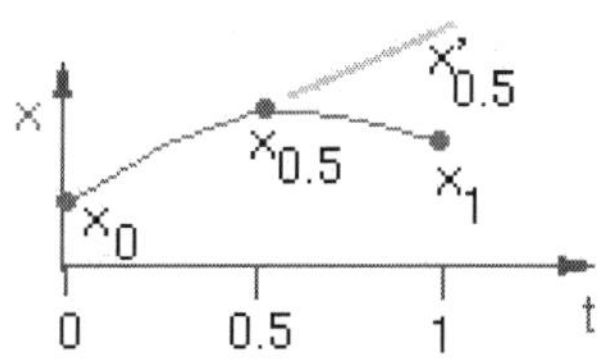

$$x(0) = [\,0\ 0\ 0\ 1\,]\ A$$
$$x(0.5) = [\,0.5^3\ \ 0.5^2\ \ 0.5^1\ \ 0.5^0\,]\ A$$
$$x'(0.5) = [\,3(0.5^2)\ \ 2(0.5)\ \ 1\ \ 0\,]\ A$$
$$x(1) = [\,1\ 1\ 1\ 1\,]\ A$$

or $G_x = B\,A$ where

$$G_x = \begin{bmatrix} x_0 \\ x_{0.5} \\ x'_{0.5} \\ x_1 \end{bmatrix} \qquad B = \begin{bmatrix} 0 & 0 & 0 & 1 \\ 0.125 & 0.25 & 0.5 & 1 \\ 0.75 & 1 & 1 & 0 \\ 1 & 1 & 1 & 1 \end{bmatrix}$$

We can solve for A using A = B_inv G_x. The final equation for x is thus:
x(t) = T B_inv G_x

The matrix B_inv is often called the basis matrix, which we shall denote by M. We can thus write x(t) = T M G_x

In this case, we have

```
     [ -4  0 -4  4 ]
M =  [  8 -4  6 -4 ]
     [ -5  4 -2  1 ]
     [  1  0  0  0 ]
```

Lastly, we can also write

x(t) = [f1(t) f2(t) f3(t) f4(t)] G_x

where f1(t) ... f4(t) are the functions obtained from the product T M. These are functions are called the blending or basis functions, because they serve to give a weighting to the various components of the geometry vector, G. In this case, these are

f1(t) = -4t^3 + 8t^2 - 5t + 1

f2(t) = -4t^2 + 4t

f3(t) = -4t^3 + 6t^2 - 2t

f4(t) = 4t^3 - 4t^2 + 1,

where f1(t) is the weighting function for P0, f2(t) is the weighting function for P0.5, f3(t) is the weighting function for T0.5, and f4(t) is the weighting function for P1. These basis functions look as follows:

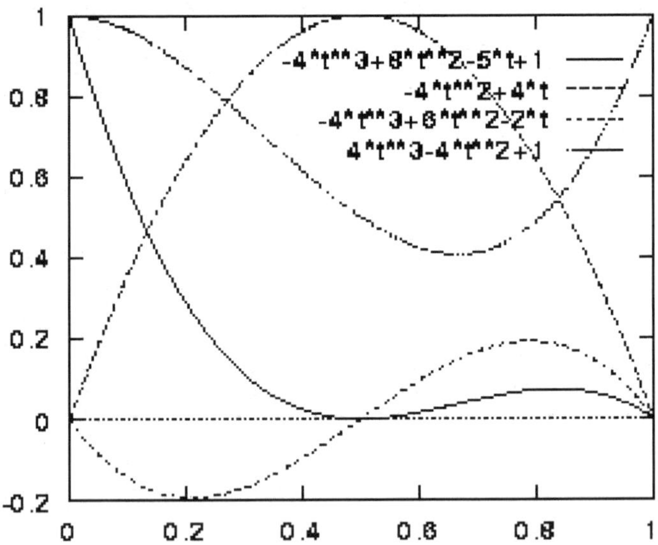

The curves for y(t) and z(t) are contructed in an analogous fashion to that for x(t). The basis matrix and basis functions thus remain the same. Only the geometry vector changes. The geometry vector for y(t) gives the y components of P0, P0.5, T0.5, and P1. In general, we can write the curve as a single vector equation

P(t) = T M G

which encompasses the following three equations:

x(t) = T M G_x

y(t) = T M G_y

z(t) = T M G_z

Hermite Curves

As a second example, let's look at Hermite curves. Hermite curves are defined by two points and two tangent vectors.

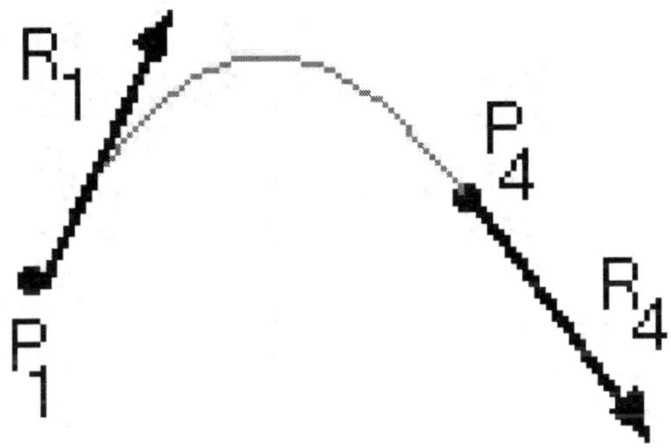

Let's derive the equation for Hermite curves using the following geometry vector:
G_h = [P1 P4 R1 R4]^T
As before, we'll express the curve as:
x(t) = T A_h
  = T M_h G_h
The constraints we'll use to define the curve are:
x(0) = P1 = [0 0 0 1] A_h
x(1) = P4 = [1 1 1 1] A_h
x'(0) = R1 = [0 0 1 0] A_h
x'(1) = R4 = [3 2 1 0] A_h
Writing these constraints in matrix form gives:
G_h = B_h A_h
A_h = inv(B_h) G_h
x(t) = T A_h
  = T inv(B_h) G_h
  = T M_h G_h
The inverse of B_h is thus defined as the basis matrix for the hermite curve.

```
      [ 2 -2  1  1 ]
M_h = [ -3  3 -2 -1 ]
      [ 0  0  1  0 ]
      [ 1  0  0  0 ]
```

As before, the basis functions are the weighting factors for the terms in the geometry vector, and are given by the product T M_h. Thus, for basis functions for Hermite curves are

f1(t) = 2t^3 - 3t^2 + 1
f2(t) = -2t^3 + 3t^2
f3(t) = t^3 - 2t^2 + t
f4(t) = t^3 - t^2

These basis functions look as follows:

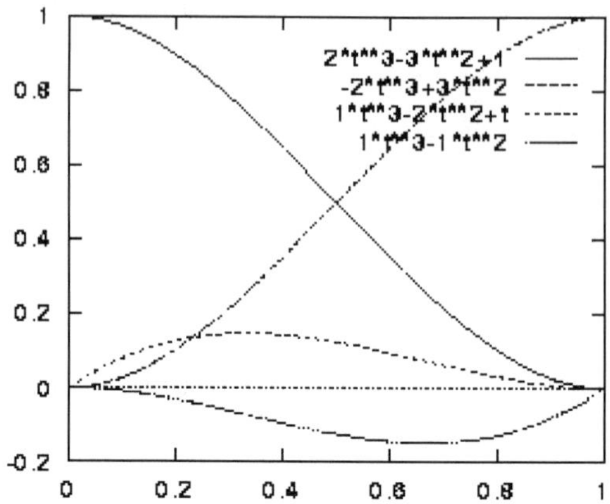

Bezier Curves

Bezier curves are a variation of the Hermite curves. They are specified by four points:

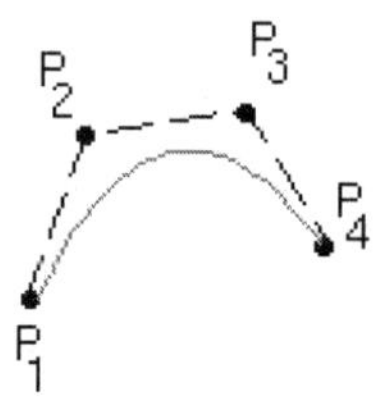

The curve is closely related to the Hermite curve:
P1_h = P1
P4_h = P4
R1_h = 3 (P2 - P1)
R4_h = 3 (P4 - P3)
We'll arrive at the basis matrix by expressing the above in matrix form:
[P1_h] [1 0 0 0] [P1]
[P4_h] = [0 0 0 1] [P2]
[R1_h] [-3 3 0 0] [P3]
[R4_h] [0 0 -3 3] [P4]
P_h = M_bh P_b
The equation for a bezier curve can thus be written as:
P(t) = T M_h M_bh P_b
P(t) = T M_b P_b
where
[-1 3 -3 1]
M_b = [3 -6 3 0]
[-3 3 0 0]
[1 0 0 0]
The Bezier basis functions are as follows:

P(t) = T M_b G_b
= f1(t) P1 + f2(t) P2 + f3(t) P3 + f4(t) P4
where
f1(t) = -t^3 + 3t^2 - 3t + 1
f2(t) = 3t^3 - 6t^2 + 3t
f3(t) = -3t^3 + 3t^2
f4(t) = t^3
or alternatively,
f1(t) = (1-t)^3
f2(t) = 3t (1-t)^2
f3(t) = 3t^2 (1-t)
f4(t) = t^3

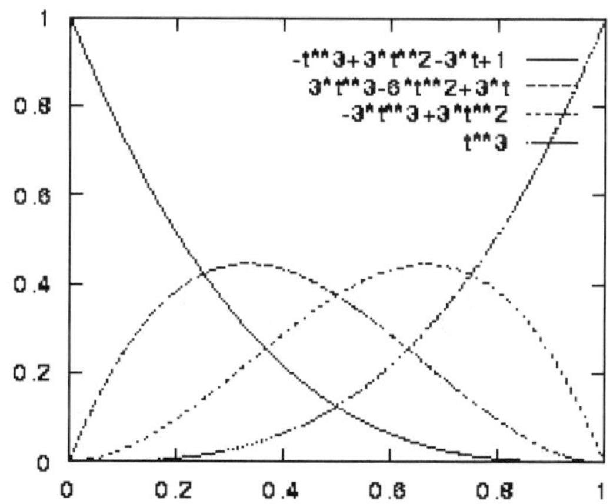

Convex hull property

The convex hull property ensures that the curve will never pass outside of the convex hull formed by the four control vertices. As such, it lends a measure of predictability to the curve. The Bezier basis functions satisfy the conditions necessary for the convex hull property, namely:

- 0 <= fi(t) <= 1 for t in [0,1].
- f1(t) + ... + fn(t) = 1

Bezier Curves of degree n

It is not per chance that the basis functions for Bezier curves have the convex hull property. How might one then go about designing a set of basis functions that sum to one and ensure that each individual basis function remain in the range [0,1]? The Bernstein polynomials have exactly the required properties. Using the Bernsteinn polynomials, we can construct a Bezier curve of arbitrary degree. For curves of higher degree than the cubic Bezier curve discussed thus far, we'll need more than four control points. The general Bezier curve of degree n is given by

$$P(t) = \sum_{i=0}^{n} B_{i,n}(t) P_i$$

$$\text{where } B_{i,n}(t) = \binom{n}{i} t^i (1-t)^{n-i}$$

The basis functions are equivalent to the terms arising from the expansion of

$$[(1-t)+t]^n$$

using the binomial expansion theorem. It is also obvious from this why the sum of all the terms is equal to one.

Piecewise Hermite and Bezier Curves

Hermite curve segments can be connected in a continuous fashion by ensuring that the end-point of one curve is the same as the starting point of another, as well as ensuring that the tangent vectors for this point have the same direction.

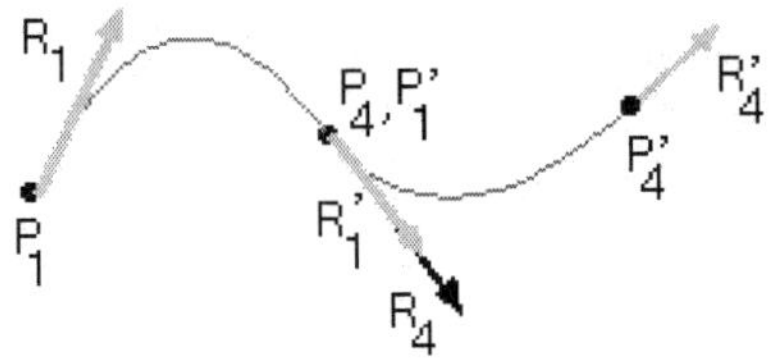

In terms of the above figure, this means P1'=P4, R1'=k R4 . For a Bezier curve, the conditions are that the the last two points of one curve and the first two points of the second curve are aligned.

Geometric and Parametric Continuity

Geometric Continuity

- G0: curves are joined
- G1: first derivatives are proportional at the join point

The curve tangents thus have the same direction, but not necessarily the same magnitude. i.e., C1'(1) = (a,b,c) and C2'(0) = (k*a, k*b, k*c).

- G2: first and second derivatives are proportional at join point

Parametric Continuity

- C0: curves are joined
- C1: first derivatives equal
- C2: first and second derivatives are equal

If t is taken to be time, this implies that the acceleration is continuous.

- Cn: nth derivatives are equal

As their names imply, geometric continuity requires the geometry to be continuous, while parametric continuity requires that the underlying parameterization be continuous as well.

Parametric continuity of order n implies geometric continuity of order n, but not vice-versa.

Summary of Bezier and Hermite Curves

- offer local control
- offer C1 continuity
- interpolates (some) control points

Splines

- splines are cubic curves which maintain C2 continuity.
- natural spline
    - o interpolates all of its control points.

- o equivalent to a thin strip of metal forced to pass through control points
- o no local control
- • B-spline
  - o local control
  - o does not interpolate control points

The following is an example of a five-segment B-spline curve (although this is simply a hand-drawn example). The points which indicate the ends of the individual curve segments and thus the join points are known as the knots.

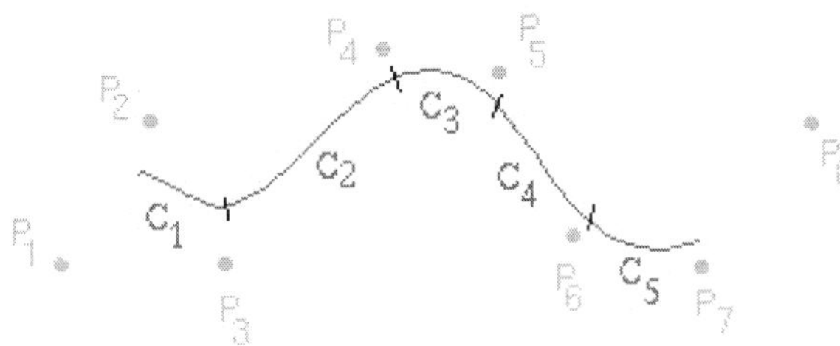

Each curve segment is determined by four control points, as follows:

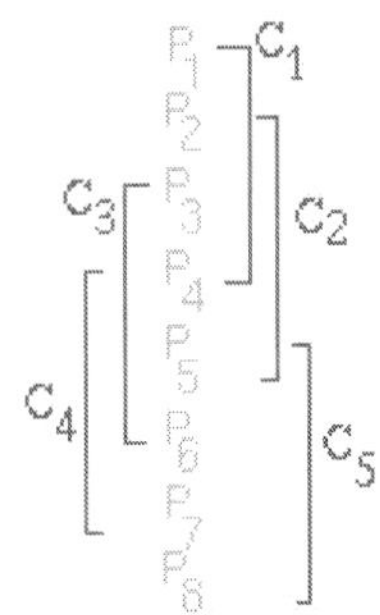

B-spline curves are defined by a basis matrix, just like the other types of cubic curves. We shall not discuss the derivation of this matrix here.

[-1 3 -3 1]
M_bspline = 1/6 [3 -6 3 0]
[-3 0 3 0]
[1 4 1 0]

where the geometry vector consists of four consecutive control points. It is easy to show that B-spline curves are C2 continuous and that they satisfy the convex-hull property. The basis functions for B-splines are as follows:

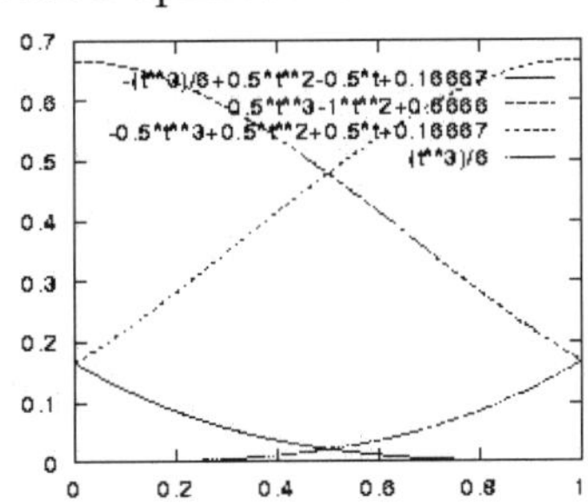

Multiple knots

Control vertices can be repeated in order to allow for reduced geometric continuity at the join points.

multiplicity	
1	G2 continuous
2	G1 continuous
3	G0 continuous, interpolates point

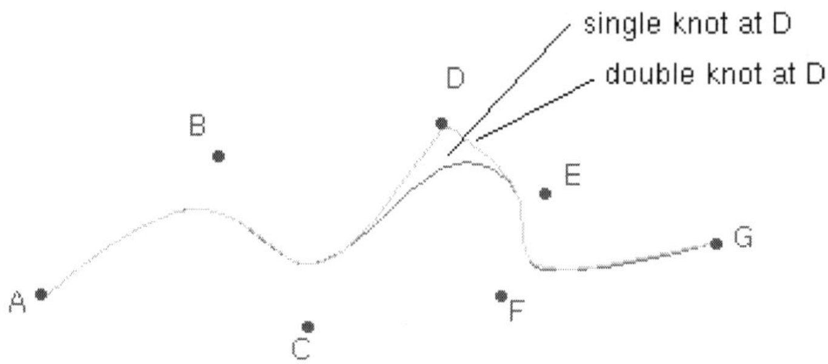

NURBS: Nonuniform Rational B-splines

One of the disadvantages of the curves discussed to date is that they cannot be used to create common conic shapes such as circles, ellipses, parabolas, etc. This can be done using rational cubic curves, however. A rational cubic curve segment in 3D can be constructed as follows

$x(t) = X(t)/W(t)$

$y(t) = Y(t)/W(t)$

$z(t) = Z(t)/W(t)$

where each of X(t), Y(t), Z(t), and W(t) are cubic polynomial curves. Defining curves as rational polynomials in this manner allows for simple exact represenations of conic sections such as circles, as well as curves which are invariant under perspective projection.

Lengths of polar subtangent and polar subnormal

Draw a line NT through the origin perpendicular to the radius vector of the point P on the curve. If PT is the tangent and PN the normal to the curve at P, then[note]

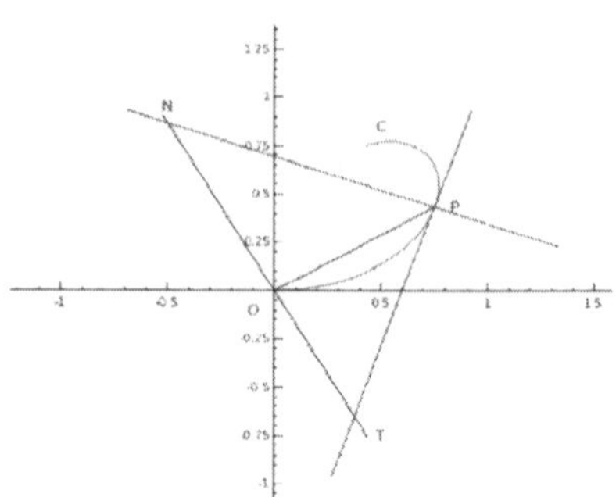

Figure6.8.1The polar subtangent and polar subnormal.

OT = length of polar subtangent,

and

ON = length of polar subnormal

of the curve at P.

In the triangle OPT, tanÈ=OTÁtanÈ=OTÁ. Therefore

OT=ÁtanÈ=Á2d¸dÁ= length of polar subtangent.(6.8.1)(6.8.1)OT=Átana È=Á2d¸dÁ= length of polar subtangent.

In the triangle OPN, tanÈ=ÁONtana È=ÁON. Therefore

ON=ÁtanÈ=dÁd¸= length of polar subnormal.(6.8.2)(6.8.2)ON=Átana È=dÁd¸= length of polar subnormal.

The length of the polar tangent (= PT) and the length of the polar normal (= PN) may be found from the figure, each being the hypotenuse of a right triangle.

Example6.8.1

{ Find lengths of polar subtangent and subnormal to the lemniscate Á2=a2cos2¸ñ2=a2cos2è.

Solution. Differentiating the equation of the curve as an implicit function with respect to ¸è, or, 2ÁdÁd¸= ″2a2sin2¸2ñdñdè= ″2a2sina 2è, dÁd¸= ″a2sin2¸Ádñdè= ″a2sina 2èñ.

Substituting in (6.8.1) and (6.8.2), we get

length of polar subtangent = ″Á3a2sin2¸″ñ3a2sina 2è,

length of polar subnormal = ″a2sin2¸Á″a2sina 2èñ.

If we wish to express the results in terms of ¸è, find Áñ in terms of ¸è from the given equation and substitute. Thus, in the above, Á=±acos2¸″″″″ ″ñ=±acosa 2è; therefore

length of polar subtangent = ±acot2¸cos2¸″″″″ ″±acota 2ècosa 2è.

13

Asymptotes

In analytic geometry, an asymptote (/ÈæsjmptoŠt/) of a curve is a line such that the distance between the curve and the line approaches zero as one or both of the x or y coordinates tends to infinity. Some sources include the requirement that the curve may not cross the line infinitely often, but this is unusual for modern authors. In projective geometry and related contexts, an asymptote of a curve is a line which is tangent to the curve at a point at infinity.

The word asymptote is derived from the Greek óýìðôùôïò (asumptôtos) which means "not falling together", from priv. + óýí "together" + ðôùô-üò "fallen". The term was introduced by Apollonius of Perga in his work on conic sections, but in contrast to its modern meaning, he used it to mean any line that does not intersect the given curve.

There are three kinds of asymptotes: horizontal, vertical and oblique asymptotes. For curves given by the graph of a function y = f(x), horizontal asymptotes are horizontal lines that the graph of the function approaches as x tends to +" or ". Vertical asymptotes are vertical lines near which the function grows without bound. An oblique asymptote has a slope that is non-zero but finite, such that the graph of the function approaches it as x tends to +" or ".

More generally, one curve is a curvilinear asymptote of another (as opposed to a linear asymptote) if the distance between the two curves tends to zero as they tend to infinity, although the term asymptote by itself is usually reserved for linear asymptotes.

Asymptotes convey information about the behavior of curves in the large, and determining the asymptotes of a function is an important step in sketching its graph. The study of asymptotes of functions, construed in a broad sense, forms a part of the subject of asymptotic analysis.

The idea that a curve may come arbitrarily close to a line without actually becoming the same may seem to counter everyday experience. The representations of a line and a curve as marks on a piece of paper or as pixels on a computer screen have a positive width. So if they were to be extended far enough they would seem to merge, at least as far as the eye could discern. But these are physical representations of the corresponding mathematical entities; the line and the curve are idealized concepts whose width is 0. Therefore, the understanding of the idea of an asymptote requires an effort of reason rather than experience.

Consider the graph of the function shown to the right. The coordinates of the points on the curve are of the form where x is a number other than 0. For example, the graph contains the points (1, 1), (2, 0.5), (5, 0.2), (10, 0.1), ... As the values ofbecome larger and larger, say 100, 1,000, 10,000 ..., putting them far to the right of the illustration, the corresponding values of {\displaystyle y}, .01, .001, .0001, ..., become infinitesimal relative to the scale shown. But no matter how largebecomes, its reciprocal {\displaystyle {\frac {1}{x}}} is never 0, so the curve never actually touches the x-axis. Similarly, as the values of {\displaystyle x} become smaller and smaller, say .01, .001, .0001, ..., making them infinitesimal relative to the scale shown, the corresponding values of {\displaystyle y}, 100, 1,000, 10,000 ..., become larger and larger. So the curve extends farther and farther upward as it comes closer and closer to the y-axis. Thus, both the x and y-axes are asymptotes of the curve. These ideas are part of the basis of concept of a limit in mathematics, and this connection is explained more fully below.

Asymptotes of functions

The asymptotes most commonly encountered in the study of calculus are of curves of the form y = *f*(x). These can be computed using limits and classified into horizontal, verticaland oblique asymptotes depending on their orientation. Horizontal asymptotes are horizontal lines that the graph of the function approaches as x tends to +" or ". As the name indicates they are parallel to the x-axis. Vertical asymptotes are vertical lines (perpendicular to the x-axis) near which the function grows without bound. Oblique asymptotes are diagonal lines such that the difference between the curve and the line approaches 0 as x tends to +" or ".

An asymptote of a curve y=f(x) that has an infinite branch is called a line such that the distance between the point (x,f(x)) lying on the curve and the line approaches zero as the point moves along the branch to infinity.

Asymptotes can be vertical, oblique (slant) and horizontal. A horizontal asymptote is often considered as a special case of an oblique asymptote.

Vertical Asymptote

The straight line x=a is a vertical asymptote of the graph of the function y=f(x) if at least one of the following conditions is true:

limx'!a"0f(x)=±",limx'!a+0f(x)=±".

In other words, at least one of the one-sided limits at the point x=a must be equal to infinity.

A vertical asymptote occurs in rational functions at the points when the denominator is zero and the numerator is not equal to zero (i.e. at the points of discontinuity of the second kind). For example, the graph of the function y=1x has the vertical asymptote x=0 (Figure 1). In this case, both one-sided limits (from the left and from the right) tend to infinity:

limx'!0"01x= ",limx'!0+01x=+".

A function which is continuous on the whole set of real numbers has no vertical asymptotes.

Oblique Asymptote

The straight line y=kx+b is called an oblique (slant) asymptote of the graph of the function y=f(x) as x′!+″ (Figure 2) if

limx′!+″[f(x)″(kx+b)]=0.

Similarly, we introduce oblique asymptotes as x′! ″.

The oblique asymptotes of the graph of the function y=f(x) may be different when x′!+″ and x′! ″.

Therefore, when finding oblique (or horizontal) asymptotes, both cases should be considered separately.

The coefficients k and b of an oblique asymptote y=kx+b are defined by the following theorem:

A straight line y=kx+b is an asymptote of a function y=f(x) as x′!+″ if and only if the following two limits are finite:

limx′!+″f(x)x=kandlimx′!+″[f(x)″kx]=b.

Proof.

Necessity. Let the straight line y=kx+b be an asymptote of the graph of y=f(x) as x′!+″.Then the following condition is true:

limx′!+″[f(x)″(kx+b)]=0

or, equivalently,

f(x)=kx+b+±(x),wherelimx′!+″±(x)=0.

Dividing both sides of the equation by x, we obtain:

f(x)x=kx+b+±(x)x,Ò!f(x)x=k+bx+±(x)x.

Consequently, in the limit as x′!+″, we have

limx′!+″f(x)x=limx′!+″[k+bx+±(x)x]=k,

limx′!+″[f(x)″kx]=limx′!+″[b+±(x)]=b.

Sufficiency. Suppose that there are finite limits

limx′!+″f(x)x=kandlimx′!+″[f(x)″kx]=b.

The second limit can be written as

limx′!+″[f(x)″(kx+b)]=0,

that meets the definition of an oblique asymptote. Thus, the straight line y=kx+b is an asymptote of the function y=f(x).

Note: Similarly we can prove the theorem for the case of x′!″.

Horizontal Asymptote

In particular, if k=0, we obtain a horizontal asymptote, which is described by the equation y=b.The theorem on necessary and sufficient conditions for the existence of a horizontal asymptote is stated as follows: A straight line y=b is an asymptote of a function y=f(x) as x′!+″, if and only if the following limit is finite:

limx′!+″f(x)=b.

The case x′! ″ is considered in the same way.

Asymptotes of a Parametrically Defined Curve

Let a plane curve be defined by the parametric equations

x=Æ(t),y=È(t).

This curve has a vertical asymptote x=a as t′!t1 if the following conditions are true:

limt′!t1Æ(t)=aandlimt′!t1È(t)=±″.

Similarly, a parametrically defined curve has a horizontal asymptote y=b as t′!t2 if

limt′!t2Æ(t)=±″andlimt′!t2È(t)=b.

Here a and b are finite quantities.

A parametrically defined curve has an oblique asymptote y=kx+b as t′!t3 if, at this value of t,both limits are infinite:

limt′!t3Æ(t)=±″,limt′!t3È(t)=±″,

and the coefficients k and b have finite values:

k=limt′!t3È(t)Æ(t),b=limt′!t3[È(t)″kÆ(t)].

Asymptote of a Polar Curve

Consider a curve given in polar coordinates by the equation

Á=Á(Æ)orF(Á,Æ)=0.

Its asymptote (if it exists) can be described by the two parameters: the distance p from the center to the asymptote (segment OA in Figure 3) and the angle ± of inclination of the asymptote to the polar axis.

These parameters ± and p are defined by the formulas

Figure 3.

±=limÁ′!″Æ,p=limÁ′!″[Ásin(±″Æ)].

Notes:

1. In the last formula, the limit transition Á′!″ can be replaced by an equivalent condition Æ′!±;
2. The parameter p can be both positive and negative.

Asymptotes of an Implicit Curve

An implicitly defined algebraic curve is described by the equation

F(x,y)=0,

where the left-hand side is a polynomial in the variables x and y.

In differential geometry, the following method for finding an oblique asymptote of an algebraic curve is used. Suppose that the asymptote is described by the equation y=kx+b. Substituting this expression for y in the equation of the curve, we obtain an algebraic equation in one variable x:

A0xn+A1xn″1+…+An″1x+An=0,

where the coefficients Ai depend on the parameters of the asymptote k and b (the coefficient A0depends only on k). The values of k and b can be determined from the condition:

{A0(k)=0A1(k,b)=0.

To find a vertical asymptote, it is necessary to substitute its equation x=a in the equation of the curve and convert the latter to the form:

B0yn+B1yn″1+…+Bn″1y+Bn=0.

A necessary condition for the existence of a vertical asymptote is the absence of the term of the highest degree B0yn in the last equation.

The value of the parameter a is determined from the condition

B1(a)=0.

The above formulas for the asymptotes of an implicit curve are valid if the curve has no singular points at infinity.

An asymptote is a line that a graph approaches without touching.

If a graph has a horizontal asymptote of y = k, then part of the graph approaches the line y = k without touching it—y is almost equal to k, but y is never exactly equal to k. The following graph has a horizontal asymptote of y = 3:

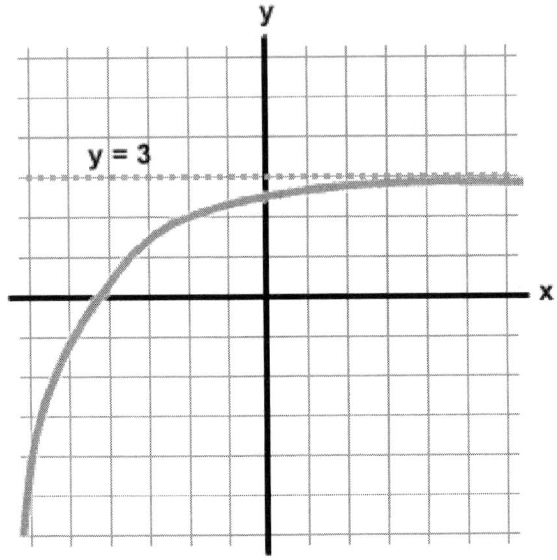

Horizontal Asymptote y = 3

The following graph has a horizontal asymptote of y = 0:

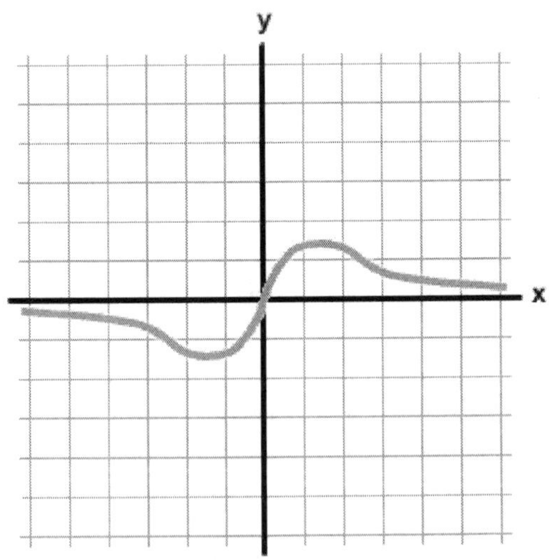

Horizontal Asymptote y = 0

If a graph has a vertical asymptote of x = h, then part of the graph approaches the line x = h without touching it—x is almost equal to h, but x is never exactly equal to h. The following graph has a vertical asymptote of x = 3:

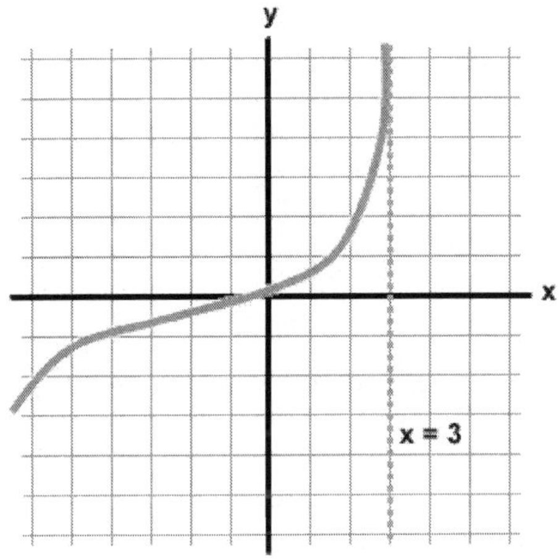

Vertical Asymptote x = 3

One reason vertical asymptotes occur is due to a zero in the denominator of a rational function. For example, if f (x) = $\frac{1}{x-5}$ then x cannot equal 5, but x can equal values very close to 5 (4.99, for example). The graph of f (x) = looks like:

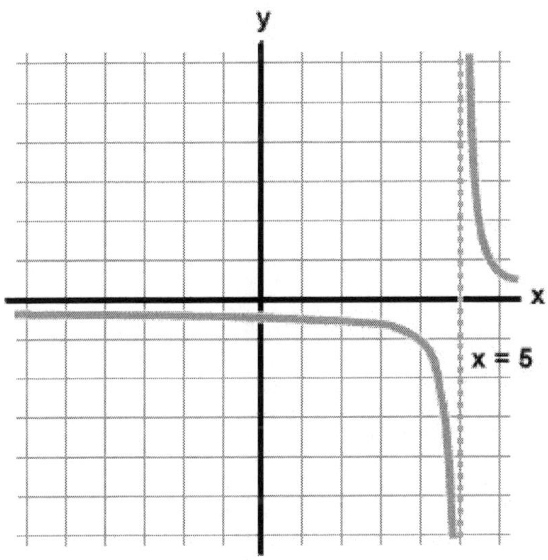

f (x) = $\frac{1}{x-5}$

Similarly, horizontal asymptotes occur because y can come close to a value, but can never equal that value. In the previous graph, there is no value of x for which y = 0 ($\frac{1}{a}$ ‴ 0), but as x gets very large or very small, y comes close to 0. Thus, f (x) = $\frac{1}{x-5}$ has a horizontal asymptote at y = 0.

The graph of a function may have several vertical asymptotes. f (x) = $\frac{1}{(x-2)(x+3)}$ has vertical asymptotes of x = 2 and x = - 3, and f (x) = $\frac{x(x-5)}{(x+4)(2x-1)^2}$ has vertical asymptotes of x= - 4 and x = $\frac{1}{2}$ In general, a vertical asymptote occurs in a rational function at any value of x for which the denominator is equal to 0, but for which the numerator is not equal to 0.

When a value of x sets both the denominator And the numerator of a rational function equal to 0, there is a hole in the graph; that is, a single point at which the function has no value. f (x) = $\frac{x^2(x-2)}{x-2}$ has a hole at x = 2:

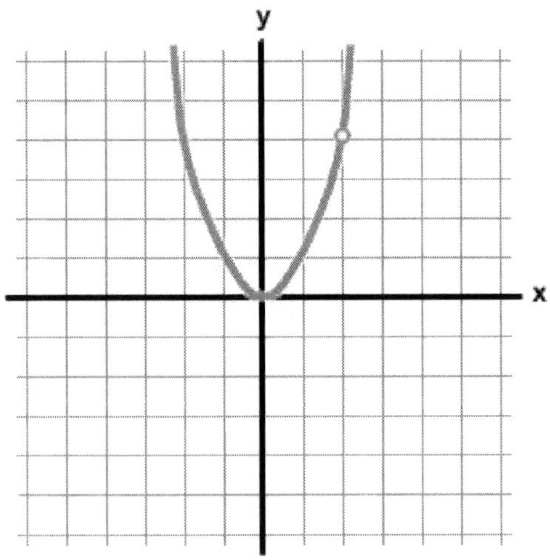

f (x) = $\frac{x^2(x-2)}{x-2}$

Examples: Name the vertical asymptotes and holes in the graphs of the following equations:

1. f (x) = $\frac{(x-2)(x+4)}{(x-3)(x+3)(x+4)}$
2. f (x) = $\frac{x^2(x+2)}{x^3(x-2)^2}$
3. f (x) = $\frac{(x-1)^2}{(x+1)(2x-3)(\frac{1}{2}x+6)}$

Solutions:

1. Asymptotes: x = 3, - 3. Holes: x = - 4.
2. Asymptotes: x = 2. Holes: x = 0.
3. Asymptotes: x = - 1, $\frac{3}{2}$ - 12.

rational function ?

A rational function is a function that can be written as a fraction of two polynomials where the denominator is not zero.

$$f(x) = \frac{p(x)}{q(x)} = \frac{a_n x^n + a_{n-1}x^{n-1} + \cdots + a_1 x + a_0}{b_m x^m + b_{m-1}x^{m-1} + \cdots + b_1 x + b_0}$$

Domain

The domain of a rational function is all real values except where the denominator, $q(x) = 0$

Example

Find the domain of :

$$f(x) = \frac{x^3 - 2x^2 + x - 7}{x^2 - 3x + 2} = \frac{x^3 - 2x^2 + x - 7}{(x-1)(x-2)}$$

Solution

Domain = {(-¥, 1) È (1, 2) È (2, +¥)}

Roots

The roots (zeros, solutions, x-intercepts) of the rational function can be found by solving:

$p(x) = 0$

This roots can be found usually by factorizing p(x). If the multiplicity of a factor (x - c) is odd, the curve cuts the x-axis at x = c. If the multiplicity of a factor is even, then the curve touches the x-axis at x = c.

We shall discuss the case in which (x – c) is both a factor of numerator and denominator.

VERTICAL ASYMPTOTES

An asymptote is a line that the curve goes nearer and nearer but does not cross. The equations of the vertical asymptotes can be found by solving $q(x) = 0$ for roots. We shall study more closely if some roots are also roots of $p(x) = 0$.

If you write p(x) in factorized form, then you can tell whether the graph is asymptotic in the same direction or in opposite directions by whether the multiplicity is even or odd.

Asymptotic in the same direction means that the curve will go up or down on both the left and right sides of the vertical asymptotes. Asymptotic in opposite direction means that the one side of the curve will go down and the other side of the curve will go up the vertical asymptotes.

Example

$$f(x) = \frac{(x+1)(x-3)^2}{(x-1)^2(x-4)}$$

(1) The curve cuts the x-axis at x = -1 since the multiplicity of (x + 1) is 1, which is odd.

The curve touches the x-axis at x = 4 since the multiplicity of (x -3) is 2, which is even.

(2) The curve is asymptotic in the same direction of x = 1, since the multiplicity of (x – 1) is 2.

The curve is asymptotic in opposite direction of x = 4, since the multiplicity of (x – 4) is 1.

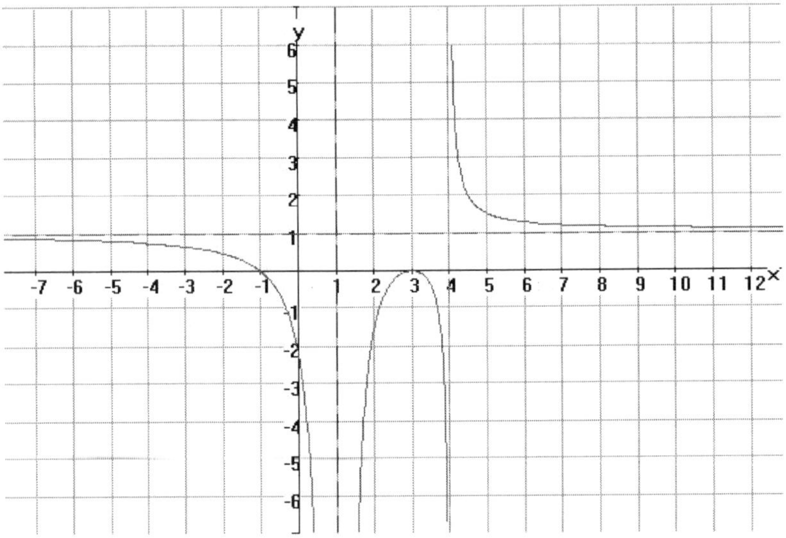

Horizontal asymptotes

A horizontal asymptote is a horizontal line which the curve approaches at far left and far right of the graph.

The location of the horizontal asymptote is found by looking at the degrees of the numerator (n) and the denominator (m).

(1) If n < m, the x-axis (or y = 0) is the horizontal asymptote.

(2) If n = m, then $y = a_n / b_m$ is the horizontal asymptote. (ratio of the leading coefficients)

(3) If n > m, then there is not horizontal asymptote.

In the above graph, the horizontal asymptote is x = 1. Note that the curve is asymptotic at opposite side of the line x = 1. It can be asymptotic in the same direction depending on the given curve. Rational function has at most one horizontal asymptote. Other function may have more than one horizontal asymptote.

Hole

Sometimes, a factor may appear in both the numerator and denominator. Let us assume that the factor $(x – c)^s$ is in the numerator and $(x – c)^t$ is in the denominator.

(1) s < t, then there will be a vertical asymptote x = c.

(2) s > t, then there will be a hole in the graph on the x-axis at x = c. There is no vertical asymptote there.

(3) s = t, then there will be a hole in the graph at x = c, but not on the x-axis. The y-value of the hole can be found by canceling the factors and substituting $x = c$ in the reduced function.

Oblique Asymptote

When the degree of the numerator is exactly one more the degree of the denominator, the graph of the rational function has an oblique asymptote. In other cases, there will be no oblique asymptote.

The equation of the oblique asymptote can be found by division.

Example

$$f(x) = \frac{(x+1)^3}{4(x-2)^2} = \frac{x^3+3x^2+3x+1}{4(x^2-4x+4)}$$

By division, we get

$$f(x) = \frac{1}{4}[x+7] + \frac{27x-27}{(x-2)^2} = \left(\frac{1}{4}x + \frac{7}{4}\right) + \frac{27}{4}\frac{x-1}{(x-2)^2}$$

$$y = \frac{1}{4}x + \frac{7}{4}$$

or $4y = x + 7$ is an oblique asymptote.

and $x = 2$ is a vertical asymptote.

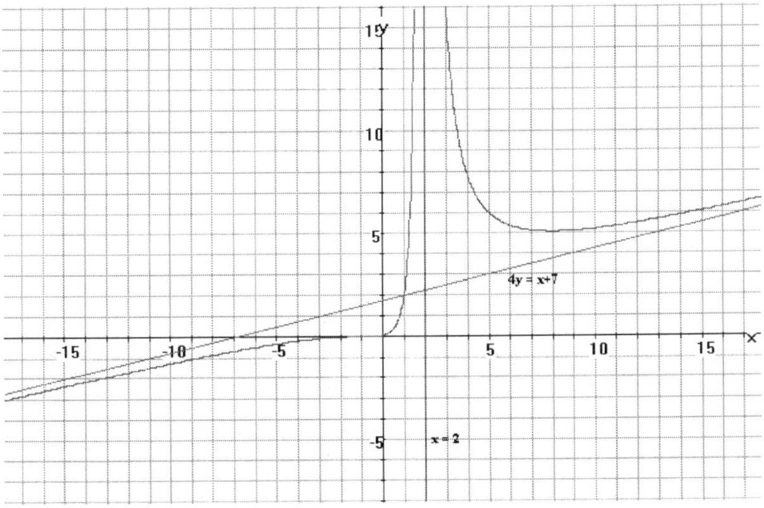

ASYMPTOTES AND RATIONAL FUNCTIONS

Suppose you're going for a walk along a trail lined with poison ivy. There is a river running next to the trail that you are trying to video as you walk along the trail. You want to get as close to the edge of the trail as possible for the best view of the river, but you don't want to touch the edge because of the poison ivy. You just walk along, getting closer and closer to the edge without actually touching it. This scenario gives us an idea of how a graph approaches an asymptote.

An asymptote is a line that the graph of a function approaches, but never actually touches. There are vertical asymptotes, horizontal asymptotes, and oblique asymptotes. Oblique asymptotes are also called slant asymptotes. Vertical and horizontal asymptotes are vertical and horizontal lines, respectively. An oblique, or slant, asymptote is an asymptote that is neither vertical nor horizontal.

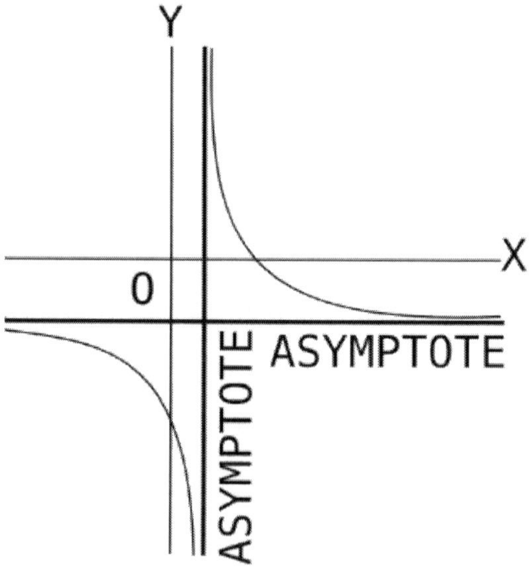

Asymptotes are a well-known characteristic of rational functions. A rational function is a function that is a quotient of functions.

$$\text{Rational Function}$$

$$R(x) = \frac{f(x)}{g(x)}$$

Let's look at how to find the asymptotes of a rational function.

Finding Asymptotes of Rational Functions

As we said, there are three types of asymptotes. First, let's look at how to find the vertical asymptotes of a rational function.

Vertical Asymptotes

To find vertical asymptotes, we want to follow these steps.

1. Set the denominator equal to zero and solve.
2. The values that make the denominator zero are where you will have vertical asymptotes.

For example, consider the function h(x) = x / (x^2 - 4). To find the vertical asymptotes, we would set the denominator equal to zero and solve.

$$x^2 - 4 = 0$$

$$(x+2)(x-2) = 0 \quad Set\ each\ factor\ equal\ to\ 0$$

$$x + 2 = 0 \quad or \quad x - 2 = 0 \quad Solve\ both\ equations$$

$$x = -2 \qquad or \quad x = 2$$

Thus, the vertical asymptotes of the function are x = -2 and x = 2.

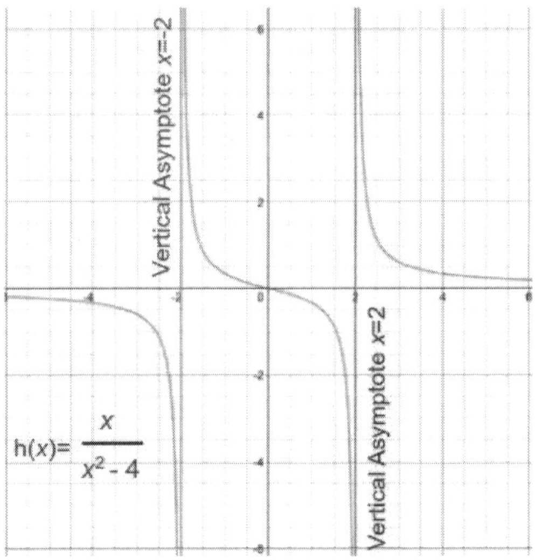

Horizontal Asymptotes

Horizontal asymptotes can be identified in a rational function by examining the degree of both the numerator and the denominator. The degree of a function is the highest exponent of the function. There are three possibilities in a rational function.

1.) The degree of the numerator is less than the degree of the denominator.

Example: $h(x) = \dfrac{x}{x^2 - 4}$

2.) The degree of the numerator is equal to the degree of the denominator.

Example: $d(x) = \dfrac{4x^2}{5x^2 - x}$

3.) The degree of the numerator is greater than the degree of the denominator.

Example: $s(x) = \dfrac{x^2 + 2x + 1}{x}$

Once we figure out which of these is true for our function, we can find the horizontal asymptotes. It is important to note that a graph can cross a horizontal asymptote. When this is the case, the graph approaches the asymptote but doesn't touch it on one part of the graph, instead crossing it somewhere else.

Let's look at each of the above mentioned cases.

The first case happens when the degree of the numerator is less than the degree of the denominator. When this is the case, $y = 0$ is a horizontal asymptote. The example given is the function we looked at when we found our vertical asymptotes, $h(x) = x/(x^2 - 4)$. In this function, the degree of the numerator is 1, and the degree of the denominator is 2. Since $1<2$, $y = 0$ is a horizontal asymptote of this function.

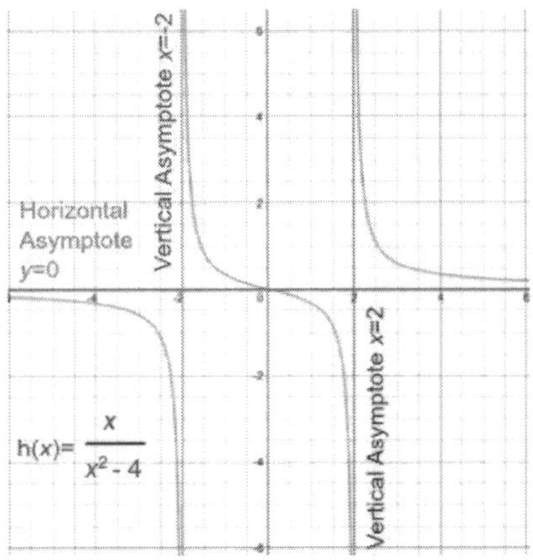

The second case applies when the degrees of the numerator and denominator are equal. When this is the case, we need to deal with lead coefficients. The lead coefficient of a function is the number in front of the term with the highest exponent. For example, in the function $r(x) = 2x^4 + 4x^3 - x + 5$, the lead coefficient is 2, the number in front of x^4.

When a rational function falls in the second category, the function has a horizontal asymptote at $y = a / b$, where a is the lead coefficient of the numerator, and b is the lead coefficient of the denominator. In the function $d(x) = 4x^2 / (5x^2 - x)$, the degree of the numerator is 2, and the degree of the denominator is 2. The degrees are equal. Thus, the function has a horizontal asymptote at $y = 4/5$, where 4 is the lead coefficient of the numerator, and 5 is the lead coefficient of the denominator.

asymptote of a curve?

An asymptote of the curve $y = f(x)$ (or in implicit form: $f(x,y) = 0$) is a straight line such that the distance between the curve and the straight line lends to zero when the points on the curve approach infinity. There are three types of asymptotes, namely, vertical, horizontal and oblique asymptotes. Here is an algebraic method for finding oblique (and also horizontal) asymptotes of algebraic curves.

The method

(1) Replace y by $mx + c$ in the equation of the curve and arrange the result in the form:

$$a_n x^n + a_{n-1} x^{n-1} + ... + a_1 x + a_0 = 0$$

(2) Solve the simultaneous equation :

The method (1) Replace y by $mx + c$ in the equation of the curve and arrange the result in the form : $a_n x^n + a_{n-1} x^{n-1} + ... + a_1 x + a_0 = 0$ (2) Solve the simultaneous equation : $\begin{cases} a_n = 0 \\ a_{n-1} = 0 \end{cases}$ 3) For each pair of solutions of m and c, write the equation of an asymptote $y = mx + c$. (4) If there is no a_{n-1} term in (1), solve : $\begin{cases} a_n = 0 \\ a_{n-2} = 0 \end{cases}$ instead of the set of simultaneous equation in (2).

Reason for (2) above

The equation $a_n x^n + a_{n-1} x^{n-1} + ... + a_1 x + a_0 = 0$
has a double root at infinity.

Put $y = \frac{1}{x}$,

The equation becomes $a_0 y^n + a_1 y^{n-1} + ... + a_{n-1} y + a_n = 0$
Then the equation has a double root at $y = 0$.
That is, the equation is of the form:

$$a_0 y^n + a_1 y^{n-1} + ... + a_{n-2} y^2 = 0$$

$$y^2 \left(a_0 y^{n-2} + a_1 y^{n-3} + ... + a_{n-2}\right) = 0$$

Therefore :

$$\begin{cases} a_n = 0 \\ a_{n-2} = 0 \end{cases}$$

Example

Find the oblique asymptotes of $y = \frac{x^3}{x^2 - 1}$

Solution

Put $y = mx + c$ in $y = \frac{x^3}{x^2 - 1}$

\ $(x^2 - 1)(mx+c) = x^3$.
\ $(m-1)x^3 - cx^2 - mx - c = 0$ (1)

Put $y = 1/x$ in (1),
\ $cy^3 + my^2 + cy - (m-1) = 0$ (2)
$y = mx + c$ is an asymptote of
iff there is a double root at infinity for (1)
iff there is a double root at 0 for (2).
\ $m - 1 = 0$ and $c = 0$.
\ $m = 1$ and $c = 0$.
\ The oblique asymptote is $y = x$.
Appendix (Calculus method)

$$m = \lim_{x\to\infty} \frac{y}{x} = \lim_{x\to\infty} \frac{1}{x}\left(\frac{x^3}{x^2-1}\right) = \frac{1}{1-\frac{1}{x^2}} = 1$$

$$c = \lim_{x\to\infty} (y - mx) = \lim_{x\to\infty}\left[\frac{x^3}{x^2-1} - (1)x\right] = \lim_{x\to\infty} \frac{x}{x^2-1} = 0$$

\ The oblique asymptote is $y = x$.

An asymptote of a curve y=f(x) that has an infinite branch is called a line such that the distance between the point (x,f(x)) lying on the curve and the line approaches zero as the point moves along the branch to infinity.

Asymptotes can be vertical, oblique (slant) and horizontal. A horizontal asymptote is often considered as a special case of an oblique asymptote.

Vertical Asymptote

The straight line x=a is a vertical asymptote of the graph of the function y=f(x) if at least one of the following conditions is true:

limx'!a"0f(x)=±",limx'!a+0f(x)=±".

In other words, at least one of the one-sided limits at the point x=a must be equal to infinity.

A vertical asymptote occurs in rational functions at the points when the denominator is zero and the numerator is not equal to zero (i.e. at the points of discontinuity of the second kind). For example, the graph of the function y=1x has the vertical asymptote x=0 (Figure 1). In this case, both one-sided limits (from the left and from the right) tend to infinity:

limx'!0"01x= ",limx'!0+01x=+".

A function which is continuous on the whole set of real numbers has no vertical asymptotes.

Oblique Asymptote

The straight line y=kx+b is called an oblique (slant) asymptote of the graph of the function y=f(x) as x'!+" (Figure 2) if limx′!+″[f(x)″(kx+b)]=0.

Similarly, we introduce oblique asymptotes as x′! ″.

The oblique asymptotes of the graph of the function y=f(x) may be different when x′!+″ and x′! ″.

Therefore, when finding oblique (or horizontal) asymptotes, both cases should be considered separately.

The coefficients k and b of an oblique asymptote y=kx+b are defined by the following theorem:

A straight line y=kx+b is an asymptote of a function y=f(x) as x′!+″ if and only if the following two limits are finite:

limx′!+″f(x)x=kandlimx′!+″[f(x)″kx]=b.

Proof.

Necessity. Let the straight line y=kx+b be an asymptote of the graph of y=f(x) as x′!+″.Then the following condition is true:

limx′!+″[f(x)″(kx+b)]=0

or, equivalently,

f(x)=kx+b+±(x),wherelimx′!+″±(x)=0.

Dividing both sides of the equation by x, we obtain:

f(x)x=kx+b+±(x)x,Ò!f(x)x=k+bx+±(x)x.

Consequently, in the limit as x′!+″, we have

limx′!+″f(x)x=limx′!+″[k+bx+±(x)x]=k,

limx′!+″[f(x)″kx]=limx′!+″[b+±(x)]=b.

Sufficiency. Suppose that there are finite limits

limx′!+″f(x)x=kandlimx′!+″[f(x)″kx]=b.

The second limit can be written as

limx′!+″[f(x)″(kx+b)]=0,

that meets the definition of an oblique asymptote. Thus, the straight line y=kx+b is an asymptote of the function y=f(x).

Note: Similarly we can prove the theorem for the case of x′!″.

Horizontal Asymptote

In particular, if k=0, we obtain a horizontal asymptote, which is described by the equation y=b.The theorem on necessary and sufficient conditions for the existence of a horizontal asymptote is stated as follows:

A straight line y=b is an asymptote of a function y=f(x) as x′!+″, if and only if the following limit is finite:

limx′!+″f(x)=b.

The case x′! ″ is considered in the same way.

Asymptotes of a Parametrically Defined Curve

Let a plane curve be defined by the parametric equations

x=Æ(t),y=È(t).

This curve has a vertical asymptote x=a as t′!t1 if the following conditions are true:

limt′!t1Æ(t)=aandlimt′!t1È(t)=±″.

Similarly, a parametrically defined curve has a horizontal asymptote y=b as t′!t2 if

limt′!t2Æ(t)=±″andlimt′!t2È(t)=b.

Here a and b are finite quantities.

A parametrically defined curve has an oblique asymptote y=kx+b as t'!t3 if, at this value of t,both limits are infinite:

limt'!t3Æ(t)=±",limt'!t3È(t)=±",

and the coefficients k and b have finite values:

k=limt'!t3È(t)Æ(t),b=limt'!t3[È(t)"kÆ(t)].

Asymptote of a Polar Curve

Consider a curve given in polar coordinates by the equation

Á=Á(Æ)orF(Á,Æ)=0.

Its asymptote (if it exists) can be described by the two parameters: the distance p from the center to the asymptote (segment OA in Figure 3) and the angle ± of inclination of the asymptote to the polar axis.

These parameters ± and p are defined by the formulas

±=limÁ'!"Æ,p=limÁ'!"[Ásin(±"Æ)].

Notes:

1. In the last formula, the limit transition Á'!" can be replaced by an equivalent condition Æ'!±;
2. The parameter p can be both positive and negative.

Asymptotes of an Implicit Curve

An implicitly defined algebraic curve is described by the equation

F(x,y)=0,

where the left-hand side is a polynomial in the variables x and y.

In differential geometry, the following method for finding an oblique asymptote of an algebraic curve is used. Suppose that the asymptote is described by the equation y=kx+b. Substituting this expression for y in the equation of the curve, we obtain an algebraic equation in one variable x:

A0xn+A1xn"1+...+An"1x+An=0,

where the coefficients Ai depend on the parameters of the asymptote k and b (the coefficient A0depends only on k). The values of k and b can be determined from the condition:

{A0(k)=0A1(k,b)=0.

To find a vertical asymptote, it is necessary to substitute its equation

x=a in the equation of the curve and convert the latter to the form:

B0yn+B1yn"1+...+Bn"1y+Bn=0.

A necessary condition for the existence of a vertical asymptote is the absence of the term of the highest degree B0yn in the last equation.

The value of the parameter a is determined from the condition

B1(a)=0.

The above formulas for the asymptotes of an implicit curve are valid if the curve has no singular points at infinity.

How to Find Asymptotes

This section is based on finding Asymptotes. Below is explained the procedure to find asymptotes:

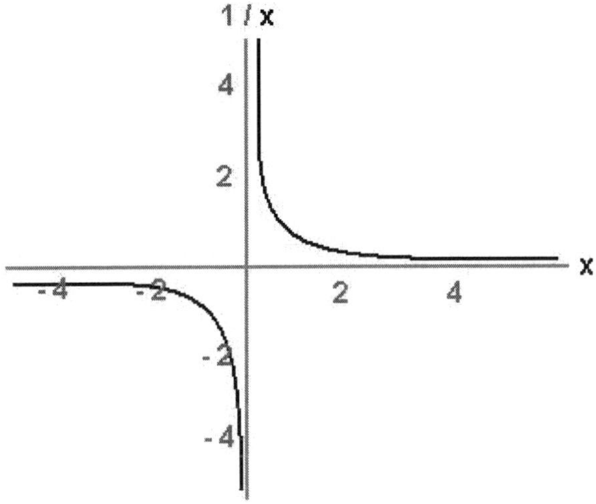

The above plot shows 1/x1/x . where as the vertical asymptote in the above figure is x=0x=0 and that of the horizontal asymptote is y=0y=0 .

Procedure to find asymptotes:

(1) Substitute mx+cmx+c = yy in the equation of the curve and equate the coefficients of the two highest powers of xx to zero.

Determine the values of mm and cc from these equations. If m1,c1,m2,c2m1,c1,m2,c2 ,...........are the values of mm and cc ,the asymptotes are y=m1x+c1,y=m2x+c2y=m1x+c1,y=m2x+c2 etc.

(2) Shorter Method:

If we put x=1x=1 , and y=my=m in the highest degree terms of the equation, we get Õn(m)Õn(m) , and Õn"1(m)Õn"1(m) is obtained by putting x=1,y=mx=1,y=m in the (n"1)(n"1) th degree terms. Equating PhinPhin (m)=0(m)=0 , we get the roots as m1,m2,m3,......mnm1,m2,m3,......mn .

The corresponding values of cc say c1,c2,c3...........cnc1,c2,c3...........cn are obtained by substituting in the formula:

cc = - Õn"1(m)Õn(m)Õn"1(m)Õn(m).

So the asymptotes are yy = m1x+c1m1x+c1 and y=m2x+c2y=m2x+c2 etc.

Curvilinear Asymptotes

Let us suppose that there be a plane parametric curve C : (a,b) '!'! R22 having coordinates C(t) = (x(t), y(t)) as defined in above section. Also, suppose that D be another curve that is unparameterized. Suppose, the curve C approaches to infinity as above discussed. Then, the curve D is defined as the curvilinear asymptote of curve C in case if the smallest of the distances from a point C(t) to any point on D tending to zero as t tends to b. The curve D can also be called simply an asymptote of C, when we want to take no risk of confusion with the curvilinear asymptote.

Asymptotes and Curve Sketching

The Asymptotes are widely used in the process of curve sketching. The asympotes of a function provide the basic guideline in telling the behavior of the curve as it tends to infinity. The curvilinear asymptotes provide a better approximations of given curve, thus they are also being used in curve sketching. Therefore, for such curve sketching, the term asymptotic curve is often used. We can say a curve an asymptotic curve if it always stays tangent to the asymptotic direction of its surface. This is also known as an asymptotic line, although it is not necessarily needed to be a line.

Asymptotes of Functions

in calculus, the common functions of the form y = f(x) do have asymptotes which can be found by encountering limits and classifying into three - horizontal, vertical and oblique.

Vertical asymptotes

For a function y = f(x), the vertical line x = c is said to be a vertical asymptote when at least one of following two statements is true:

limx′!c+f(x)=±″limx′!c+f(x)=±″

limx′!c″f(x)=±″limx′!c″f(x)=±″

Horizontal asymptotes

For given function y = f(x), the horizontal line y = b is said to be horizontal asymptote if

limx′!+″f(x)=blimx′!+″f(x)=b

Here, y = b is an asymptote of f as x tends to +infinity.

or

limx′! ″f(x)=blimx′! ″f(x)=b

Here, y = b is an asymptote of f as x tends to -infinity.

Oblique Asymptote

An oblique or slant asymptote is formed when a linear asymptote is not horizontal or vertical, i.e. it is not parallel to x-axis or y-axis.

An oblique asymptope of a function f(x) is given by a straight line y = mx + c (where m is not equal to zero) if

limx′!+″[f(x)″(mx+c)]=0limx′!+″[f(x)″(mx+c)]=0

Here, y = mx + n is an oblique asymptote of function f when x tends to +infinity.

or

limx′! ″[f(x)″(mx+c)]=0limx′! ″[f(x)″(mx+c)]=0

Here, y = mx + n is an oblique asymptote of f when x tends to -infinity.

Asymptotes parallel to the cordinate axes

Consider a curve given by the equation f(x,y)=0f(x,y)=0, where f(x,y)f(x,y) is a polynomial of degree nn (in xxand yy). If it contains the term ynyn (with some non-zero coefficient), then it has no asymptote parallel to the yy-axis. Otherwise, equate the total coefficient of the highest degree term of yy to 0 and solve (if possible). This gives all asymptotes parallel to the yy-axis.

Example

x2"xy″y+1=0x2"xy″y+1=0

The degree of the equation is 22, and y2y2 is not present, which means the curve might have an asymptote parallel to the yy-axis. The highest degree term of yy is yy, and its total coefficient is "(x+1)"(x+1) ("total" here meaning we include not only constant coefficients, but also any xx-term that is multiplied with yy). Equating this to zero, "(x+1)=0Ò!x="1"(x+1)=0Ò!x="1 is an asymptote parallel to the yy-axis.

Similarly, we can find the asymptotes parallel to the xx-axis as well, but that is unnecessary as the procedure for finding the "oblique" asymptotes does this job as well. Suppose there are asymptotes of the form y=mx+cy=mx+c. Substitute y=mx+cy=mx+c in f(x,y)f(x,y) to get a polynomial in xx. Equate the coefficients of the highest and next highest degree terms of xx to zero. This gives two equations in mmand cc (in general). If these can be solved simultaneously, each solution pair (m,c)(m,c) corresponds to an asymptote y=mx+cy=mx+c.

The method

(1) Replace y by mx + c in the equation of the curve and arrange the result in the form:

$$a_n x^n + a_{n-1} x^{n-1} + ... + a_1 x + a_0 = 0$$

(2) Solve the simultaneous equation :

$$\begin{cases} a_n = 0 \\ a_{n-1} = 0 \end{cases}$$

(3) For each pair of solutions of m and c, write the equation of an asymptote $y = mx + c$.

(4) If there is no a_{n-1} term in (1), solve :

$$\begin{cases} a_n = 0 \\ a_{n-2} = 0 \end{cases}$$

instead of the set of simultaneous equation in (2).
instead of the set of simultaneous equation in (2).
Reason for (2) above

The equation $a_n x^n + a_{n-1} x^{n-1} + ... + a_1 x + a_0 = 0$ has a double root at infinity.

Put $y = \frac{1}{x}$,

The equation becomes

$$a_0 y^n + a_1 y^{n-1} + ... + a_{n-1} y + a_n = 0$$

Then the equation has a double root at $y = 0$.
That is, the equation is of the form:

$$a_0 y^n + a_1 y^{n-1} + ... + a_{n-2} y^2 = 0$$

$$y^2\left(a_0 y^{n-2} + a_1 y^{n-3} + \dots + a_{n-2}\right) = 0$$

Therefore :

$$\begin{cases} a_n = 0 \\ a_{n-2} = 0 \end{cases}$$

Example

Find the oblique asymptotes of

$$y = \frac{x^3}{x^2 - 1}$$

Solution

Put $y = mx + c$ in $y = \dfrac{x^3}{x^2 - 1}$

\ $(x^2 - 1)(mx+c) = x^3$.

\ $(m-1)x^3 - cx^2 - mx - c = 0$ (1)

Put $y = 1/x$ in (1),

\ $cy^3 + my^2 + cy - (m-1) = 0$ (2)

$y = mx + c$ is an asymptote of

iff there is a double root at infinity for (1)

iff there is a double root at 0 for (2).

\ $m - 1 = 0$ and $c = 0$.

\ $m = 1$ and $c = 0$.

\ The oblique asymptote is $y = x$.

Appendix (Calculus method)

$$m = \lim_{x\to\infty} \frac{y}{x} = \lim_{x\to\infty} \frac{1}{x}\left(\frac{x^3}{x^2-1}\right) = \frac{1}{1 - \frac{1}{x^2}} = 1$$

$$c = \lim_{x\to\infty} (y - mx) = \lim_{x\to\infty}\left[\frac{x^3}{x^2-1} - (1)x\right] = \lim_{x\to\infty} \frac{x}{x^2-1} = 0$$

\ The oblique asymptote is $y = x$.

RATIONAL ALGEBRAIC CURVE

In mathematics, a plane real algebraic curve is the set of points on the Euclidean plane whose coordinates are zeros of some polynomial in two variables. More generally an algebraic curve is similar but may be embedded in a higher dimensional space or defined over some more general field. For example, the unit circle is a real algebraic curve, being the set of zeros of the polynomial $x^2 + y^2 - 1$.

Various technical considerations result in the complex zeros of a polynomial being considered as belonging to the curve. Also, the notion of algebraic curve has been generalized to allow the coefficients of the defining polynomial and the coordinates of the points of the curve to belong to any field, leading to the following definition.

In algebraic geometry, a plane affine algebraic curve defined over a field k is the set of points of K^2 whose coordinates are zeros of some bivariate polynomial with coefficients in k, where K is some algebraically closed extension of k. The points of the curve with coordinates in kare the k-points of the curve and, all together, are the k part of the curve.

For example, (2,""3) is a point of the curve defined by $x^2 + y^2$ " $1 = 0$ and the usual unit circle is the real part of this curve. The term "unit circle" may refer to all the complex points as well as to only the real points, the exact meaning usually being clear from the context. The equation $x^2 + y^2 + 1 = 0$ defines an algebraic curve, whose real part is empty.

More generally, one may consider algebraic curves that are not contained in the plane, but in a space of higher dimension. A curve that is not contained in some plane is called a skew curve. The simplest example of a skew algebraic curve is the twisted cubic. One may also consider algebraic curves contained in the projective space and even algebraic curves that are defined independently to any embedding in an affine or projective space. This leads to the most general definition of an algebraic curve:

In algebraic geometry, an algebraic curve is an algebraic variety of dimension one.

In Euclidean geometry

An algebraic curve in the Euclidean plane is the set of the points whose coordinates are the solutions of a bivariate polynomial equation $p(x, y) = 0$. This equation is often called the implicit equation of the curve, in contrast to the curves that are the graph of a function defining explicitly y as a function of x.

With a curve given by such an implicit equation, the first problems are to determine the shape of the curve and to draw it. These problems are not as easy to solve as in the case of the graph of a function, for which y may easily be computed for various values of x. The fact that the defining equation is a polynomial implies that the curve has some structural properties that may help in solving these problems.

Every algebraic curve may be uniquely decomposed into a finite number of smooth monotone arcs (also called branches) sometimes connected by some points sometimes called "remarkable points", and possibly a finite number of isolated points called acnodes. A smooth monotone arc is the graph of a smooth function which is defined and monotone on an open interval of the x-axis. In each direction, an arc is either unbounded (usually called an infinite arc) or has an end point which is either a singular point (this will be defined below) or a point with a tangent parallel to one of the coordinate axes.

For example, for the Tschirnhausen cubic, there are two infinite arcs having the origin (0,0) as end point. This point is the only singular point of the curve. There are also two arcs having this singular point as one end point and having a second end point with a horizontal tangent. Finally, there are two other arcs each having one of these points with horizontal tangent as the first end point and having the unique point with vertical tangent as the second end point. In contrast, the sinusoid is certainly not an algebraic curve, having an infinite number of monotone arcs.

To draw an algebraic curve, it is important to know the remarkable points and their tangents, the infinite branches and their asymptotes (if any) and the way in which the arcs connect them. It is also useful to consider the inflection points as remarkable points. When all this information is drawn on a sheet of paper, the shape of the curve usually appears rather clearly. If not, it suffices to add a few other points and their tangents to get a good description of the curve.

The methods for computing the remarkable points and their tangents are described below, after the section Projective curves.

Plane projective curves

It is often desirable to consider curves in the projective space. An algebraic curve in the projective plane or plane projective curve is the set of the points in a projective plane whose projective coordinates are zeros of a homogeneous polynomial in three variables P(x, y, z).

Every affine algebraic curve of equation p(x, y) = 0 may be completed into the projective curve of equation where is the result of the homogenization of p. Conversely, if P(x, y, z) = 0 is the homogeneous equation of a projective curve, then P(x, y, 1) = 0 is the equation of an affine curve, which consists of the points of the projective curve whose third projective coordinate is not zero. These two operations are reciprocal one to the other, as {\displaystyle ^{h}p(x,y,1)=p(x,y)} and, if p is defined by {\displaystyle p(x,y)=P(x,y,1)}, then {\displaystyle ^{h}p(x,y,z)=P(x,y,z),}, as soon as the homogeneous polynomial P is not divisible by z.

For example, the projective curve of equation $x^2 + y^2$ " z^2 is the projective completion of the unit circle of equation $x^2 + y^2$ " 1 = 0.

This implies that an affine curve and its projective completion are the same curve, or, more precisely that the affine curve is a part of the projective curve that is large enough to well define the "complete" curve. This point of view is commonly expressed by calling "points at infinity" of the affine curve the points (in finite number) of the projective completion that do not belong to the affine part.

Remarkable points of a plane curve

In this section, we consider a plane algebraic curve defined by a bivariate polynomial p(x, y) and its projective completion, defined by the homogenization of p.

Intersection with a line

Knowing the points of intersection of a curve with a given line is frequently useful. The intersection with the axes of coordinates and the asymptotes are useful to draw the curve. Intersecting with a line parallel to the axes allows one to find at least a point in each branch of the curve. If an efficient root-finding algorithm is available, this allows to draw the curve by plotting the intersection point with all the lines parallel to the y-axis and passing through each pixel on the x-axis.

If the polynomial defining the curve has degree d, any line cuts the curve in at most d points. Bézout's theorem asserts that this number is exactly d, if the points are searched in the projective plane over an algebraically closed field (for example the complex numbers), and counted with their multiplicity. The method of computation that follows proves again this theorem, in this simple case.

To compute the intersection of the curve defined by the polynomial p with the line of equation ax+by+c = 0, one solves the equation of the line for x (or for y if a = 0). Substituting the result in p, one gets a univariate equation q(y) = 0 (or q(x) = 0, if the equation of the line has been solved in y), each of whose roots is one coordinate of an intersection point. The other coordinate is deduced from the equation of the line. The multiplicity of an intersection point is the multiplicity of the corresponding root. There is an intersection point at infinity if the degree of q is lower than the degree of p; the multiplicity of such an intersection point at infinity is the difference of the degrees of p and q.

This implies that the number of singular points is finite as long as p(x,y) or P(x,y,z) is square free. Bézout's theorem implies thus that the number of singular points is at most $(d''1)^2$, but this bound is not sharp because the system of equations is overdetermined. If reducible polynomials are allowed, the sharp bound is d(d"1)/ 2, this value being reached when the polynomial factors in linear factors, that is if the curve is the union of d lines. For irreducible curves and polynomials, the number of singular points is at most (d"1)(d"2)/2, because of the formula expressing the genus in term of the singularities. The maximum is reached by the curves of genus zero whose all singularities have multiplicity two and distinct tangents.

The equation of the tangents at a singular point is given by the nonzero homogeneous part of lowest degree in the Taylor series of the polynomial at the singular point. When one changes the coordinates to put the singular point at the origin, the equation of the tangents at the singular point is thus the nonzero homogeneous part of lowest degree of the polynomial, and the multiplicity of the singular point is the degree of this homogeneous part.

ANALYTIC STRUCTURE

The study of the analytic structure of an algebraic curve in the neighborhood of a singular point provides an accurate information of the topology of singularities. In fact, near a singular point, a real algebraic curve is the union of a finite number of branches that intersect only at the singular point, and look either as a cusp or as a smooth curve.

Near a regular point, one of the coordinates of the curve may be expressed as an analytic function of the other coordinate. This is a corollary of the analytic implicit function theorem, and implies that the curve is smooth near the point. Near a singular point, the situation is more complicated, and involves Puiseux series, which provide analytic parametric equationsof the branches.

For describing a singularity, it is worth to translate the curve for having the singularity at the origin. This consists of a change of variable of the form where are the coordinates of the singular point. In the following, the singular point under consideration is always supposed to be at the origin.

The equation of an algebraic curve is {\displaystyle f(x,y)=0,} where f is a polynomial in x and y. This polynomial may be considered as a polynomial in y, with coefficients in the algebraically closed field of the Puiseux series in x. Thus f may be factored in factors of the form {\displaystyle y-P(x),} where P is a Puiseux series. These factors are all different if f is an irreducible polynomial, because this implies that f is square-free, a property which is independent of the field of coefficients.

Algebraic function fields

The study of algebraic curves can be reduced to the study of irreducible algebraic curves: those curves that cannot be written as the union of two smaller curves. Up to birationalequivalence, the irreducible curves over a field F are categorically equivalent to algebraic function fields in one variable over F. Such an algebraic function field is a field extension Kof F that contains an element x which is transcendental over F, and such that K is a finite algebraic extension of F(x), which is the field of rational functions in the indeterminate xover F.

For example, consider the field C of complex numbers, over which we may define the field C(x) of rational functions in C. If $y^2 = x^3$ ″ x ″ 1, then the field C(x, y) is an elliptic function field. The element x is not uniquely determined; the field can also be regarded, for instance, as an extension of C(y). The algebraic curve corresponding to the function field is simply the set of points (x, y) in C^2 satisfying $y^2 = x^3$ ″ x ″ 1.

If the field F is not algebraically closed, the point of view of function fields is a little more general than that of considering the locus of points, since we include, for instance, "curves" with no points on them. For example, if the base field F is the field R of real numbers, then $x^2 + y^2 =$ ″1 defines an algebraic extension field of R(x), but the corresponding curve considered as a subset of R^2 has no points. The equation $x^2 + y^2 =$ ″1 does define an irreducible algebraic curve over R in the scheme sense (an integral, separated one-dimensional schemes of finite type over R). In this sense, the one-to-one correspondence between irreducible algebraic curves over F (up to birational equivalence) and algebraic function fields in one variable over F holds in general.

Two curves can be birationally equivalent (i.e. have isomorphic function fields) without being isomorphic as curves. The situation becomes easier when dealing with nonsingularcurves, i.e. those that lack any singularities. Two nonsingular projective curves over a field are isomorphic if and only if their function fields are isomorphic.

Tsen's theorem is about the function field of an algebraic curve over an algebraically closed field.

Complex curves and real surfaces

A complex projective algebraic curve resides in n-dimensional complex projective space CP^n. This has complex dimension n, but topological dimension, as a real manifold, 2n, and is compact, connected, and orientable. An algebraic curve over C likewise has topological dimension two; in other words, it is a surface.

The topological genus of this surface, that is the number of handles or donut holes, is equal to the geometric genus of the algebraic curve that may be computed by algebraic means. In short, if one consider a plane projection of a nonsingular curve that has degree d and only ordinary singularities (singularities of multiplicity two with distinct tangents), then the genus is (d " 1)(d " 2)/2 " k, where k is the number of these singularities.

Compact Riemann surfaces

A Riemann surface is a connected complex analytic manifold of one complex dimension, which makes it a connected real manifold of two dimensions. It is compact if it is compact as a topological space.

There is a triple equivalence of categories between the category of smooth irreducible projective algebraic curves over C (with non-constant regular maps as morphisms), the category of compact Riemann surfaces (with non-constant holomorphic maps as morphisms), and the opposite of the category of algebraic function fields in one variable over C (with field homomorphisms that fix C as morphisms). This means that in studying these three subjects we are in a sense studying one and the same thing. It allows complex analytic methods to be used in algebraic geometry, and algebraic-geometric methods in complex analysis, and field-theoretic methods to be used in both. This is characteristic of a much wider class of problems in algebraic geometry.

See also algebraic geometry and analytic geometry for more general theory.

Singularities

Using the intrinsic concept of tangent space, points P on an algebraic curve C are classified as smooth (synonymous: non-singular), or else singular. Given n"1 homogeneous polynomials in n+1 variables, we may find the Jacobian matrix as the (n"1)×(n+1) matrix of the partial derivatives. If the rank of this matrix is n"1, then the polynomials define an algebraic curve (otherwise they define an algebraic variety of higher dimension). If the rank remains n"1 when the Jacobian matrix is evaluated at a point P on the curve, then the point is a smooth or regular point; otherwise it is a singular point.

Classification of singularities

Singular points include multiple points where the curve crosses over itself, and also various types of cusp, for example that shown by the curve with equation $x^3 = y^2$ at (0,0).

A curve C has at most a finite number of singular points. If it has none, it can be called smooth or non-singular. Commonly, this definition is understood over an algebraically closed field and for a curve C in projective space (i.e., complete in the sense of algebraic geometry). For example, the curve of equation {\displaystyle y-x^{3}=0} is considered as singular, as having a singular point (a cusp) at infinity.

The singular points are classified by means of several invariants. The multiplicity m is defined as the maximum integer such that the derivatives of f to all orders up to m – 1 vanish (also the minimal intersection number between the curve and a straight line at P). Intuitively, a singular point has delta invariant ä if it concentrates ä ordinary double points at P. To make this precise, the blow up process produces so-called infinitely near points, and summing m(m"1)/2 over the infinitely near points, where m is their multiplicity, produces ä. For an irreducible and reduced curve and a point P we can define ä algebraically as the length of {\displaystyle {\widetilde {{\mathcal {O}}_{P}}}/{\mathcal {O}}_{P}} where {\displaystyle {\mathcal {O}}_{P}} is the local ring at P and {\displaystyle {\widetilde {{\mathcal {O}}_{P}}}} is its integral closure.

The Milnor number ì of a singularity is the degree of the mapping grad f(x,y)/ | grad f(x,y) | on the small sphere of radius å, in the sense of the topological degree of a continuous mapping, where grad f is the (complex) gradient vector field of f. It is related to ä and r by the Milnor-Jung formula,

ì = 2ä “ r + 1.

Here, the branching number r of P is the number of locally irreducible branches at P. For example, r = 1 at an ordinary cusp, and r = 2 at an ordinary double point. Note that m is at least r, and that P is singular if and only if m is at least 2. Moreover, ä is at least m(m-1)/2.

Rational curves

A rational curve, also called a unicursal curve, is any curve which is birationally equivalent to a line, which we may take to be a projective line; accordingly, we may identify the function field of the curve with the field of rational functions in one indeterminate F(x). If F is algebraically closed, this is equivalent to a curve of genus zero; however, the field of all real algebraic functions defined on the real algebraic variety x^2+y^2 = “1 is a field of genus zero which is not a rational function field.

Concretely, a rational curve embedded in an affine space of dimension n over F can be parameterized (except for isolated exceptional points) by means of n rational functions of a single parameter t; by reducing these rational functions to the same denominator, the n+1 resulting polynomials define a polynomial parametrization of the projective completion of the curve in the projective space. An example is the rational normal curve, where all these polynomials are monomials.

Any conic section defined over F with a rational point in F is a rational curve. It can be parameterized by drawing a line with slope t through the rational point, and intersection with the plane quadratic curve; this gives a polynomial with F-rational coefficients and one F-rational root, hence the other root is F-rational (i.e., belongs to F) also. Elliptic curves carry the structure of an abelian group with the distinguished point as the identity of the group law. In a plane cubic model three points sum to zero in the group if and only if they are collinear. For an elliptic curve defined over the complex numbers the group is isomorphic to the additive group of the complex plane modulo the period lattice of the corresponding elliptic functions.

The intersection of two quadric surfaces is in general a nonsingular curve of genus one and degree four, and thus an elliptic curve, if it has a rational point. In special cases, the intersection either may be a rational singular quartic, or is decomposed in curves of smaller degrees which are not always distinct (either a cubic curve and a line, or two conics, or a conic and two lines, or four lines).

Curves of genus greater than one

Curves of genus greater than one differ markedly from both rational and elliptic curves. Such curves defined over the rational numbers, by Faltings's theorem, can have only a finite number of rational points, and they may be viewed as having a hyperbolic geometry structure. Examples are the hyperelliptic curves, the Klein quartic curve, and the Fermat curve$x^n + y^n = z^n$ when n is greater than three.

Asymptotes and Other Things to Look For

A vertical asymptote is a place where the function becomes infinite, typically because the formula for the function has a denominator that becomes zero. For example, the reciprocal function f(x)=1/xf(x)=1/x has a vertical asymptote at x=0x=0, and the function tanxtana x has a vertical asymptote at x=À/2x=ð/2 (and also at x="À/2x="ð/2, x=3À/2x=3ð/2, etc.). Whenever the formula for a function contains a denominator it is worth looking for a vertical asymptote by checking to see if the denominator can ever be zero, and then checking the limit at such points. Note that there is not always a vertical asymptote where the denominator is zero: f(x)=(sinx)/xf(x)=(sina x)/xhas a zero denominator at x=0x=0, but since limx′!0(sinx)/x=1limx′!0(sina x)/x=1 there is no asymptote there.

A horizontal asymptote is a horizontal line to which f(x)f(x) gets closer and closer as xxapproaches ″″ (or as xx approaches ″″ ″). For example, the reciprocal function has the xx-axis for a horizontal asymptote. Horizontal asymptotes can be identified by computing the limits limx′!″f(x)limx′!″f(x) and limx′!″″f(x)limx′! ″f(x). Since limx′!″1/x=limx′!″″1/x=0limx′!″1/x=limx′! ″1/x=0, the line y=0y=0 (that is, the xx-axis) is a horizontal asymptote in both directions.

Some functions have asymptotes that are neither horizontal nor vertical, but some other line. Such asymptotes are somewhat more difficult to identify and we will ignore them.

If the domain of the function does not extend out to infinity, we should also ask what happens as xx approaches the boundary of the domain. For example, the function y=f(x)=1/r2″x2″″″″″″″″″y=f(x)=1/r2″x2 has domain ″r<x<r″r<x<r, and yy becomes infinite as xx approaches either rr or ″r″r. In this case we might also identify this behavior because when x=±rx=±r the denominator of the function is zero.

If there are any points where the derivative fails to exist (a cusp or corner), then we should take special note of what the function does at such a point.

Finally, it is worthwhile to notice any symmetry. A function f(x)f(x) that has the same value for ″x″x as for xx, i.e., f(″x)=f(x)f(″x)=f(x), is called an "even function." Its graph is symmetric with respect to the yy-axis. Some examples of even functions are: xnxn when nn is an even number, cosxcosa x, and sin2xsin2a x. On the other hand, a function that satisfies the property f(″x)=″f(x)f(″x)=″f(x) is called an "odd function." Its graph is symmetric with respect to the origin. Some examples of odd functions are: xnxn when nn is an odd number, sinxsina x, and tanxtana x. Of course, most functions are neither even nor odd, and do not have any particular symmetry.

An asymptote of a curve y=f(x) that has an infinite branch is called a line such that the distance between the point (x,f(x)) lying on the curve and the line approaches zero as the point moves along the branch to infinity.

Asymptotes can be vertical, oblique (slant) and horizontal. A horizontal asymptote is often considered as a special case of an oblique asymptote.

Circular Asymptotes

Could an asymptote be defined as a point, or a circle? I assume it would be in a rather exotic topology, or a very trivial one. Furthermore, can we define each of these as the other's asymptote? The points would probably have to turn into lines(functions) by virtue of an extra dimension in that topology. What I'm basically trying to figure out is how a generic attractive force, coupled with a generic mass-like property(i.e. the points can't overlap; passive repelling force) could be defined mathematically, without using vectors. I first thought of setting x,y as each other's limit(x->y and y->x), with x ‴ ‴ y, but that leads to x=y, due to the nature of the limit operation. With asymptotes, I'm guessing an infinity should show up.

14

Change of the Independent Variables

As with ODEs (ordinary differential equations), a PDE (partial differential equation, or more accurately, the initial-boundary value problem (IBVP) as a whole) may be made more amenable with the help of some kind of modification of variables. So far, we've dealt only with boundary conditions that specify the value of u, which represented fluid velocity, as zero at the boundaries. Though fluid mechanics can get more complicated than that (understatement of the millennium), let's look at heat transfer now for the sake of variety.

As hinted previously, the one dimensional diffusion equation can also describe heat flow in one dimension. Think of how heat could flow in one dimension: one possibility is a rod that's completely laterally insulated, so that the heat will flow only along the rod and not across it (be aware, though, it is possible to consider heat loss/gain along the rod without going two dimensional).

If this rod has finite length, heat could flow in and out of the uninsulated ends. A 1D rod can have at most two ends (it can also have one or zero: the rod could be modeled as "very long"), and the boundary conditions could specify what happens at these ends. For example, the temperature could be specified at a boundary, or maybe the flow of heat, or maybe some combination of the two.

ADDED VARIABLES, SAME TECHNIQUES

In the real world, it is very difficult to explain behavior as a function of only one variable, and economics is no different. More specific economic interpretations will be discussed in the next section, but for now, we'll just concentrate on developing the techniques we'll be using.

First, to define the functions themselves. We want to describe behavior where a variable is dependent on two or more variables. Every rule and notation described from now on is the same for two variables, three variables, four variables, and so on, so we'll use the simplest case; a function of two independent variables. Conventionally, z is the dependent variable (like y in univariate functions) and x and y are the independent variables (like x in univariate functions):

$$z = f(x,y)$$

For example, suppose that the following function describes some behavior:

$$z = 3x^2 + 2xy - y^2$$

Differentiating this function still means the same thing—still we are looking for functions that give us the slope, but now we have more than one variable, and more than one slope.

Visualize this by recalling from graphing what a function with two independent variables looks like. Whereas a 2-dimensional picture can represent a univariate function, our z function above can be represented as a 3-dimensional shape. Think of the x and y variables as being measured along the sides of a chessboard. Then every combination of x and y would map onto a square somewhere on the chessboard. For example, suppose x=1 and y=1. Start at one of the corners of the chessboard. Then move one square in on the x side for x=1, and one square up into the board to represent y=1. Now, calculate the value of z.

$$\begin{aligned} z &= 3(1)^2 + 2(1)(1) - (1)^2 \\ &= 3 + 2 - 1 = 4 \end{aligned}$$

The function z takes on a value of 4, which we graph as a height of 4 over the square that represents x=1 and y=1. Map out the entire function this way, and the result will be a shape, usually looking like a mountain peak in typical economic analysis problems.

Now back to slope. Imagine standing on the mountain shape, facing parallel to the x side of the chessboard. If you allow x to increase, while holding y constant, then you would move forward in a straight line along the mountain shape. We define the slope in this direction as the change in the z variable, or a change in the height of the shape, in response to a movement along the chessboard in one direction, or a change in the variable x, holding y constant.

Formally, the definition is: the partial derivative of z with respect to x is the change in z for a given change in x, holding y constant. Notation, like before, can vary. Here are some common choices:

$$\frac{\partial z}{\partial x} \qquad Z_x \qquad f_x(x, y)$$

Now go back to the mountain shape, turn 90 degrees, and do the same experiment. Now, we define a second slope as the change in the height of the z function in response to a movement forward on the chessboard (perpendicular to the movement measured by the first slope calculation), or a change in the y variable, holding the x variable constant. Typical notation for this operation would be

$$\frac{\partial z}{\partial y} \qquad Z_y \qquad f_y(x,y)$$

Therefore, calculus of multivariate functions begins by taking partial derivatives, in other words, finding a separate formula for each of the slopes associated with changes in one of the independent variables, one at a time. Before we discuss economic applications, let's review the rules of partial differentiation.

Basic rules of partial differentiation

The rules of partial differentiation follow exactly the same logic as univariate differentiation. The only difference is that we have to decide how to treat the other variable. Recall that in the previous section, slope was defined as a change in z for a given change in x or y, holding the other variable constant. There's our clue as to how to treat the other variable. If we hold it constant, that means that no matter what we call it or what variable name it has, we treat it as a constant. Suppose, for example, we have the following equation:

$$z = 2xy$$

If we are taking the partial derivative of z with respect to x, then y is treated as a constant. Since it is multiplied by 2 and x and is constant, it is also defined as a coefficient of x. Therefore,

$$\frac{\partial z}{\partial x} = 2y$$

Therefore, once all other variables are held constant, then the partial derivative rules for dealing with coefficients, simple powers of variables, constants, and sums/ differences of functions remain the same, and are used to determine the function of the slope for each independent variable. Let's use the function from the previous section to illustrate.

$$z = 3x^2 + 2xy - y^2$$

First, differentiate with respect to x, holding y constant:

$$\frac{\partial z}{\partial x} = 6x + 2y - 0 = 6x + 2y$$

Note that there were no y variables in the first term, so differentiation was exactly like the univariate process; in the last term there were no x variables, therefore the derivative is zero, according to the constant rule, since y is treated as a constant.

Now, take the partial derivative with respect to y, holding x constant:

$$\frac{\partial z}{\partial y} = 0 + 2x - 2y = 2x + 2y$$

Again, note that the first term had no "variables" in it, since x is being treated as a constant, therefore the derivative of that term is 0.

To make sure you have a clear picture of more than one slope in a function, let's evaluate the two partial derivatives at the point on the function where x = 1 and y = 2:

$$
\begin{aligned}
z &= 3x^2 + 2xy - y^2 \\
&= 3(1)^2 + 2(1)(2) - (2)^2 \\
&= 3
\end{aligned}
$$

$$\frac{\partial z}{\partial x} = 6x + 2y = 6(1) + 2(2) = 10$$

$$\frac{\partial z}{\partial x} = 2x + 2y = 2(1) + 2(2) = 6$$

How do we interpret this information? First, note that when x = 1 and y = 2, then the function z takes on a value of 3. At this point on our "mountain' or 3 dimensional shape, we can evaluate the change in the function z in 2 different directions. First, the change in z with respect to x is 10. In other words, the slope in a direction parallel to the x-axis is 10. Now turn 90 degrees. The slope in a direction perpendicular to our previous slope is 6, therefore not quite as steep. Also, note that although each slope depends on the change in only one variable, the position or fixed value of the other variable does matter; since you need both x and y to actually calculate the numerical values of slope. We'll come back to this in the next section, and look at the economic meaning behind this relatedness. But first, back to the rules.

The product and quotient of functions rules follow exactly the same logic: hold all variables constant except for the one that is changing in order to determine the slope of the function with respect to that variable. To illustrate the product rule, first let's redefine the rule, using partial differentiation notation:

$$\text{Given } Z = f(x,y)\,g(x,y)$$

$$\frac{\partial Z}{\partial x} = \frac{\partial f}{\partial x}\cdot g(x,y) + \frac{\partial g}{\partial x}\cdot f(x,y)$$

$$\frac{\partial Z}{\partial y} = \frac{\partial f}{\partial y}\cdot g(x,y) + \frac{\partial g}{\partial y}\cdot f(x,y)$$

Now use the product rule to determine the partial derivatives of the following function:

$$z = (4x+9)(8x+5y)$$

$$
\begin{aligned}
\frac{\partial z}{\partial x} &= (4)(8x+5y) + 8(4x+9) \\
&= 32x + 20y + 32x + 72 \\
&= 64x + 20y + 72
\end{aligned}
$$

$$
\begin{aligned}
\frac{\partial z}{\partial y} &= (0)(8x+5y) + (5)(4x+9) \\
&= 20x + 45
\end{aligned}
$$

To illustrate the quotient rule, first redefine the rule using partial differentiation notation:

$$\text{Given } Z = \frac{f(x,y)}{g(x,y)}$$

$$\frac{\partial Z}{\partial x} = \frac{\frac{\partial f}{\partial x} \cdot g(x,y) - \frac{\partial g}{\partial x} \cdot f(x,y)}{\left[g(x,y)\right]^2}$$

$$\frac{\partial Z}{\partial y} = \frac{\frac{\partial f}{\partial y} \cdot g(x,y) - \frac{\partial g}{\partial y} \cdot f(x,y)}{\left[g(x,y)\right]^2}$$

Use the new quotient rule to take the partial derivatives of the following function:

$$z = \frac{(2x+y)}{(x+y)}$$

$$\frac{\partial z}{\partial x} = \frac{(2)(x+y)-(1)(2x+y)}{[x+y]^2}$$

$$= \frac{2x+2y-2x-y}{[x+y]^2}$$

$$= \frac{y}{[x+y]^2}$$

$$\frac{\partial z}{\partial y} = \frac{(1)(x+y)-(y)(2x+y)}{[x+y]^2}$$

$$= \frac{x+y-2xy-y^2}{\left[x+y^2\right]^2}$$

Not-so-basic rules of partial differentiation

Just as in the previous univariate section, we have two specialized rules that we now can apply to our multivariate case.

First, the generalized power function rule. Again, we need to adjust the notation, and then the rule can be applied in exactly the same manner as before.

When a multivariate function takes the following form:

$$z = \left[f(x,y)\right]^n$$

Then the rule for taking the derivative is:

$$\frac{\partial z}{\partial x} = n\left[f(x,y)\right]^{n-1} \cdot \frac{\partial f}{\partial x}$$
$$\frac{\partial z}{\partial y} = n\left[f(x,y)\right]^{n-1} \cdot \frac{\partial f}{\partial y}$$

Use the power rule on the following function to find the two partial derivatives:

$$z = \left(2x + y^2\right)^3$$
$$\frac{\partial z}{\partial x} = 3\left[2x + y^2\right]^2 \cdot (2) = 6\left(2x + y^2\right)^2$$
$$\frac{\partial z}{\partial y} = 3\left[2x + y^2\right]^2 \cdot (2y) = 6y\left(2x + y^2\right)^2$$

The composite function chain rule notation can also be adjusted for the multivariate case:

$$\text{Given } z = f(u) \text{ and } u = g(x,y)$$
$$\text{such that } z = f\left[g(x,y)\right]$$

Then the partial derivatives of z with respect to its two independent variables are defined as:

$$\frac{\partial z}{\partial x} = \frac{dy}{du} \cdot \frac{\partial u}{\partial x}$$
$$\frac{\partial z}{\partial y} = \frac{dy}{du} \cdot \frac{\partial u}{\partial y}$$

Let's do the same example as above, this time using the composite function notation where functions within the z function are renamed. Note that either rule could be used for this problem, so when is it necessary to go to the trouble of presenting the more formal composite function notation? As problems become more complicated, renaming parts of a composite function is a better way to keep track of all parts of the problem. It is slightly more time consuming, but mistakes within the problem are less likely.

$$\text{Given } z = \left(2x + y^2\right)^3$$
$$\text{let } z = f(u) = u^3 \text{ and } u = g(x,y) = 2x + y^2$$
$$\text{then } \frac{\partial z}{\partial x} = \frac{dy}{du} \cdot \frac{\partial u}{\partial x} = \left(3u^2\right)(2) = 6u^2$$
$$\text{and } \frac{\partial z}{\partial y} = \frac{dy}{du} \cdot \frac{\partial u}{\partial y} = \left(3u^2\right)(2y) = (6y)u^2$$

The final step is the same, replace u with function g:

$$\frac{\partial z}{\partial x} = 6u^2 = 6\left(2x+y^2\right)^2$$
$$\frac{\partial z}{\partial y} = 6y(u)^2 = (6y)\left(2x+y^2\right)^2$$

Special cases in multivariate functions

The last two special cases in multivariate differentiation also follow the same logic as their univariate counterparts.

The rule for differentiating multivariate natural logarithmic functions, with appropriate notation changes is as follows:

$$\text{Given } z = f(u) = \ln(u) \text{ and } u = g(x,y)$$
$$\text{such that } z = \ln g(x,y)$$

Then the partial derivatives of z with respect to its independent variables are defined as:

$$\frac{\partial z}{\partial x} = \frac{1}{u}\cdot\frac{\partial g}{\partial x} = \frac{1}{g(x,y)}\cdot\frac{\partial g}{\partial x}$$
$$\frac{\partial z}{\partial y} = \frac{1}{u}\cdot\frac{\partial g}{\partial y} = \frac{1}{g(x,y)}\cdot\frac{\partial g}{\partial y}$$

Let's do an example. Find the partial derivatives of the following function:

$$z = \ln\left(2x^2+4y^2\right)$$
$$\frac{\partial z}{\partial x} = \frac{1}{g(x,y)}\cdot\frac{\partial g}{\partial x} = \frac{1}{\left(2x^2+4y^2\right)}\cdot(4x) = \frac{4x}{\left(2x^2+4y^2\right)}$$
$$\frac{\partial z}{\partial y} = \frac{1}{g(x,y)}\cdot\frac{\partial g}{\partial y} = \frac{1}{\left(2x^2+4y^2\right)}\cdot(8y) = \frac{4y}{\left(x^2+2y^2\right)}$$

The rule for taking partials of exponential functions can be written as:

$$\text{Given } z = f(u) = e^u \text{ and } u = g(x, y)$$
$$\text{such that } z = e^{g(x,y)}$$

Then the partial derivatives of z with respect to its independent variables are defined as:

$$\frac{\partial z}{\partial x} = e^{g(x,y)} \cdot \frac{\partial g}{\partial x}$$
$$\frac{\partial z}{\partial y} = e^{g(x,y)} \cdot \frac{\partial g}{\partial y}$$

One last time, we look for partial derivatives of the following function using the exponential rule:

$$z = e^{(3xy^2)}$$
$$\frac{\partial z}{\partial x} = e^{g(x,y)} \cdot \frac{\partial g}{\partial x} = e^{(3xy^2)} \cdot (3y^2) = (3y^2)e^{(3xy^2)}$$
$$\frac{\partial z}{\partial y} = e^{g(x,y)} \cdot \frac{\partial g}{\partial y} = e^{(3xy^2)} \cdot (6xy) = (6xy \cdot)e^{(3xy^2)}$$

Higher order partial and cross partial derivatives

The story becomes more complicated when we take higher order derivatives of multivariate functions. The interpretation of the first derivative remains the same, but there are now two second order derivatives to consider.

First, there is the direct second-order derivative. In this case, the multivariate function is differentiated once, with respect to an independent variable, holding all other variables constant. Then the result is differentiated a second time, again with respect to the same independent variable. In a function such as the following:

$$z = f(x,y)$$

There are 2 direct second-order partial derivatives, as indicated by the following examples of notation:

$$f_{xx} = \frac{\partial}{\partial x}\left(\frac{\partial z}{\partial x}\right) = \frac{\partial^2 z}{\partial x^2} = Z_{xx}$$
$$f_{yy} = \frac{\partial}{\partial y}\left(\frac{\partial z}{\partial y}\right) = \frac{\partial^2 z}{\partial y^2} = Z_{yy}$$

These second derivatives can be interpreted as the rates of change of the two slopes of the function z.

Now the story gets a little more complicated. The cross-partials, f_{xy} and f_{yx} are defined in the following way. First, take the partial derivative of z with respect to x. Then take the derivative again, but this time, take it with respect to y, and hold the x constant. Spatially, think of the cross partial as a measure of how the slope (change in z with respect to x) changes, when the y variable changes. The following are examples of notation for cross-partials:

$$f_{xy} = \frac{\partial}{\partial y}\left(\frac{\partial z}{\partial x}\right) = \frac{\partial^2 z}{\partial y \partial x}$$

$$f_{yx} = \frac{\partial}{\partial x}\left(\frac{\partial z}{\partial y}\right) = \frac{\partial^2 z}{\partial x \partial y}$$

We'll discuss economic meaning further in the next section, but for now, we'll just show an example, and note that in a function where the cross-partials are continuous, they will be identical. For the following function:

$$z = 2x^3 + 3xy + 2y^2$$

Take the first and second partial derivatives.

$$z = 2x^3 + 3xy + 2y^2$$

$$z_x = 6x^2 + 3y \qquad z_y = 3x + 4y$$

$$z_{xx} = 12x \qquad z_{yy} = 4$$

Now, starting with the first partials, find the cross partial derivatives:

$$z_x = 6x^2 + 3y \qquad z_y = 3x + 4y$$

$$z_{xy} = 3 \qquad z_{yx} = 3$$

Note that the cross partials are indeed identical, a fact that will be very useful to us in future optimization sections.

Dependent and Independent Variables

In mathematical modeling, statistical modeling and experimental sciences, the values of dependent variables depend on the values of independent variables. The dependent variables represent the output or outcome whose variation is being studied. The independent variables, also known in a statistical context as regressors, represent inputs or causes, that is, potential reasons for variation. In an experiment, any variable that the experimenter manipulates can be called an independent variable. Models and experiments test the effects that the independent variables have on the dependent variables. Sometimes, even if their influence is not of direct interest, independent variables may be included for other reasons, such as to account for their potential confounding effect.

In mathematics, a function is a rule for taking an input (in the simplest case, a number or set of numbers) and providing an output (which may also be a number). A symbol that stands for an arbitrary input is called an independent variable, while a symbol that stands for an arbitrary output is called a dependent variable. The most common symbol for the input is x, and the most common symbol for the output is y; the function itself is commonly written.

It is possible to have multiple independent variables and/or multiple dependent variables. For instance, in multivariable calculus, one often encounters functions of the form, where z is a dependent variable and x and y are independent variables. Functions with multiple outputs are often referred to as vector-valued functions.

In set theory, a function between a set X and a set Y is a subset of the Cartesian product such that every element of X appears in an ordered pair with exactly one element of Y. In this situation, a symbol representing an element of X may be called an independent variable and a symbol representing an element of Y may be called a dependent variable, such as when X is a manifold and the symbol x represents an arbitrary point in the manifold. However, many advanced textbooks do not distinguish between dependent and independent variables.

Statistics

In an experiment, a variable, manipulated by an experimenter, is called an independent variable. The dependent variable is the event expected to change when the independent variable is manipulated.

In data mining tools (for multivariate statistics and machine learning), the dependent variable is assigned a role as target variable (or in some tools as label attribute), while an independent variable may be assigned a role as regular variable. Known values for the target variable are provided for the training data set and test data set, but should be predicted for other data. The target variable is used in supervised learning algorithms but not in unsupervised learning.

Modeling

In mathematical modeling, the dependent variable is studied to see if and how much it varies as the independent variables vary. In the simple stochastic linear model is the i th value of the dependent variable and is the i th value of the independent variable. The term {\displaystyle e_{i}} is known as the "error" and contains the variability of the dependent variable not explained by the independent variable.

With multiple independent variables, the model is {\displaystyle y_+...+bx_{n,i}+e_{i}}, where n is the number of independent variables.

Simulation

In simulation, the dependent variable is changed in response to changes in the independent variables.

Statistics synonyms

Depending on the context, an independent variable is sometimes called a "predictor variable", regressor, covariate, "controlled variable", "manipulated variable", "explanatory variable", exposure variable, "risk factor", "feature" (in machine learning and pattern recognition) or "input variable." In econometrics, the term "control variable" is usually used instead of "covariate".

Depending on the context, a dependent variable is sometimes called a "response variable", "regressand", "criterion", "predicted variable", "measured variable", "explained variable", "experimental variable", "responding variable", "outcome variable", "output variable" or "label".

"Explanatory variable" is preferred by some authors over "independent variable" when the quantities treated as independent variables may not be statistically independent or independently manipulable by the researcher. If the independent variable is referred to as an "explanatory variable" then the term "response variable" is preferred by some authors for the dependent variable.

"Explained variable" is preferred by some authors over "dependent variable" when the quantities treated as "dependent variables" may not be statistically dependent. If the dependent variable is referred to as an "explained variable" then the term "predictor variable" is preferred by some authors for the independent variable.

Variables may also be referred to by their form: continuous, binary/dichotomous, nominal categorical, and ordinal categorical, among others.

An example is provided by the analysis of trend in sea level by Woodworth (1987). Here the dependent variable (and variable of most interest) was the annual mean sea level at a given location for which a series of yearly values were available. The primary independent variable was time. Use was made of a covariate consisting of yearly values of annual mean atmospheric pressure at sea level. The results showed that inclusion of the covariate allowed improved estimates of the trend against time to be obtained, compared to analyses which omitted the covariate.

Other variables

A variable may be thought to alter the dependent or independent variables, but may not actually be the focus of the experiment. So that variable will be kept constant or monitored to try to minimize its effect on the experiment. Such variables may be designated as either a "controlled variable", "control variable", or "extraneous variable".

Extraneous variables, if included in a regression analysis as independent variables, may aid a researcher with accurate response parameter estimation, prediction, and goodness of fit, but are not of substantive interest to the hypothesis under examination. For example, in a study examining the effect of post-secondary education on lifetime earnings, some extraneous variables might be gender, ethnicity, social class, genetics, intelligence, age, and so forth. A variable is extraneous only when it can be assumed (or shown) to influence the dependent variable. If included in a regression, it can improve the fit of the model. If it is excluded from the regression and if it has a non-zero covariance with one or more of the independent variables of interest, its omission will bias the regression's result for the effect of that independent variable of interest. This effect is called confounding or omitted variable bias; in these situations, design changes and/or controlling for a variable statistical control is necessary.

Extraneous variables are often classified into three types:

1. Subject variables, which are the characteristics of the individuals being studied that might affect their actions. These variables include age, gender, health status, mood, background, etc.
2. Blocking variables or experimental variables are characteristics of the persons conducting the experiment which might influence how a person behaves. Gender, the presence of racial discrimination, language, or other factors may qualify as such variables.

3. Situational variables are features of the environment in which the study or research was conducted, which have a bearing on the outcome of the experiment in a negative way. Included are the air temperature, level of activity, lighting, and the time of day.

In modelling, variability that is not covered by the independent variable is designated by {\displaystyle e_{I}} and is known as the "residual", "side effect", "error", "unexplained share", "residual variable", or "tolerance".

Examples

- Effect of fertilizer on plant growth

In a study measuring the influence of different quantities of fertilizer on plant growth, the independent variable would be the amount of fertilizer used. The dependent variable would be the growth in height or mass of the plant. The controlled variables would be the type of plant, the type of fertilizer, the amount of sunlight the plant gets, the size of the pots, etc.

- Effect of drug dosage on symptom severity

In a study of how different doses of a drug affect the severity of symptoms, a researcher could compare the frequency and intensity of symptoms when different doses are administered. Here the independent variable is the dose and the dependent variable is the frequency/intensity of symptoms.

- Effect of temperature on pigmentation

In measuring the amount of color removed from beetroot samples at different temperatures, temperature is the independent variable and amount of pigment removed is the dependent variable.

15

Maxima and Minima

In mathematical analysis, the maxima and minima (the respective plurals of maximum and minimum) of a function, known collectively as extrema (the plural of extremum), are the largest and smallest value of the function, either within a given range (the local or relativeextrema) or on the entire domain of a function (the global or absolute extrema). Pierre de Fermat was one of the first mathematicians to propose a general technique, adequality, for finding the maxima and minima of functions.

As defined in set theory, the maximum and minimum of a set are the greatest and least elements in the set, respectively. Unbounded infinite sets, such as the set of real numbers, have no minimum or maximum.

DEFINITION

A real-valued function f defined on a domain X has a global (or absolute) maximum point at x″ if f(x″) e″ f(x) for all x in X. Similarly, the function has a global (or absolute) minimum point at x″ if f(x″) d″ f(x) for all x in X. The value of the function at a maximum point is called the maximum value of the function and the value of the function at a minimum point is called the minimum value of the function.

If the domain X is a metric space then f is said to have a local (or relative) maximum point at the point x″ if there exists some å > 0 such that f(x″) e″ f(x) for all x in X within distance å of x″. Similarly, the function has a local minimum point at x″ if f(x″) d″ f(x) for all x in X within distance å of x″. A similar definition can be used when X is a topological space, since the definition just given can be rephrased in terms of neighbourhoods.

In both the global and local cases, the concept of a strict extremum can be defined. For example, x″ is a strict global maximum point if, for all x in X with x ‴ x″, we have f(x″) > f(x), and x″ is a strict local maximum point if there exists some å > 0 such that, for all x in X within distance å of x″ with x ‴ x″, we have f(x″) > f(x). Note that a point is a strict global maximum point if and only if it is the unique global maximum point, and similarly for minimum points.

A continuous real-valued function with a compact domain always has a maximum point and a minimum point. An important example is a function whose domain is a closed (and bounded) interval of real numbers.

Finding Functional Maxima and Minima

Finding global maxima and minima is the goal of mathematical optimization. If a function is continuous on a closed interval, then by the extreme value theorem global maxima and minima exist. Furthermore, a global maximum (or minimum) either must be a local maximum (or minimum) in the interior of the domain, or must lie on the boundary of the domain. So a method of finding a global maximum (or minimum) is to look at all the local maxima (or minima) in the interior, and also look at the maxima (or minima) of the points on the boundary, and take the largest (or smallest) one.

Local extrema of differentiable functions can be found by Fermat's theorem, which states that they must occur at critical points. One can distinguish whether a critical point is a local maximum or local minimum by using the first derivative test, second derivative test, or higher-order derivative test, given sufficient differentiability.

For any function that is defined piecewise, one finds a maximum (or minimum) by finding the maximum (or minimum) of each piece separately, and then seeing which one is largest (or smallest).

Examples

- The function $x^?$ has a unique global minimum at x = 0.
- The function x^3 has no global minima or maxima. Although the first derivative ($3x^2$) is 0 at x = 0, this is an inflection point.
- The function has a unique global maximum at x = e. (See figure at right)
- The function $x^{''x}$ has a unique global maximum over the positive real numbers at x = 1/e.
- The function x^3/3 " x has first derivative x^2 " 1 and second derivative 2x. Setting the first derivative to 0 and solving for x gives stationary points at "1 and +1. From the sign of the second derivative we can see that "1 is a local maximum and +1 is a local minimum. Note that this function has no global maximum or minimum.
- The function |x| has a global minimum at x = 0 that cannot be found by taking derivatives, because the derivative does not exist at x = 0.
- The function cos(x) has infinitely many global maxima at 0, ±2ð, ±4ð, ..., and infinitely many global minima at ±ð, ±3ð, ...
- The function 2 cos(x) " x has infinitely many local maxima and minima, but no global maximum or minimum.
- The function cos(3ðx)/x with 0.1 d" x d" 1.1 has a global maximum at x = 0.1 (a boundary), a global minimum near x = 0.3, a local maximum near x = 0.6, and a local minimum near x = 1.0. (See figure at top of page.)
- The function $x^3 + 3x^2$ " 2x + 1 defined over the closed interval (segment) ["4,2] has a local and maximum at x = "1""15/3, a local minimum at x = "1+"15/3, a global maximum at x = 2 and a global minimum at x = "4.

Functions of more than one Variable

For functions of more than one variable, similar conditions apply. For example, in the (enlargeable) figure at the right, the necessary conditions for a local maximum are similar to those of a function with only one variable. The first partial derivatives as to z (the variable to be maximized) are zero at the maximum (the glowing dot on top in the figure). The second partial derivatives are negative. These are only necessary, not sufficient, conditions for a local maximum because of the possibility of a saddle point. For use of these conditions to solve for a maximum, the function z must also be differentiable throughout. The second partial derivative test can help classify the point as a relative maximum or relative minimum. In contrast, there are substantial differences between functions of one variable and functions of more than one variable in the identification of global extrema. For example, if a bounded differentiable function f defined on a closed interval in the real line has a single critical point, which is a local minimum, then it is also a global minimum (use the intermediate value theorem and Rolle's theorem to prove this by reductio ad absurdum). In two and more dimensions, this argument fails, as the function

IN RELATION TO SETS

Maxima and minima can also be defined for sets. In general, if an ordered set S has a greatest element m, m is a maximal element. Furthermore, if S is a subset of an ordered set Tand m is the greatest element of S with respect to order induced by T, m is a least upper bound of S in T. The similar result holds for least element, minimal element and greatest lower bound.

In the case of a general partial order, the least element (smaller than all other) should not be confused with a minimal element (nothing is smaller). Likewise, a greatest element of a partially ordered set (poset) is an upper bound of the set which is contained within the set, whereas a maximal element m of a poset A is an element of A such that if m d" b (for any b in A) then m = b. Any least element or greatest element of a poset is unique, but a poset can have several minimal or maximal elements. If a poset has more than one maximal element, then these elements will not be mutually comparable.

In a totally ordered set, or chain, all elements are mutually comparable, so such a set can have at most one minimal element and at most one maximal element. Then, due to mutual comparability, the minimal element will also be the least element and the maximal element will also be the greatest element. Thus in a totally ordered set we can simply use the terms minimum and maximum. If a chain is finite then it will always have a maximum and a minimum. If a chain is infinite then it need not have a maximum or a minimum. For example, the set of natural numbers has no maximum, though it has a minimum. If an infinite chain S is bounded, then the closure Cl(S) of the set occasionally has a minimum and a maximum, in such case they are called the greatest lower bound and the least upper bound of the set S, respectively.

Minimum And Maximum Values

Many of our applications in this chapter will revolve around minimum and maximum values of a function. While we can all visualize the minimum and maximum values of a function we want to be a little more specific in our work here.

In particular, we want to differentiate between two types of minimum or maximum values. The following definition gives the types of minimums and/or maximums values that we'll be looking at.

Note that when we say an "open interval aroundx=cx=c" we mean that we can find some interval (a,b)(a,b), not including the endpoints, such that a<c<ba<c<b. Or, in other words, cc will be contained somewhere inside the interval and will not be either of the endpoints.

Also, we will collectively call the minimum and maximum points of a function the extrema of the function. So, relative extrema will refer to the relative minimums and maximums while absolute extrema refer to the absolute minimums and maximums.

Now, let's talk a little bit about the subtle difference between the absolute and relative in the definition above.

We will have an absolute maximum (or minimum) at x=cx=c provided f(c)f(c) is the largest (or smallest) value that the function will ever take on the domain that we are working on. Also, when we say the "domain we are working on" this simply means the range of xx's that we have chosen to work with for a given problem. There may be other values of xx that we can actually plug into the function but have excluded them for some reason.

A relative maximum or minimum is slightly different. All that's required for a point to be a relative maximum or minimum is for that point to be a maximum or minimum in some interval of xx's around x=cx=c. There may be larger or smaller values of the function at some other place, but relative to x=cx=c, or local to x=cx=c, f(c)f(c) is larger or smaller than all the other function values that are near it.

Note as well that in order for a point to be a relative extrema we must be able to look at function values on both sides of x=cx=c to see if it really is a maximum or minimum at that point. This means that relative extrema do not occur at the end points of a domain. They can only occur interior to the domain.

There is actually some debate on the preceding point. Some folks do feel that relative extrema can occur on the end points of a domain. However, in this class we will be using the definition that says that they can't occur at the end points of a domain. This will be discussed in a little more detail at the end of the section once we have a relevant fact taken care of.

It's usually easier to get a feel for the definitions by taking a quick look at a graph.

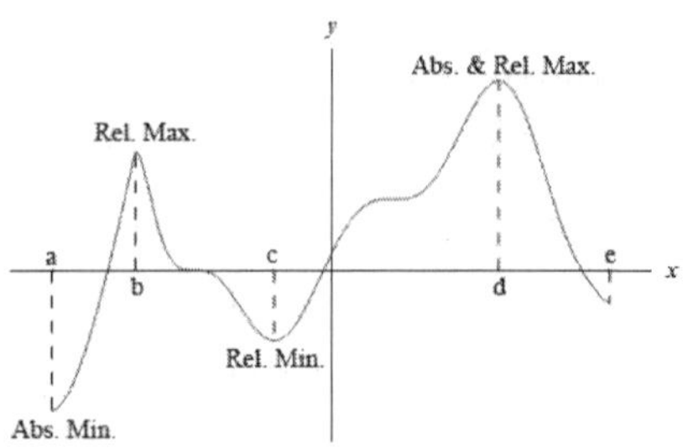

For the function shown in this graph we have relative maximums at x=bx=b and x=dx=d. Both of these points are relative maximums since they are interior to the domain shown and are the largest point on the graph in some interval around the point. We also have a relative minimum at x=cx=c since this point is interior to the domain and is the lowest point on the graph in an interval around it. The far-right end point, x=ex=e, will not be a relative minimum since it is an end point.

The function will have an absolute maximum at x=dx=d and an absolute minimum at x=ax=a. These two points are the largest and smallest that the function will ever be. We can also notice that the absolute extrema for a function will occur at either the endpoints of the domain or at relative extrema. We will use this idea in later sections so it's more important than it might seem at the present time.

Let's take a quick look at some examples to make sure that we have the definitions of absolute extrema and relative extrema straight.

So, if we have a continuous function on an interval [a,b][a,b] then we are guaranteed to have both an absolute maximum and an absolute minimum for the function somewhere in the interval. The theorem doesn't tell us where they will occur or if they will occur more than once, but at least it tells us that they do exist somewhere. Sometimes, all that we need to know is that they do exist.

This theorem doesn't say anything about absolute extrema if we aren't working on an interval. We saw examples of functions above that had both absolute extrema, one absolute extrema, and no absolute extrema when we didn't restrict ourselves down to an interval.

The requirement that a function be continuous is also required in order for us to use the theorem. Consider the case of

f(x)=1x2on["1,1]f(x)=1x2on["1,1]

Here's the graph.

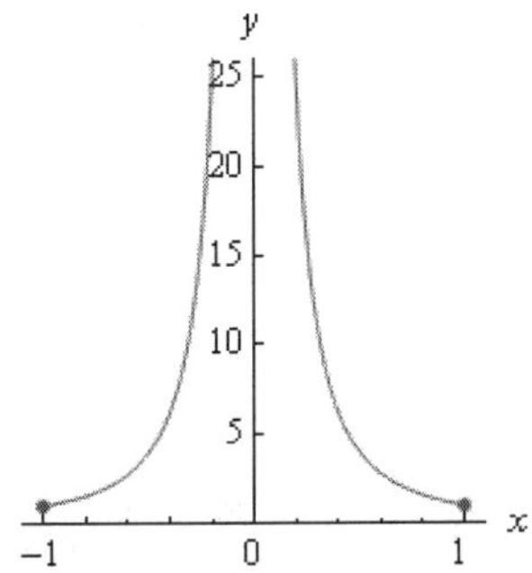

This function is not continuous at x=0x=0 as we move in towards zero the function is approaching infinity. So, the function does not have an absolute maximum. Note that it does have an absolute minimum however. In fact the absolute minimum occurs twice at both x="1x="1 and x=1x=1.

If we changed the interval a little to say,

f(x)=1x2on[12,1]f(x)=1x2on[12,1]

the function would now have both absolute extrema. We may only run into problems if the interval contains the point of discontinuity. If it doesn't then the theorem will hold.

We should also point out that just because a function is not continuous at a point that doesn't mean that it won't have both absolute extrema in an interval that contains that point. Below is the graph of a function that is not continuous at a point in the given interval and yet has both absolute extrema.

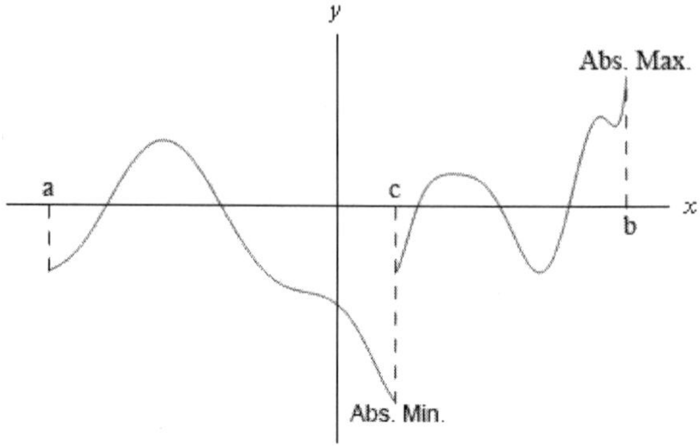

This graph is not continuous at x=cx=c, yet it does have both an absolute maximum (x=bx=b) and an absolute minimum (x=cx=c). Also note that, in this case one of the absolute extrema occurred at the point of discontinuity, but it doesn't need to. The absolute minimum could just have easily been at the other end point or at some other point interior to the region. The point here is that this graph is not continuous and yet does have both absolute extrema

The point of all this is that we need to be careful to only use the Extreme Value Theorem when the conditions of the theorem are met and not misinterpret the results if the conditions aren't met.

In order to use the Extreme Value Theorem we must have an interval that includes its endpoints, often called a closed interval, and the function must be continuous on that interval. If we don't have a closed interval and/or the function isn't continuous on the interval then the function may or may not have absolute extrema.

We need to discuss one final topic in this section before moving on to the first major application of the derivative that we're going to be looking at in this chapter.

Fermat's Theorem

If f(x)f(x) has a relative extrema at x=cx=c and f2 (c)f2 (c) exists then x=cx=c is a critical point of f(x)f(x). In fact, it will be a critical point such that f2 (c)=0f2 (c)=0.

To see the proof of this theorem see the Proofs From Derivative Applications section of the Extras chapter.

Also note that we can say that f2 (c)=0f2 (c)=0 because we are also assuming that f2 (c)f2 (c) exists.

This theorem tells us that there is a nice relationship between relative extrema and critical points. In fact, it will allow us to get a list of all possible relative extrema. Since a relative extrema must be a critical point the list of all critical points will give us a list of all possible relative extrema.

Consider the case of f(x)=x2f(x)=x2. We saw that this function had a relative minimum at x=0x=0 in several earlier examples. So according to Fermat's theorem x=0x=0 should be a critical point. The derivative of the function is,

f2 (x)=2xf2 (x)=2x

Sure enough x=0x=0 is a critical point.

Be careful not to misuse this theorem. It doesn't say that a critical point will be a relative extrema. To see this, consider the following case.

f(x)=x3f2 (x)=3x2f(x)=x3f2 (x)=3x2

Clearly x=0x=0 is a critical point. However, we saw in an earlier example this function has no relative extrema of any kind. So, critical points do not have to be relative extrema.

Also note that this theorem says nothing about absolute extrema. An absolute extrema may or may not be a critical point.

Before we leave this section we need to discuss a couple of issues.

First, Fermat's Theorem only works for critical points in which f2 (c)=0f2 (c)=0. This does not, however, mean that relative extrema won't occur at critical points where the derivative does not exist. To see this consider f(x)= | x | f(x)= | x |. This function clearly has a relative minimum at x=0x=0 and yet in a previous section we showed in an example that f(0)f(0) does not exist.

What this all means is that if we want to locate relative extrema all we really need to do is look at the critical points as those are the places where relative extrema may exist.

Finally, recall that at that start of the section we stated that relative extrema will not exist at endpoints of the interval we are looking at. The reason for this is that if we allowed relative extrema to occur there it may well (and in fact most of the time) violate Fermat's Theorem. There is no reason to expect end points of intervals to be critical points of any kind. Therefore, we do not allow relative extrema to exist at the endpoints of intervals.

The turning points of a graph

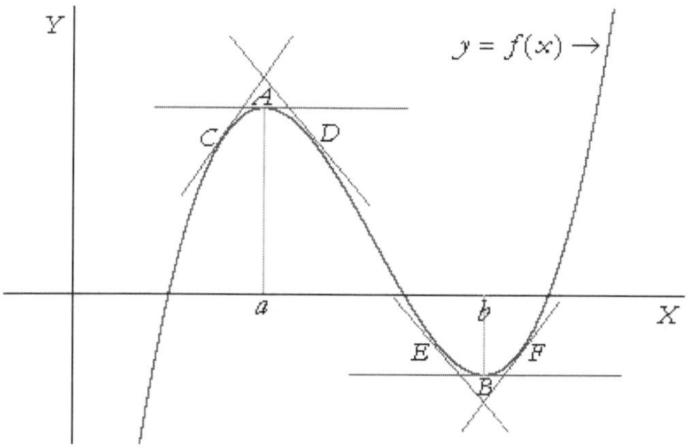

WE SAY THAT A FUNCTION f(x) has a relative maximum value at x = a, if f(a) is greater than any value immediately preceding or follwing.

We call it a "relative" maximum because other values of the function may in fact be greater.

We say that a function f(x) has a relative minimum value at x = b, if f(b) is less than any value immediately preceding or follwing.

Again, other values of the function may in fact be less. With that understanding, then, we will drop the term relative.

The value of the function, the value of y, at either a maximum or a minimum is called an extreme value.

Now, what characterizes the graph at an extreme value?

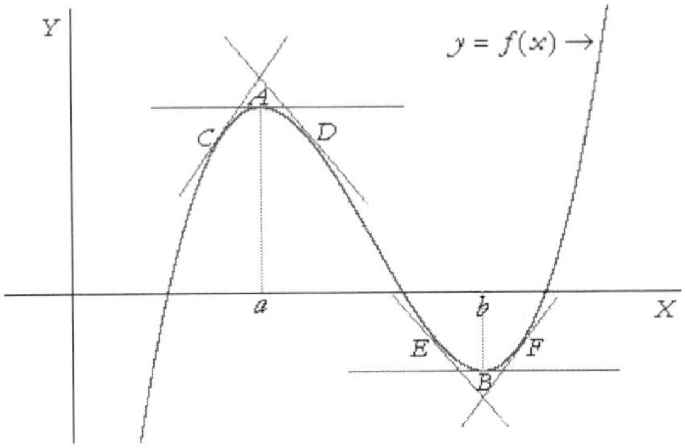

The tangent to the curve is horizontal. We see this at the points A and B. The slope of each tangent line — the derivative when evaluated at a or b — is 0.

$f'(x) = 0$.

Moreover, at points immediately to the left of a maximum — at a point C — the slope of the tangent is positive: $f'(x) > 0$. While at points immediately to the right — at a point D — the slope is negative: $f'(x) < 0$.

In other words, at a maximum, $f'(x)$ changes sign from + to “ .

At a minimum, $f'(x)$ changes sign from “ to + . We can see that at the points E and F.

We can also observe that at a maximum, at A, the graph is concave downward. (Topic 14 of Precalculus.) While at a minimum, at B, it is concave upward.

A value of x at which the function has either a maximum or a minimum is called a critical value. In the figure —

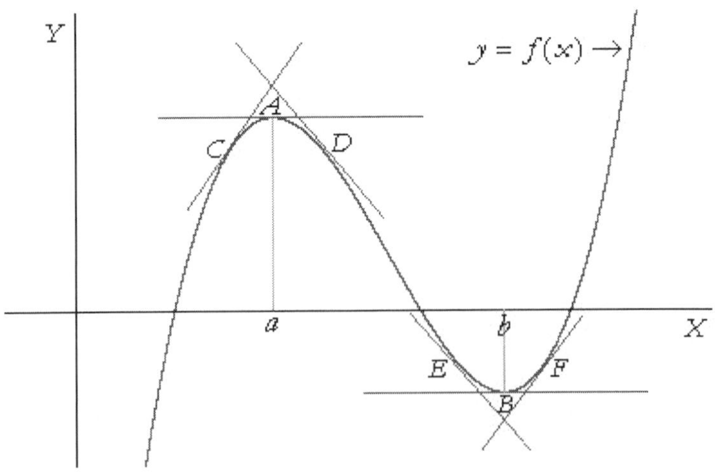

— the critical values are x = a and x = b.

The critical values determine turning points, at which the tangent is parallel to the x-axis. The critical values — if any — will be the solutionsto $f'(x) = 0$.

Example 1. Let f(x) = x^2 ″ 6x + 5.

Are there any critical values — any turning points? If so, do they determine a maximum or a minimum? And what are the coördinates on the graph of that maximum or minimum?

Solution. f ′(x) = 2x ″ 6 = 0 implies x = 3. (Lesson 9 of Algebra.)

x = 3 is the only critical value. It is the x-coördinate of the turning point. To determine the y-coördinate, evaluate f at that critical value — evaluate f(3):

$f(x)$	=	$x^2 - 6x + 5$
$f(3)$	=	$3^2 - 6 \cdot 3 + 5$
	=	$-4.$

The extreme value is ″4. To see whether it is a maximum or a minimum, in this case we can simply look at the graph.

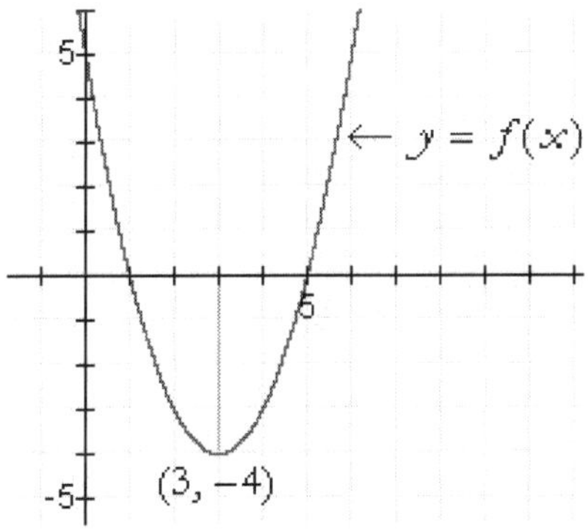

f(x) is a parabola, and we can see that the turning point is a minimum.

By finding the value of x where the derivative is 0, then, we have discovered that the vertex of the parabola is at (3, ″4).

But we will not always be able to look at the graph. The algebraic condition for a minimum is that f ′(x) changes sign from ″ to + . We see this at the points E, B, F above. The value of the slope is increasing.

Now to say that the slope is increasing, is to say that, at a critical value, the second derivative (Lesson 9) — which is rate of change of the slope — is positive.

Again, here is f(x):

$f(x)$	=	$x^2 - 6x + 5.$
$f'(x)$	=	$2x - 6.$
$f''(x)$	=	$2.$

f ″ evaluated at the critical value 3 — f ″(3) = 2 — is positive. This tells us algebraically that the critical value 3 determines a minimum.

Sufficient conditions

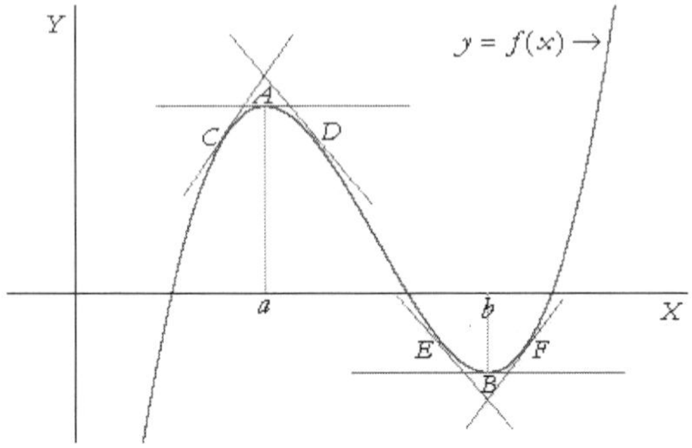

We can now state these sufficient conditions for extreme values of a function at a critical value a:

The function has a minimum value at x = a if $f'(a) = 0$ and $f''(a)$ = a positive number.

The function has a maximum value at x = a if $f'(a) = 0$ and $f''(a)$ = a negative number.

In the case of the maximum, the slope of the tangent is decreasing — it is going from positive to negative. We can see that at the points C, A, D.

Example 2. Let f(x) = $2x^{3}$" $9x^{2}$ + 12x ″ 3.

Are there any extreme values? First, are there any critical values — solutions to $f'(x) = 0$ — and do they determine a maximum or a minimum? And what are the coördinates on the graph of that maximum or minimum? Where are the turning points?

Solution. $f'(x) = 6x^2 - 18x + 12$	=	$6(x^2 - 3x + 2)$
	=	$6(x - 1)(x - 2)$
	=	0

implies:

x = 1 or x = 2.

(Lesson 37 of Algebra.)

Those are the critical values. Does each one determine a maximum or does it determine a minimum? To answer, we must evaluate the second derivative at each value.

$f'(x)$	=	$6x^2 - 18x + 12.$
$f''(x)$	=	$12x - 18.$
$f''(1)$	=	$12 - 18 = -6.$

The second derivative is negative. The function therefore has a maximum at x = 1.

To find the y-coördinate — the extreme value — at that maximum we evaluate f(1):

$f(x)$	=	$2x^3 - 9x^2 + 12x - 3$
$f(1)$	=	$2 - 9 + 12 - 3$
	=	2.

The maximum occurs at the point (1, 2).
Next, does x = 2 determine a maximum or a minimum?

$f''(x)$	=	$12x - 18.$
$f''(2)$	=	$24 - 18 = 6.$

The second derivative is positive. The function therefore has a minimum at x = 2.

To find the y-coördinate — the extreme value — at that minimum, we evaluate f(2):

$f(x)$	=	$2x^3 - 9x^2 + 12x - 3.$
$f(2)$	=	$16 - 36 + 24 - 3$
	=	1.

The minimum occurs at the point (2, 1).
Here in fact is the graph of f(x):

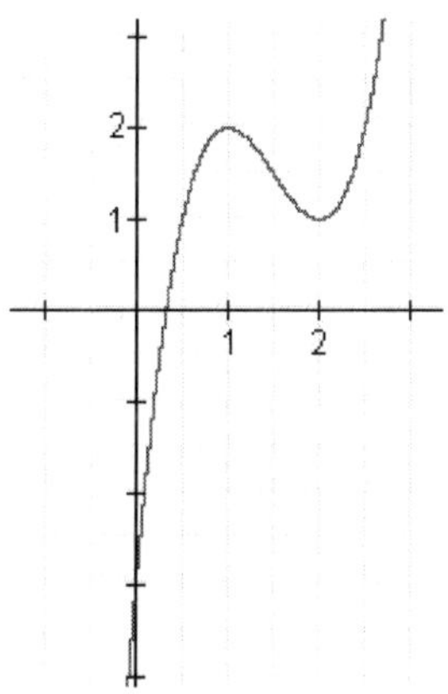

Solutions to $f''(x) = 0$ indicate a point of inflection at those solutions, not a maximum or minimum. An example is $y = x^3$. $y'' = 6x = 0$ implies $x = 0$. But $x = 0$ is a point of inflection in the graph of $y = x^3$, not a maximum or minimum. Another example is y = sin x. The solutions to $y'' = 0$ are the multiplies of ő, which are points of inflection.

Problem 1. Find the coördinates of the vertex of the parabola,

$y = x^2$ ″ 8x + 1.

To see the answer, pass your mouse over the colored area.

To cover the answer again, click "Refresh" ("Reload").

Do the problem yourself first!

Problem 2. Examine each function for maxima and minima.

a) $y = x^3$ d $3x^2 + 2$.

$y' = 3x^2$ d 6x = 3x(x d 2) = 0 implies

y(2) = 2^3 ″ $3\cdot 2^2 + 2 = 8$ ″ 12 + 2 = ″2.

b) y = ″$2x^3$ ″ $3x^2$ + 12 x + 10.

At x = 1 there is a maximum of y = 17.

At x = ″2 there is a minimum of y = ″10.

c) $y = 2x^3 + 3x^2 + 12$ x ″ 4.

Since f ′(x) = 0 has no real solutions, there are no extreme values.

d) y = $3x^{4"}$ $4x^3$ ″ $12x^2 + 2$.

At x = 0 there is a maximum of y = 2.

At x = ″1 there is a minimum of y = ″3.

At x = 2 there is a minimum of y = ″30.

A local maximum point on a function is a point (x,y)(x,y) on the graph of the function whose yycoordinate is larger than all other yy coordinates on the graph at points "close to" (x,y)(x,y). More precisely, (x,f(x))(x,f(x)) is a local maximum if there is an interval (a,b)(a,b) with a<x<ba<x<b and f(x)e″f(z)f(x)e″f(z) for every zz in (a,b)(a,b). Similarly, (x,y)(x,y) is a local minimum point if it has locally the smallest yy coordinate. Again being more precise: (x,f(x))(x,f(x)) is a local minimum if there is an interval (a,b)(a,b) with a<x<ba<x<b and f(x)d″f(z)f(x)d″f(z) for every zz in (a,b)(a,b). A local extremumis either a local minimum or a local maximum.

Local maximum and minimum points are quite distinctive on the graph of a function, and are therefore useful in understanding the shape of the graph. In many applied problems we want to find the largest or smallest value that a function achieves (for example, we might want to find the minimum cost at which some task can be performed) and so identifying maximum and minimum points will be useful for applied problems as well. Some examples of local maximum and minimum points are shown in figure 5.1.1.

AA BB	AA AA BB

Figure 5.1.1. Some local maximum points (AA) and minimum points (BB).

If (x,f(x))(x,f(x)) is a point where f(x)f(x) reaches a local maximum or minimum, and if the derivative of ff exists at xx, then the graph has a tangent line and the tangent line must be horizontal. This is important enough to state as a theorem, though we will not prove it.

Theorem 5.1.1 (Fermat's Theorem) If f(x)f(x) has a local extremum at x=ax=a and ff is differentiable at aa, then f2 (a)=0f2 (a)=0.

Thus, the only points at which a function can have a local maximum or minimum are points at which the derivative is zero, as in the left hand graph in figure 5.1.1, or the derivative is undefined, as in the right hand graph. Any value of xx for which f2 (x)f2 (x) is zero or undefined is called a critical value for ff. When looking for local maximum and minimum points, you are likely to make two sorts of mistakes: You may forget that a maximum or minimum can occur where the derivative does not exist, and so forget to check whether the derivative exists everywhere. You might also assume that any place that the derivative is zero is a local maximum or minimum point, but this is not true. A portion of the graph of f(x)=x3f(x)=x3 is shown in figure 5.1.2. The derivative of ff is f2 (x)=3x2f2 (x)=3x2, and f2 (0)=0f2 (0)=0, but there is neither a maximum nor minimum at (0,0)(0,0).

Figure 5.1.2. No maximum or minimum even though the derivative is zero.

Since the derivative is zero or undefined at both local maximum and local minimum points, we need a way to determine which, if either, actually occurs. The most elementary approach, but one that is often tedious or difficult, is to test directly whether the yy coordinates "near" the potential maximum or minimum are above or below the yy coordinate at the point of interest. Of course, there are too many points "near" the point to test, but a little thought shows we need only test two provided we know that ff is continuous (recall that this means that the graph of ff has no jumps or gaps).

Suppose, for example, that we have identified three points at which f2 f2 is zero or nonexistent: (x1,y1)(x1,y1), (x2,y2)(x2,y2), (x3,y3)(x3,y3), and x1<x2<x3x1<x2<x3. Suppose that we compute the value of f(a)f(a) for x1<a<x2x1<a<x2, and that f(a)<f(x2)f(a)<f(x2). What can we say about the graph between aa and x2x2? Could there be a point (b,f(b))(b,f(b)), a<b<x2a<b<x2 with f(b)>f(x2)f(b)>f(x2)? No: if there were, the graph would go up from (a,f(a))(a,f(a)) to (b,f(b))(b,f(b)) then down to (x2,f(x2))(x2,f(x2)) and somewhere in between would have a local maximum point. (This is not obvious; it is a result of the Extreme Value Theorem, theorem 6.1.2.) But at that local maximum point the derivative of ff would be zero or nonexistent, yet we already know that the derivative is zero or nonexistent only at x1x1, x2x2, and x3x3. The upshot is that one computation tells us that (x2,f(x2))(x2,f(x2)) has the largest yy coordinate of any point on the graph near x2x2 and to the left of x2x2. We can perform the same test on the right. If we find that on both sides of x2x2 the values are smaller, then there must be a local maximum at (x2,f(x2))(x2,f(x2)); if we find that on both sides of x2x2 the values are larger, then there must be a local minimum at (x2,f(x2))(x2,f(x2)); if we find one of each, then there is neither a local maximum or minimum at x2x2.

x1x1

aa

bb

x2x2

x3x3

Figure 5.1.3. Testing for a maximum or minimum.

It is not always easy to compute the value of a function at a particular point. The task is made easier by the availability of calculators and computers, but they have their own drawbacks—they do not always allow us to distinguish between values that are very close together. Nevertheless, because this method is conceptually simple and sometimes easy to perform, you should always consider it.

Example 5.1.2 Find all local maximum and minimum points for the functionf(x)=x3"xf(x)=x3"x. The derivative is f2 (x)=3x2"1f2 (x)=3x2"1. This is defined everywhere and is zero at x=±3–"/3x=±3/3. Looking first at x=3–"/3x=3/3, we see that f(3–"/3)="23–"/9f(3/3)="23/9. Now we test two points on either side of x=3–"/3x=3/3, making sure that neither is farther away than the nearest critical value; since 3–"<33<3, 3–"/3<13/3<1 and we can use x=0x=0 and x=1x=1. Since f(0)=0>"23–"/9f(0)=0>"23/9 and f(1)=0>"23–"/9f(1)=0>"23/9, there must be a local minimum atx=3–"/3x=3/3. For x="3–"/3x="3/3, we see that f("3–"/3)=23–"/9f("3/3)=23/9. This time we can use x=0x=0 and x="1x="1, and we find that f("1)=f(0)=0<23–"/9f("1)=f(0)=0<23/9, so there must be a local maximum at x="3–"/3x="3/3.

Of course this example is made very simple by our choice of points to test, namely x="1x="1, 00, 11. We could have used other values, say "5/4"5/4, 1/31/3, and 3/43/4, but this would have made the calculations considerably more tedious.

Example 5.1.3 Find all local maximum and minimum points for f(x)=sinx+cosxf(x)=sina x+cosa x. The derivative is f2 (x)=cosx"sinxf2 (x)=cosa x"sina x. This is always defined and is zero whenever cosx=sinxcosa x=sina x. Recalling that the cosxcosa x and sinxsina x are the xx and yy coordinates of points on a unit circle, we see that cosx=sinxcosa x=sina x when xx is À/4õ/4, À/4±Àõ/4±õ, À/4±2Àõ/4±2õ, À/4±3Àõ/4±3õ, etc. Since both sine and cosine have a period of 2À2õ, we need only determine the status of x=À/4x=õ/4 and x=5À/4x=5õ/4. We can use 00 and À/2õ/2 to test the critical value x=À/4x=õ/4. We find that f(À/4)=2–"f(õ/4)=2, f(0)=1<2–"f(0)=1<2 and f(À/2)=1f(õ/2)=1, so there is a local maximum when x=À/4x=õ/4 and also when x=À/4±2Àx=õ/4±2õ, À/4±4Àõ/4±4õ, etc. We can summarize this more neatly by saying that there are local maxima at À/4±2kÀõ/4±2kõ for every integer kk.

We use Àõ and 2À2õ to test the critical value x=5À/4x=5õ/4. The relevant values are f(5À/4)="2–"f(5õ/4)="2, f(À)="1>"2–"f(õ)="1>"2, f(2À)=1>"2–"f(2õ)=1>"2, so there is a local minimum at x=5À/4x=5õ/4, 5À/4±2À5õ/4±2õ, 5À/4±4À5õ/4±4õ, etc. More succinctly, there are local minima at 5À/4±2kÀ5õ/4±2kõ for every integer kk.

MAXIMA AND MINIMA OF FUNCTIONS OF TWO VARIABLES

The problem of determining the maximum or minimum of function is encountered in geometry, mechanics, physics, and other fields, and was one of the motivating factors in the development of the calculus in the seventeenth century.

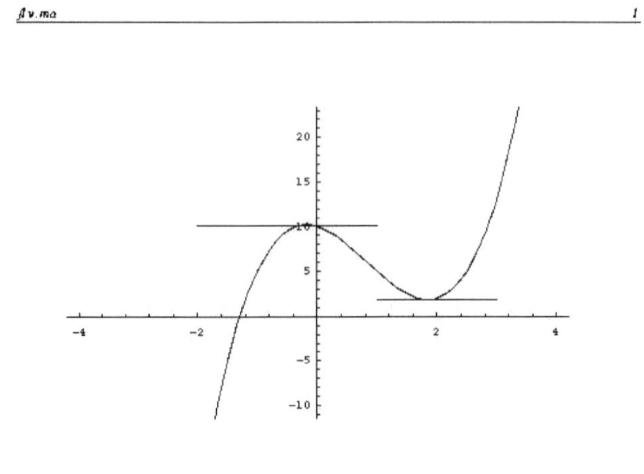

Let us recall the procedure for the case of a function of one variable y=f(x). First, we determine points x_c where f'(x)=0. These points are called critical points. At critical points the tangent line is horizontal. This is shown in the figure below.

The second derivative test is employed to determine if a critical point is a relative maximum or a relative minimum. If f''(x_c)>0, then x_c is a relative minimum. If f''(x_c)<0, then x_c is a maximum. If f''(x_c)=0, then the test gives no information.

The notions of critical points and the second derivative test carry over to functions of two variables. Let z=f(x,y). Critical points are points in the xy-plane where the tangent plane is horizontal.

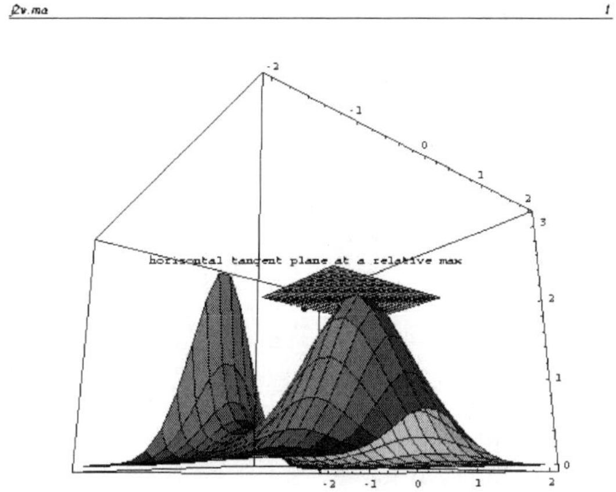

Since the normal vector of the tangent plane at (x,y) is given by

$$f_x(x,y)\mathbf{i} + f_y(x,y)\mathbf{j} - \mathbf{k}$$

The tangent plane is horizontal if its normal vector points in the z direction. Hence, critical points are solutions of the equations:

$$f_x(x,y) = 0 \quad and \quad f_y(x,y) = 0$$

because horizontal planes have normal vector parallel to z-axis. The two equations above must be solved simultaneously.

Example

Let us find the critical points of

$$z = f(x,y) = \exp\left(-\frac{1}{3}x^3 + x - y^2\right)$$

The partial derivatives are

$$f_x(x,y) = (-x^2 + 1)\exp\left(-\frac{1}{3}x^3 + x - y^2\right)$$

$$f_y(x,y) = -2y\exp\left(-\frac{1}{3}x^3 + x - y^2\right)$$

f_x=0 if 1-x^2=0 or the exponential term is 0. f_y=0 if -2y=0 or the exponential term is 0. The exponential term is not 0 except in the degenerate case. Hence we require 1-x^2=0 and -2y=0, implying x=1 or x=-1 and y=0. There are two critical points (-1,0) and (1,0).

The Second Derivative Test for Functions of Two Variables

How can we determine if the critical points found above are relative maxima or minima? We apply a second derivative test for functions of two variables.

Let (x_c,y_c) be a critical point and define

$$D(x_c, y_c) = f_{xx}(x_c, y_c) f_{yy}(x_c, y_c) - [f_{xy}(x_c, y_c)]^2.$$

We have the following cases:

- If D>0 and f_xx(x_c,y_c)<0, then f(x,y) has a relative maximum at (x_c,y_c).
- If D>0 and f_xx(x_c,y_c)>0, then f(x,y) has a relative minimum at (x_c,y_c).
- If D<0, then f(x,y) has a saddle point at (x_c,y_c).
- If D=0, the second derivative test is inconclusive.

An example of a saddle point is shown in the example below.

Example: Continued

For the example above, we have

$$f_{xx}(x,y) = (-2x + (1-x^2)^2) exp\left(-\frac{1}{3}x^3 + x - y^2\right),$$

$$f_{yy}(x,y) = (-2 + 4y^2) exp\left(-\frac{1}{3}x^3 + x - y^2\right),$$

$$f_{xy}(x,y) = -2y(1-x^2) exp\left(-\frac{1}{3}x^3 + x - y^2\right),$$

For x=1 and y=0, we have D(1,0)=4exp(4/3)>0 with f_xx(1,0)=-2exp(2/3)<0. Hence, (1,0) is a relative maximum. For x=-1 and y=0, we have D(-1,0)=-4exp(-4/3)<0. Hence, (-1,0) is a saddle point.

The figure below plots the surface z=f(x,y).

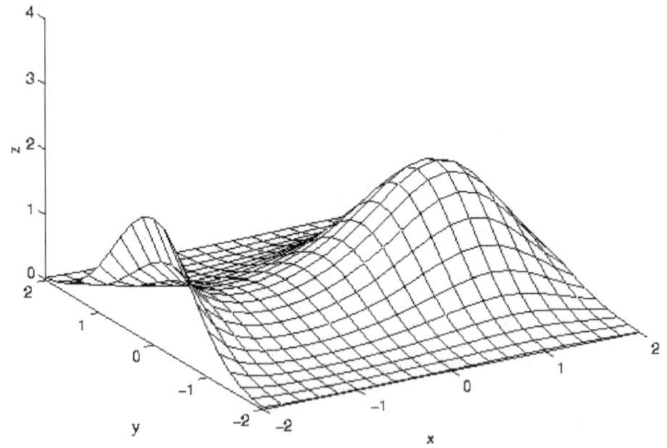

Notice the relative maximum at (x=1,y=0). (x=-1,y=0) is a relative maximum if one travels in the y direction and a relative minimum if one travels in the x-direction. Near (-1,0) the surface looks like a saddle, hence the name.

Maxima and Minima in a Bounded Region

Suppose that our goal is to find the global maximum and minimum of our model function above in the square -2<=x<=2 and -2<=y<=2? There are three types of points that can potentially be global maxima or minima:

1. Relative extrema in the interior of the square.
2. Relative extrema on the boundary of the square.

3. Corner Points.

We have already done step 1. There are extrema at (1,0) and (-1,0). The boundary of square consists of 4 parts. Side 1 is y=-2 and -2<=x<=2. On this side, we have

$$z = f(x, -2) = exp\left(-\frac{1}{3}x^3 + x - (-2)^2\right) = g(x)$$

The original function of 2 variables is now a function of x only. We set g'(x)=0 to determine relative extrema on Side 1. It can be shown that x=1 and x=-1 are the relative extrema. Since y=-2, the relative extrema on Side 1 are at (1,-2) and (-1,-2).

On Side 2 (x=-2 and -2<=y<=2)

$$z = f(-2, y) = exp\left(-\frac{1}{3}(-2)^3 - 2 - y^2\right) = h(y).$$

We set h'(y)=0 to determine the relative extrema. It can be shown that y=0 is the only critical point, corresponding to (-2,0).

We play the same game to determine the relative extrema on the other 2 sides. It can be shown that they are (2,0), (1,2), and (-1,2).

Finally, we must include the 4 corners (-2,-2), (-2,2), (2,-2), and (2,2). In summary, the candidates for global maximum and minimum are (-1,0), (1,0), (1,-2), (-1,-2), (-2,0), (2,0), (1,2), (-1,2), (-2,-2), (-2,2), (2,-2), and (2,2). We evaluate f(x,y) at each of these points to determine the global max and min in the square. The global maximum occurs (-2,0) and (1,0). This can be seen in the figure above. The global minimum occurs at 4 points: (-1,2), (-1,-2), (2,2), and (2,-2).

Example: Maxima and Minima in a Disk

Another example of a bounded region is the disk of radius 2 centered at the origin. We proceed as in the previous example, determining in the 3 classes above. (1,0) and (-1,0) lie in the interior of the disk. The boundary of the disk is the circle x^2+y^2=4. To find extreme points on the disk we parameterize the circle. A natural parameterization is x=2cos(t) and y=2sin(t) for 0<=t<=2*pi. We substitute these expressions into z=f(x,y) and obtain

$$z = f(x, y) = f(\cos(t), \sin(t)) = exp\left(-\frac{8}{3}\cos^3 t + 2\cos t - 4\sin^2 t\right) = g(t)$$

On the circle, the original functions of 2 variables is reduced to a function of 1 variable. We can determine the extrema on the circle using techniques from calculus of on variable.

In this problem there are not any corners. Hence, we determine the global max and min by considering points in the interior of the disk and on the circle. An alternative method for finding the maximum and minimum on the circle is the method of Lagrange multipliers.

Maxima and Minima for Functions of More than 2 Variables

The notion of extreme points can be extended to functions of more than 2 variables. Suppose z=f(x_1,x_2,...,x_n). (a_1,a_2,...,a_n) is extreme point if it satisfies the n equations

$$\frac{\partial f}{\partial x_i}(x_1, x_2, \ldots, x_n) = 0 \quad i = 1, 2, \ldots, n.$$

There is not a general second derivative test to determine if a point is a relative maximum or minimum for functions of more than two variables.

Minimum & Maximum

Imagine you own a company, and your company's profit can be modeled by the function $P(x) = -10x^2 + 1760x - 50000$, where P(x) is your company's profit, and x is the number of products sold. Of course, you want your profit to be as large as possible, so you want to know how many products you should sell to make this happen. In other words, you want to know how many products you need to sell in order to maximize your profit. To find that maximum profit and to solve problems similar to this one, we need to be familiar with maximum and minimum points of a function.

A maximum point of a function is the highest point on the graph of a function, or the point that takes on the largest y-value. The minimum point of a function is the lowest point on the graph of a function, or the point that takes on the smallest y-value. The maximum or minimum point in a given interval of x-values is called a local maximum or local minimum, respectively. The maximum or minimum point of the whole function is called the global maximum or global minimum, respectively.

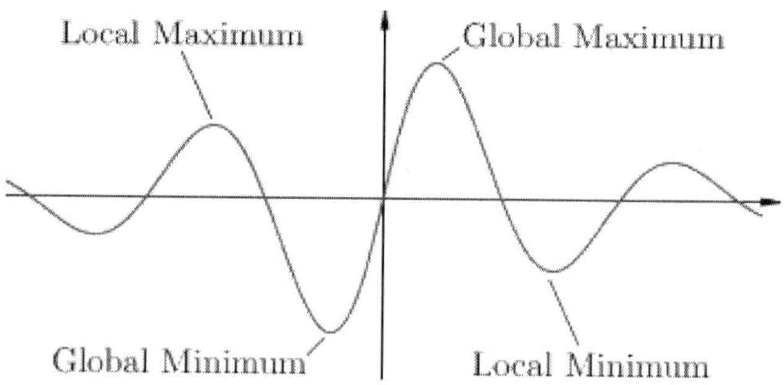

Local and Global Maxima and Minima

Looking back at our profit example, since we want to maximize profit, we want to find the global maximum of the function. This point is shown in the following image of the graph of the function $P(x) = -10x^2 + 1760x-50000$.

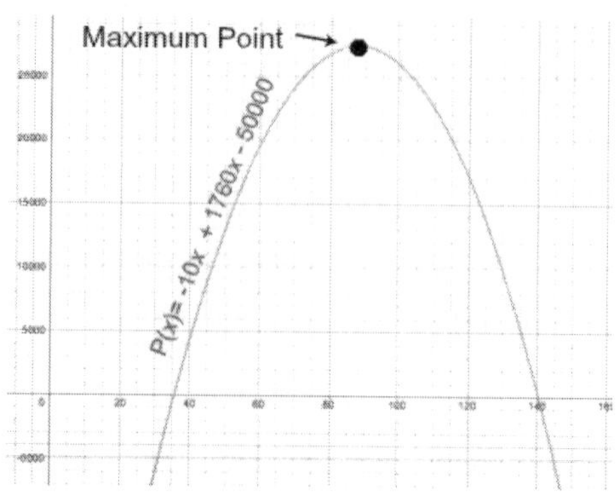

Finding Maxima & Minima Using Derivatives

Let's take a look at our earlier image of local and global maxima and minima.

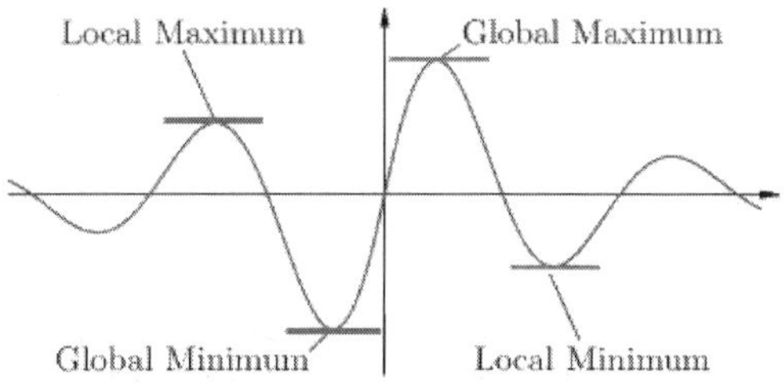

What do you notice about where the function has a maximum or minimum? Do you see how the graph flattens out at these points? If not, you should see it now that it's been pointed out, and that's great since this is the key to finding the maxima and minima of a function using derivatives.

The derivative of a function tells us the slope of the function at any given point. Since the graph flattens out at maximum and minimum points, let's think about what the slope of the function would be at these points. Ah-ha! The slope of the function would be 0, since the slope of a flat line is 0. This tells us that the derivative of the function would be zero at any maximum or minimum point. This information gives us the following steps to finding maxima and minima using derivatives.

1. Find the derivative of the function.
2. Set the derivative equal to 0, and solve for x. This gives you the x-values of the maximum and minimum points.
3. Plug those x-values back into the function to find the corresponding y-values. This will give you your maximum and minimum points of the function.

Let's take our profit example through these steps to solve the problem. The first thing we want to do is find the derivative of $P(x) = -10x^2 + 1760x - 50000$, which is $P'(x) = -20x + 1760$. Next, we set this equal to zero and solve for x.

10.5

$-20x + 1760 = 0$	Subtract 1760 from both sides.
$-20x = -1760$	Divide both sides by -20.
$x = 88$	This is the x-value of your maximum point.

We now know that we need to sell 88 products to maximize your profit. The last step is to plug x = 88 into the function to find the corresponding y-value, or in this instance, the maximum profit.

Stationary point

In mathematics, particularly in calculus, a stationary point of a differentiable function of one variable is a point on the graph of the function where the function's derivative is zero. Informally, it is a point where the function "stops" increasing or decreasing (hence the name). Some authors treat this term as synonymous with "critical point," but for other authors the latter is a broader term.

For a differentiable function of several real variables, a stationary (critical) point is a point on the surface of the graph where all its partial derivatives are zero (equivalently, the gradient is zero).

Stationary points are easy to visualize on the graph of a function of one variable: they correspond to the points on the graph where the tangent is horizontal (i.e., parallel to the x-axis). For a function of two variables, they correspond to the points on the graph where the tangent plane is parallel to the xy plane.

Turning points

A turning point is a point at which the derivative changes sign. A turning point may be either a relative maximum or a relative minimum (also known as local minimum and maximum). If the function is differentiable, then a turning point is a stationary point; however not all stationary points are turning points. If the function is twice differentiable, the stationary points that are not turning points are horizontal inflection points. For example, the functionhas a stationary point at x=0, which is also an inflection point, but is not a turning point.

Classification

Isolated stationary points of a real valued function \mathbb are classified into four kinds, by the first derivative test:

- a local minimum (minimal turning point or relative minimum) is one where the derivative of the function changes from negative to positive;
- a local maximum (maximal turning point or relative maximum) is one where the derivative of the function changes from positive to negative;
- a rising point of inflection (or inflexion) is one where the derivative of the function is positive on both sides of the stationary point; such a point marks a change in concavity;
- a falling point of inflection (or inflexion) is one where the derivative of the function is negative on both sides of the stationary point; such a point marks a change in concavity.

The first two options are collectively known as "local extrema". Similarly a point that is either a global (or absolute) maximum or a global (or absolute) minimum is called a global (or absolute) extremum. The last two options—stationary points that are not local extremum—are known as saddle points.

By Fermat's theorem, global extrema must occur on the boundary or at stationary points.

Curve sketching

Determining the position and nature of stationary points aids in curve sketching of differentiable functions. Solving the equation $f'(x) = 0$ returns the x-coordinates of all stationary points; the y-coordinates are trivially the function values at those x-coordinates. The specific nature of a stationary point at x can in some cases be determined by examining the second derivative $f''(x)$:

- If $f''(x) < 0$, the stationary point at x is concave down; a maximal extremum.
- If $f''(x) > 0$, the stationary point at x is concave up; a minimal extremum.
- If $f''(x) = 0$, the nature of the stationary point must be determined by way of other means, often by noting a sign change around that point.

A more straightforward way of determining the nature of a stationary point is by examining the function values between the stationary points (if the function is defined and continuous between them).

A simple example of a point of inflection is the function $f(x) = x^3$. There is a clear change of concavity about the point x = 0, and we can prove this by means of calculus. The second derivative of f is the everywhere-continuous 6x, and at x = 0, f2 2 = 0, and the sign changes about this point. So x = 0 is a point of inflection.

More generally, the stationary points of a real valued function {\displaystyle f\colon \mathbb {R} ^{n}\to \mathbb {R} } are those points x_0 where the derivative in every direction equals zero, or equivalently, the gradient is zero.

Example

For the function $f(x) = x^4$ we have $f'(0) = 0$ and $f''(0) = 0$. Even though $f''(0) = 0$, this point is not a point of inflection. The reason is that the sign of $f'(x)$ changes from negative to positive.

For the function f(x) = sin(x) we have f′(0) ‘“ 0 and $f''(0) = 0$. But this is not a stationary point, rather it is a point of inflection. This is because the concavity changes from concave downwards to concave upwards and the sign of $f'(x)$ does not change; it stays positive.

For the function $f(x) = x^3$ we have $f'(0) = 0$ and $f''(0) = 0$. This is both a stationary point and a point of inflection. This is because the concavity changes from concave downwards to concave upwards and the sign of $f'(x)$ does not change; it stays positive.

A function f(x) is said to have a local maximum at x = a if the value of f(a) is greater than all the values of f(x) in a small neighbourhood of x = a. Mathematically, $f(a) > f(a - h)$ and $f(a) > f(a + h)$ where $h > 0$, then a is called the point of local maximum.

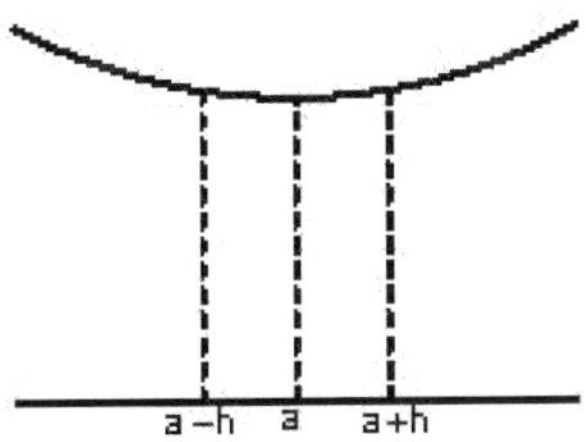

A function f(x) is said to have a local minimum at x = a, if the value of the function at x = a is less than the value of the function at the neighboring points of x = a. Mathematically, $f(a) < f(a-h)$ and $f(a) < f(a+h)$ where $h > 0$, then a is called the point of local minimum.

- A point of local maximum or a local minimum is also called a point of local extremum.
- A point where the graph of function is continuous and has a tangent line and where the concavity changes is called point of inflexion.
- At the point of inflexion, either $y'' = 0$ and changes sign or y'' fails to exist.
- At the point of inflexion, the curve crosses its tangent at that point.
- A function cannot have point of inflexion and extrema at the same point.
- Working rules to find points of local maxima and local minima:

1. First Derivative Test: If $f'(a) = 0$ and $f'(x)$ changes its sign while passing through the point $x = a$, then

- $f(x)$ would have a local maximum at $x = a$ if $f'(a-0) > 0$ and $f'(a+0) < 0$. It means that $f'(x)$ should change its sign from positive to negative.
- $f(x)$ would have local minimum at $x = a$ if $f'(a-0) < 0$ and $f'(a+0) > 0$. It means that $f'(x)$ should change its sign from negative to positive.
- If $f(x)$ doesn't change its sign while passing through $x = a$, then $f(\dot{x})$ would have neither a maximum nor minimum at $x = a$. e.g. $f(x) = x^3$ doesn't have any local maxima or minima at $x = 0$.

2. Second Derivative Test:

- Let $f(x)$ be a differentiable function on a given interval and let f'' be continuous at stationary point. Find $f'(x)$ and solve the equation $f'(x) = 0$ given let $x = a, b, \ldots$ be solutions.
- There can be two cases:
  Case (i): If $f''(a) < 0$ then $f(a)$ is maximum.
  Case (ii): If $f''(a) > 0$ then $f(a)$ is minimum.
  In case, $f''(a) = 0$ the second derivatives test fails and then one has to go back and apply the first derivative test.

If $f''(a) = 0$ and a is neither a point of local maximum nor local minimum then a is a point of inflection.

3. n^{th} Derivative Test for Maxima and Minima: Also termed as the generalization of the second derivative test, it states that if the n derivatives i.e. $f'(a) = f''(a) = f'''(a) = \ldots\ldots\ldots = f^n(a) = 0$ and $f^{n+1}(a)$ `" 0 (all derivatives of the function up to order 'n' vanish and $(n+1)$th order derivative does not vanish at $x = a$), then $f(x)$ would have a local maximum or minimum at $x = a$ iff n is odd natural number and that $x = a$ would be a point of local maxima if $f^{n+1}(a) < 0$ and would be a point of local minima if $f^{n+1}(a) > 0$.

- In some questions involving determination of maxima and minima, it might become difficult to decide whether $f(x)$ actually changes its sign while passing through $x = a$ and here, n^{th} derivative test can be applied.
- Global Minima & Maxima of $f(x)$ in $[a, b]$ is the least or the greatest value of the function $f(x)$ in interval $[a, b]$.

1. The function $f(x)$ has a global maximum at the point 'a' in the interval I if $f(a)$ e" $f(x)$, for all x " I.

2. Function $f(x)$ has a global minimum at the point 'a' if $f(a)$ d" $f(x)$, for all x " I.

- Global Maxima Minima always occur either at the critical points of $f(x)$ within $[a, b]$ or at the end points of the interval.
- Computation of Global Maxima and minima in maxima minima problems:

1. Compute the critical points of $f(x)$ in (a, b). Let the various critical points be $C_1, C_2, \ldots, C_n$.

2. Next, compute the value of the function at these critical points along with the end points of the domain. Let us denote these values by $f(C_1), f(C_2)\ldots\ldots\ldots f(C_n)$.

3. Now, compute $M^* = \max\{f(a), f(C_1), f(C_2)\ldots\ldots\ldots f(C_n), f(b)\}$ and $M^{**} = \min\{f(a), f(C_1), f(C_2)\ldots\ldots\ldots f(C_n), f(b)\}$. Now M^* is the maximum value of $f(x)$ in $[a, b]$ and M^{**} is the minimum value of $f(x)$ in $[a, b]$.

- In order to find global maxima or minima in open interval (a, b) proceed as told above and after the first two steps, compute
  $M_1 = \max\{f(C_1), f(C_2)..........f(C_n)\}$ and
  $M_2 = \min\{f(C_1),f(C_2)..........f(C_n)\}$.
- Now if x approaches a^- or if x approaches b^-, the limit of $f(x) > M_1$ or its limit $f(x) < M_1$ would not have global maximum (or global minimum) in (a, b) but if as x approaches a^- and x approaches b^-, $\lim f(x) < M_1$ and $\lim f(x) > M_2$, then M_1 and M_2 will respectively be the global maximum and global minimum of f(x) in (a,b).
- If f(x) is a continuous function on a closed bounded interval [a,b], then f(x) will have a global maximum and a global minimum on [a,b]. On the other hand, if the interval is not bounded or closed, then there is no guarantee that a continuous function f(x) will have global extrema.
- If f(x) is differentiable on the interval I, then every global extremum is a local extremum or an end point extremum.
  How do we find them?
  1. Given f(x), we differentiate once to find f ′(x).
  2. Set f ′(x)=0 and solve for x. Using our above observation, the x values we find are the 'x-coordinates' of our maxima and minima.
  3. Substitute these x-values back into f(x). This gives the corresponding 'y-coordinates' of our maxima and minima.

  Which of these points are maxima and which are minima?
  Here we may apply a simple test. Assume we've found a stationary point (a,b):
  1. Differentiate f ′(x) once more to give f ″(x), the second derivative.
  2. Calculate f ″(a). If f ″(a)<0 then (a,b) is a local maximum.
     If f ″(a)>0 then (a,b) is a local minimum.

To see why this works, imagine moving gradually towards our point (a,b), plotting the slope of our graph as we move. If our point is a local maximum, we can that this slope starts off positive, decreases to zero at the point, then becomes negative as we move through and past the point. Our slope, f ′(x), is decreasing throughout this movement, so we must have that f ″(a)<0.

The exact reverse is true if (a,b) is a local minimum. Our slope is increasing through the same movement, so here we have that f ″(a)>0.

An example

Find the maxima and minima of $f(x)=x^3+x^2$.

First, we find f ′(x). Using the rules of differentiation, we find $f'(x)=3x^2+2x$.

Now let's set f ′(x)=0: $3x^2+2x=0$

$x(3x+2)=0$ (factorising)

$x=0$ or $x=-2/3$

Substitute these values back in so that we can find our 'y-coordinates': $f(0)=(0)^3+(0)^2=0$ $f(-2/3)=(-2/3)^3+(-2/3)^2=4/27$

Hence our stationary points are (0,0) and (-2/3,4/27).

Finally, we use our test: $f''(x)=6x+2$

$f''(0)=2$ (substituting x=0)

$f''(-2/3)=-2$ (substituting x=-2/3)

2>0, so (0,0) is a local minimum of f(x).

-2<0, so (-2/3,4/27) is a local maximum of f(x).

FUNCTIONS OF SEVERAL VARIABLES

In single-variable calculus we were concerned with functions that map the real numbers RR to RR, sometimes called "real functions of one variable", meaning the "input" is a single real number and the "output" is likewise a single real number. In the last chapter we considered functions taking a real number to a vector, which may also be viewed as functions f:R'!R3f:R'!R3, that is, for each input value we get a position in space. Now we turn to functions of several variables, meaning several input variables, functions f:Rn'!Rf:Rn'!R. We will deal primarily with n=2n=2 and to a lesser extent n=3n=3; in fact many of the techniques we discuss can be applied to larger values of nn as well.

A function f:R2'!Rf:R2'!R maps a pair of values (x,y)(x,y) to a single real number. The three-dimensional coordinate system we have already used is a convenient way to visualize such functions: above each point (x,y)(x,y) in the xx-yy plane we graph the point (x,y,z)(x,y,z), where of course z=f(x,y)z=f(x,y).

Example 14.1.1 Consider f(x,y)=3x+4y"5f(x,y)=3x+4y"5. Writing this as z=3x+4y"5z=3x+4y"5 and then 3x+4y"z=53x+4y"z=5 we recognize the equation of a plane. In the form f(x,y)=3x+4y"5f(x,y)=3x+4y"5 the emphasis has shifted: we now think of xx and yy as independent variables and zz as a variable dependent on them, but the geometry is unchanged.

Example 14.1.2 We have seen that x2+y2+z2=4x2+y2+z2=4 represents a sphere of radius 2. We cannot write this in the form f(x,y)f(x,y), since for each xx and yy in the disk x2+y2<4x2+y2<4 there are two corresponding points on the sphere. As with the equation of a circle, we can resolve this equation into two functions, f(x,y)=4"x2"y2""""""""" • "f(x,y)=4"x2"y2 and f(x,y)="4"x2"y2""""""""" • "f(x,y)="4"x2"y2, representing the upper and lower hemispheres. Each of these is an example of a function with a restricted domain: only certain values of xx and yy make sense (namely, those for which x2+y2d"4x2+y2d"4) and the graphs of these functions are limited to a small region of the plane.

Example 14.1.3 Consider f=x" "+y"f=x+y. This function is defined only when both xx and yyare non-negative. When y=0y=0 we get f(x,y)=x" "f(x,y)=x, the familiar square root function in the xx-zz plane, and when x=0x=0 we get the same curve in the yy-zz plane. Generally speaking, we see that starting from f(0,0)=0f(0,0)=0 this function gets larger in every direction in roughly the same way that the square root function gets larger. For example, if we restrict attention to the line x=yx=y, we get f(x,y)=2x" "f(x,y)=2x and along the line y=2xy=2x we have f(x,y)=x" "+2x" "=(1+2–d)x" "f(x,y)=x+2x=(1+2)x.

Figure 14.1.1. f(x,y)=x" "+y"f(x,y)=x+y

A computer program that plots such surfaces can be very useful, as it is often difficult to get a good idea of what they look like. Still, it is valuable to be able to visualize relatively simple surfaces without such aids. As in the previous example, it is often a good idea to examine the function on restricted subsets of the plane, especially lines. It can also be useful to identify those points (x,y)(x,y) that share a common zz-value.

Example 14.1.4 Consider f(x,y)=x2+y2f(x,y)=x2+y2. When x=0x=0 this becomes f=y2f=y2, a parabola in the yy-zz plane; when y=0y=0 we get the "same" parabola f=x2f=x2 in the xx-zz plane. Now consider the line y=kxy=kx. If we simply replace yy by kxkx we get f(x,y)=(1+k2)x2f(x,y)=(1+k2)x2which is a parabola, but it does not really "represent" the cross-section along y=kxy=kx, because the cross-section has the line y=kxy=kx where the horizontal axis should be. In order to pretend that this line is the horizontal axis, we need to write the function in terms of the distance from the origin, which is x2+y2"""""""""=x2+k2x2"""""""""e"x2+y2=x2+k2x2. Now f(x,y)=x2+k2x2=(x2+k2x2"""""""""e")2f(x,y)=x2+k2x2=(x2+k2x2)2. So the cross-section is the "same" parabola as in the xx-zz and yy-zz planes, namely, the height is always the distance from the origin squared. This means that f(x,y)=x2+y2f(x,y)=x2+y2 can be formed by starting with z=x2z=x2 and rotating this curve around the zz axis.

Finally, picking a value z=kz=k, at what points does f(x,y)=kf(x,y)=k? This means x2+y2=kx2+y2=k, which we recognize as the equation of a circle of radius k" "k. So the graph of f(x,y)f(x,y) has parabolic cross-sections, and the same height everywhere on concentric circles with center at the origin. This fits with what we have already discovered.

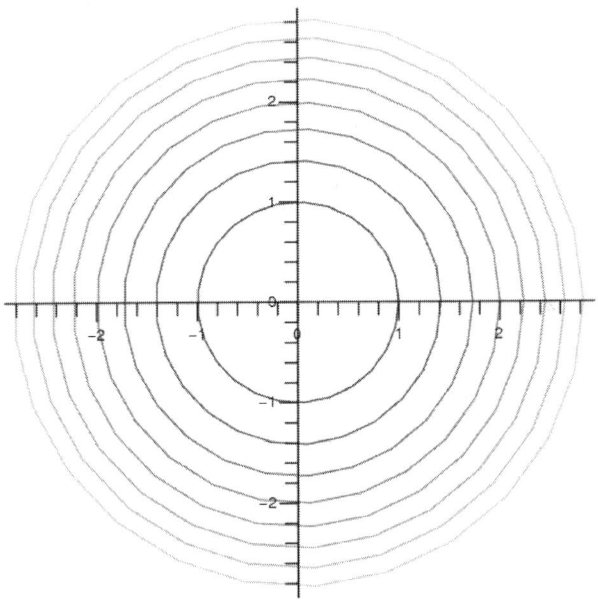

As in this example, the points (x,y)(x,y) such that f(x,y)=kf(x,y)=k usually form a curve, called a level curve of the function. A graph of some level curves can give a good idea of the shape of the surface; it looks much like a topographic map of the surface. In figure 14.1.2 both the surface and its associated level curves are shown. Note that, as with a topographic map, the heights corresponding to the level curves are evenly spaced, so that where curves are closer together the surface is steeper.

Functions f:Rn'!Rf:Rn'!R behave much like functions of two variables; we will on occasion discuss functions of three variables. The principal difficulty with such functions is visualizing them, as they do not "fit" in the three dimensions we are familiar with. For three variables there are various ways to interpret functions that make them easier to understand. For example, f(x,y,z)f(x,y,z) could represent the temperature at the point (x,y,z)(x,y,z), or the pressure, or the strength of a magnetic field. It remains useful to consider those points at which f(x,y,z)=kf(x,y,z)=k, where kk is some constant value. If f(x,y,z)f(x,y,z) is temperature, the set of points (x,y,z)(x,y,z) such that f(x,y,z)=kf(x,y,z)=k is the collection of points in space with temperature kk; in general this is called a level set; for three variables, a level set is typically a surface, called alevel surface.

Example 14.1.5 Suppose the temperature at (x,y,z)(x,y,z) is T(x,y,z)=e″(x2+y2+z2)T(x,y,z)=e″(x2+y2+z2). This function has a maximum value of 1 at the origin, and tends to 0 in all directions. If kk is positive and at most 1, the set of points for which T(x,y,z)=kT(x,y,z)=k is those points satisfying x2+y2+z2=″lnkx2+y2+z2=″lna k, a sphere centered at the origin. The level surfaces are the concentric spheres centered at the origin.

You can use Sage to graph surfaces to check your work:

Ex 14.1.1 Let f(x,y)=(x″y)2f(x,y)=(x″y)2. Determine the equations and shapes of the cross-sections when x=0x=0, y=0y=0, x=yx=y, and describe the level curves. Use a three-dimensional graphing tool to graph the surface. (answer)

Ex 14.1.2 Let f(x,y)=|x|+|y|f(x,y)=|x|+|y|. Determine the equations and shapes of the cross-sections when x=0x=0, y=0y=0, x=yx=y, and describe the level curves. Use a three-dimensional graphing tool to graph the surface. (answer)

Ex 14.1.3 Let f(x,y)=e″(x2+y2)sin(x2+y2)f(x,y)=e″(x2+y2)sina (x2+y2). Determine the equations and shapes of the cross-sections when x=0x=0, y=0y=0, x=yx=y, and describe the level curves. Use a three-dimensional graphing tool to graph the surface. (answer)

Ex 14.1.4 Let f(x,y)=sin(x″y)f(x,y)=sina (x″y). Determine the equations and shapes of the cross-sections when x=0x=0, y=0y=0, x=yx=y, and describe the level curves. Use a three-dimensional graphing tool to graph the surface. (answer)

Ex 14.1.5 Let f(x,y)=(x2"y2)2f(x,y)=(x2"y2)2. Determine the equations and shapes of the cross-sections when x=0x=0, y=0y=0, x=yx=y, and describe the level curves. Use a three-dimensional graphing tool to graph the surface. (answer)

Ex 14.1.6 Find the domain of each of the following functions of two variables:

a. 9"x2""""""""+y2"4""""""""9"x2+y2"4

b. arcsin(x2+y2"2)arcsina (x2+y2"2)

c. 16"x2"4y2"""""""""""""""""16"x2"4y2

(answer)

Ex 14.1.7 Below are two sets of level curves. One is for a cone, one is for a paraboloid. Which is which? Explain.

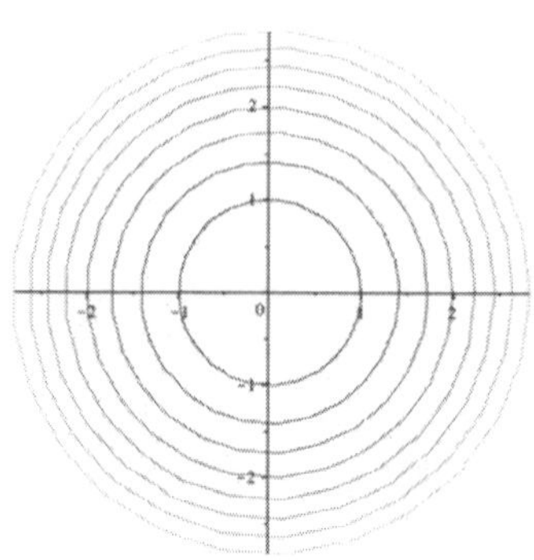

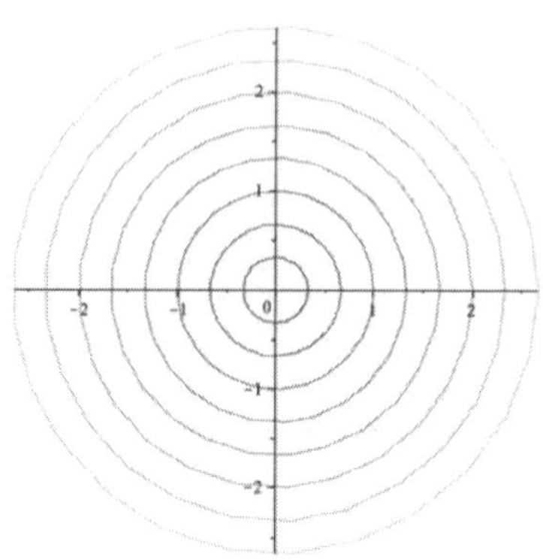

Functions Of Several Variables

In this section we want to go over some of the basic ideas about functions of more than one variable.

First, remember that graphs of functions of two variables, z=f(x,y)z=f(x,y) are surfaces in three dimensional space. For example, here is the graph of z=2x2+2y2"4z=2x2+2y2"4.

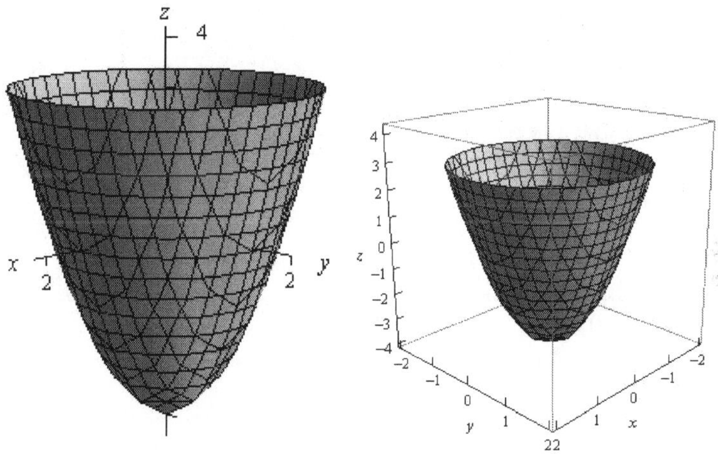

This is an elliptic paraboloid and is an example of a quadric surface. We saw several of these in the previous section. We will be seeing quadric surfaces fairly regularly later on in Calculus III.

Another common graph that we'll be seeing quite a bit in this course is the graph of a plane. We have a convention for graphing planes that will make them a little easier to graph and hopefully visualize.

Recall that the equation of a plane is given by

ax+by+cz=dax+by+cz=d

or if we solve this for zz we can write it in terms of function notation. This gives,

f(x,y)=Ax+By+Df(x,y)=Ax+By+D

To graph a plane we will generally find the intersection points with the three axes and then graph the triangle that connects those three points. This triangle will be a portion of the plane and it will give us a fairly decent idea on what the plane itself should look like. For example, let's graph the plane given by,

f(x,y)=12"3x"4yf(x,y)=12"3x"4y

For purposes of graphing this it would probably be easier to write this as,

z=12"3x"4yÒ!3x+4y+z=12z=12"3x"4yÒ!3x+4y+z=12

Now, each of the intersection points with the three main coordinate axes is defined by the fact that two of the coordinates are zero. For instance, the intersection with the zz-axis is defined by x=y=0x=y=0. So, the three intersection points are,

x"axis : (4,0,0)y"axis : (0,3,0)z"axis : (0,0,12)x"axis : (4,0,0)y"axis : (0,3,0)z"axis : (0,0,12)

Here is the graph of the plane.

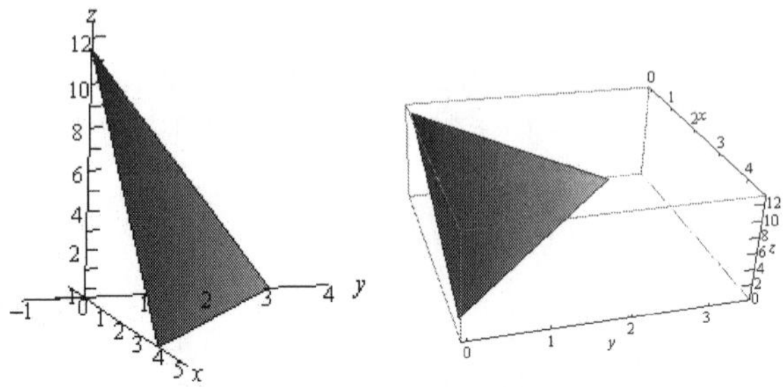

Now, to extend this out, graphs of functions of the form w=f(x,y,z)w=f(x,y,z) would be four dimensional surfaces. Of course, we can't graph them, but it doesn't hurt to point this out.

We next want to talk about the domains of functions of more than one variable. Recall that domains of functions of a single variable, y=f(x)y=f(x), consisted of all the values of xx that we could plug into the function and get back a real number. Now, if we think about it, this means that the domain of a function of a single variable is an interval (or intervals) of values from the number line, or one dimensional space.

The domain of functions of two variables, z=f(x,y)z=f(x,y), are regions from two dimensional space and consist of all the coordinate pairs, (x,y)(x,y), that we could plug into the function and get back a real number.

16

Singular Point

In the mathematical field of algebraic geometry, a singular point of an algebraic variety V is a point P that is 'special' (so, singular), in the geometric sense that at this point the tangent space at the variety may not be regularly defined. In case of varieties defined over the reals, this notion generalizes the notion of local non-flatness. A point of an algebraic variety which is not singular is said to be regular. An algebraic variety which has no singular point is said to be non-singular or smooth.

While behavior of ODEs at singular points is more complicated, certain singular points are not especially diï¬ƒcult to solve. Let us look at some examples before giving a general method. We may be lucky and obtain a power series solution using the method of the previous section, but in general we may have to try other things.

EXAMPLES

Example Let us ï¬ rst look at a simple ï¬ rst order equation
Note that $x = 0$ is a singular point. If we only try to plug in
we obtain

$$0 = 2xy' - y = 2x\left(\sum_{k=1}^{\infty} ka_k x^{k-1}\right) - \left(\sum_{k=0}^{\infty} a_k x^k\right)$$

$$= a_0 + \sum_{k=1}^{\infty}(2ka_k - a_k)x^k.$$

First, . Next, the only way to solve for is for for all k. Therefore we only get the trivial solution $y = 0$. We need a nonzero solution to get the general solution.

Let us try $y = x^r$ for some real number r. Consequently our solutionâ€"if we can ï¬ nd oneâ€"may only make sense for positive x. Then . So

Therefore , or in other words . Multiplying by a constant, the general solution for positive x is

If then the derivative of the solution â€œblows upâ€ at $x = 0$ (the singular point). There is only one solution that is diï¬€erentiable at and thatâ€™s the trivial solution .

Not every problem with a singular point has a solution of the form , of course. But perhaps we can combine the methods. What we will do is to try a solution of the form

where $f(x)$ is an analytic function.

Example 7.3.2: Suppose that we have the equation

and again note that $x = 0$ is a singular point.

Let us try

where r is a real number, not necessarily an integer. Again if such a solution exists, it may only exist for positive x. First let us ï¬ nd the derivatives

$$y' = \sum_{k=0}^{\infty}(k+r)a_k x^{k+r-1},$$

$$y'' = \sum_{k=0}^{\infty}(k+r)(k+r-1)a_k x^{k+r-2}.$$

Plugging into our equation we obtain

$$\begin{aligned}
0 &= 4x^2y'' - 4x^2y' + (1-2x)y \\
&= 4x^2\left(\sum_{k=0}^{\infty}(k+r)(k+r-1)a_kx^{k+r-2}\right) - 4x^2\left(\sum_{k=0}^{\infty}(k+r)a_kx^{k+r-1}\right) + (1-2x)\left(\sum_{k=0}^{\infty}a_kx^{k+r}\right) \\
&= \left(\sum_{k=0}^{\infty}4(k+r)(k+r-1)a_kx^{k+r}\right) - \left(\sum_{k=0}^{\infty}4(k+r)a_kx^{k+r+1}\right) + \left(\sum_{k=0}^{\infty}a_kx^{k+r}\right) - \left(\sum_{k=0}^{\infty}2a_kx^{k+r+1}\right) \\
&= \left(\sum_{k=0}^{\infty}4(k+r)(k+r-1)a_kx^{k+r}\right) - \left(\sum_{k=1}^{\infty}4(k+r-1)a_{k-1}x^{k+r}\right) + \left(\sum_{k=0}^{\infty}a_kx^{k+r}\right) - \left(\sum_{k=1}^{\infty}2a_{k-1}x^{k+r}\right) \\
&= 4r(r-1)a_0x^r + a_0x^r + \sum_{k=1}^{\infty}\left(4(k+r)(k+r-1)a_k - 4(k+r-1)a_{k-1} + a_k - 2a_{k-1}\right)x^{k+r} \\
&= (4r(r-1)+1)a_0x^r + \sum_{k=1}^{\infty}\left((4(k+r)(k+r-1)+1)a_k - (4(k+r-1)+2)a_{k-1}\right)x^{k+r}.
\end{aligned}$$

To have a solution we must ï¬ rst have . Supposing that we obtain

This equation is called the indicial equation. This particular indicial equation has a double root at .

OK, so we know what r has to be. That knowledge we obtained simply by looking at the coeï¬ƒcient of . All other coeï¬ƒcients of also have to be zero so

$$(4(k+r)(k+r-1)+1)a_k - (4(k+r-1)+2)a_{k-1} = 0.$$

If we plug in and solve for a_k we get

Let us set . Then

$$a_1 = \frac{1}{1}a_0 = 1, \qquad a_2 = \frac{1}{2}a_1 = \frac{1}{2}, \qquad a_3 = \frac{1}{3}a_2 = \frac{1}{3\cdot 2}, \qquad a_4 = \frac{1}{4}a_3 = \frac{1}{4\cdot 3\cdot 2}, \qquad \dots$$

Extrapolating, we notice that

In other words,

$$y = \sum_{k=0}^{\infty}a_kx^{k+r} = \sum_{k=0}^{\infty}\frac{1}{k!}x^{k+1/2} = x^{1/2}\sum_{k=0}^{\infty}\frac{1}{k!}x^k = x^{1/2}e^x.$$

That was lucky! In general, we will not be able to write the series in terms of elementary functions.

We have one solution, let us call it . But what about a second solution? If we want a general solution, we need two linearly independent solutions. Picking a_0 to be a diï¬€erent constant only gets us a constant multiple of y_1, and we do not have any other r to try; we only have one solution to the indicial equation. Well, there are powers of x ï¬‚oating around and we are taking derivatives, perhaps the logarithm (the antiderivative of) is around as well. It turns out we want to try for another solution of the form

which in our case is

We now diï¬€erentiate this equation, substitute into the diï¬€erential equation and solve for . A long computation ensues and we obtain some recursion relation for . The reader can (and should) try this to obtain for example the ï¬ rst three terms

We then ï¬ x and obtain a solution y_2. Then we write the general solution as .

THE METHOD OF FROBENIUS

Before giving the general method, let us clarify when the method applies. Let

$$p(x)y'' + q(x)y' + r(x)y = 0$$

be an ODE. As before, if , then x_0 is a singular point. If, furthermore, the limits

$$\lim_{x \to x_0} (x - x_0)\frac{q(x)}{p(x)} \quad \text{and} \quad \lim_{x \to x_0} (x - x_0)^2 \frac{r(x)}{p(x)}$$

both exist and are ï¬ nite, then we say that x_0 is a regular singular point.

Example 7.3.3: Often, and for the rest of this section, $x_0 = 0$. Consider

Write

$$\lim_{x \to 0} x\frac{q(x)}{p(x)} = \lim_{x \to 0} x\frac{x(1+x)}{x^2} = \lim_{x \to 0} (1+x) = 1,$$

$$\lim_{x \to 0} x^2\frac{r(x)}{p(x)} = \lim_{x \to 0} x^2\frac{(\pi + x^2)}{x^2} = \lim_{x \to 0} (\pi + x^2) = \pi.$$

So $x = 0$ is a regular singular point.

On the other hand if we make the slight change

then

$$\lim_{x \to 0} x\frac{q(x)}{p(x)} = \lim_{x \to 0} x\frac{(1+x)}{x^2} = \lim_{x \to 0} \frac{1+x}{x} = \text{DNE}.$$

Here DNE stands for does not exist. The point is a singular point, but not a regular singular point.

Let us now discuss the general Method of Frobenius[4]. Let us only consider the method at the point $x = 0$ for simplicity. The main idea is the following theorem.

Theorem 7.3.1 (Method of Frobenius). Suppose that

$$p(x)y'' + q(x)y' + r(x)y = 0$$

has a regular singular point at $x = 0$, then there exists at least one solution of the form

A solution of this form is called a Frobenius-type solution.

The method usually breaks down like this.

We seek a Frobenius-type solution of the form

We plug this y into equation (7.3). We collect terms and write everything as a single series.

The obtained series must be zero. Setting the ï¬ rst coeï¬ƒcient (usually the coeï¬ƒcient of) in the series to zero we obtain the indicial equation, which is a quadratic polynomial in r.

If the indicial equation has two real roots and such that is not an integer, then we have two linearly independent Frobenius-type solutions. Using the ï¬ rst root, we plug in

$$y_1 = x^{r_1} \sum_{k=0}^{\infty} a_k x^k,$$

and we solve for all a_k to obtain the ï¬ rst solution. Then using the second root, we plug in

and solve for all to obtain the second solution.

If the indicial equation has a doubled root r, then there we ï¬ nd one solution and then we obtain a new solution by plugging into equation (7.3) and solving for the constants .

If the indicial equation has two real roots such that is an integer, then one solution is

$$y_1 = x^{r_1} \sum_{k=0}^{\infty} a_k x^k,$$

and the second linearly independent solution is of the form

The main idea is to ï¬ nd at least one Frobenius-type solution. If we are lucky and ï¬ nd two, we are done. If we only get one, we either use the ideas above or even a diï¬€erent method such as reduction of order (Exercise 2.1.8) to obtain a second solution.

Bessel functions

An important class of functions that arises commonly in physics are the Bessel functions[6]. For example, these functions appear when solving the wave equation in two and three dimensions. First we have Besselâ€™s equation of order p:

$$y_1 = x^p \sum_{k=0}^{\infty} \frac{(-1)^k x^{2k}}{2^{2k} k!(k+p)(k-1+p)\cdots(2+p)(1+p)},$$

$$y_2 = x^{-p} \sum_{k=0}^{\infty} \frac{(-1)^k x^{2k}}{2^{2k} k!(k-p)(k-1-p)\cdots(2-p)(1-p)}.$$

Exercise 7.3.1: a) Verify that the indicial equation of Besselâ€™s equation of order p is . b) Suppose that is not an integer. Carry out the computation to obtain the solutions and above.

Bessel functions will be convenient constant multiples of and . First we must deï¬ ne the gamma function

Notice that . The gamma function also has a wonderful property

From this property, one can show that when n is an integer, so the gamma function is a continuous version of the factorial. We compute:

$$\Gamma(k+p+1) = (k+p)(k-1+p)\cdots(2+p)(1+p)\Gamma(1+p),$$
$$\Gamma(k-p+1) = (k-p)(k-1-p)\cdots(2-p)(1-p)\Gamma(1-p).$$

Exercise 7.3.2: Verify the above identities using .

We deï¬ ne the Bessel functions of the ï¬ rst kind of order p and as

$$J_p(x) = \frac{1}{2^p\Gamma(1+p)} y_1 = \sum_{k=0}^{\infty} \frac{(-1)^k}{k!\Gamma(k+p+1)} \left(\frac{x}{2}\right)^{2k+p},$$
$$J_{-p}(x) = \frac{1}{2^{-p}\Gamma(1-p)} y_2 = \sum_{k=0}^{\infty} \frac{(-1)^k}{k!\Gamma(k-p+1)} \left(\frac{x}{2}\right)^{2k-p}.$$

As these are constant multiples of the solutions we found above, these are both solutions to Besselâ€™s equation of order p. The constants are picked for convenience.

When is not an integer, and are linearly independent. When is an integer we obtain

$$J_n(x) = \sum_{k=0}^{\infty} \frac{(-1)^k}{k!(k+n)!} \left(\frac{x}{2}\right)^{2k+n}.$$

In this case it turns out that

and so we do not obtain a second linearly independent solution. The other solution is the so-called Bessel function of second kind. These make sense only for integer orders n and are deï¬ ned as limits of linear combinations of and as p approaches in the following way:

$$Y_n(x) = \lim_{p \to n} \frac{\cos(p\pi) J_p(x) - J_{-p}(x)}{\sin(p\pi)}.$$

As each linear combination of and is a solution to Besselâ€™s equation of order p, then as we take the limit as goes to , is a solution to Besselâ€™s equation of order . It also turns out that and are linearly independent. Therefore when is an integer, we have the general solution to Besselâ€™s equation of order for arbitrary constants and . Note that goes to negative inï¬ nity at . Many mathematical software packages have these functions and deï¬ ned, so they can be used just like say and . In fact, they have some similar properties. For example, is a derivative of , and in general the derivative of can be written as a linear combination of and . Furthermore, these functions oscillate, although they are not periodic. See Figure 7.4 for graphs of Bessel functions.

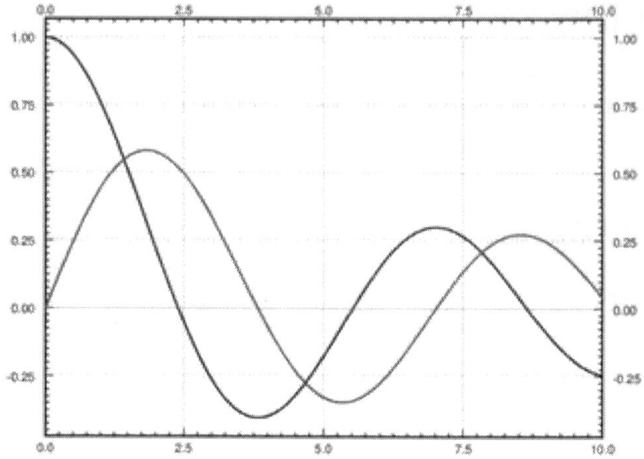

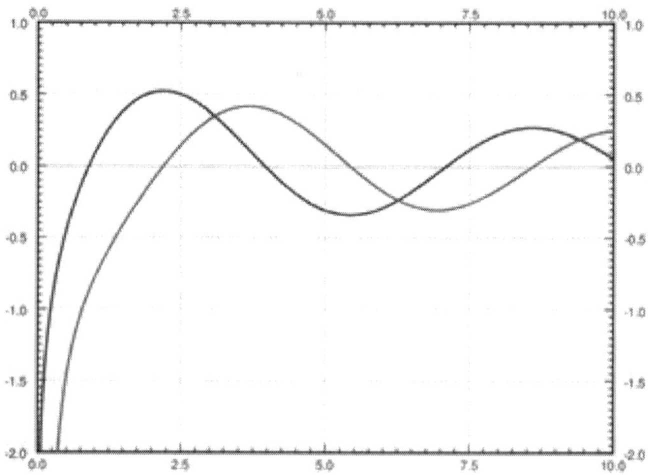

Example 7.3.4: Other equations can sometimes be solved in terms of the Bessel functions. For example, given a positive constant λ,

can be changed to . Then changing variables we obtain via chain rule the equation in y and t:

which can be recognized as Besselâ€™s equation of order 0. Therefore the general solution is , or in terms of x:

This equation comes up for example when ï¬ nding fundamental modes of vibration of a circular drum, but we digress.

Concave and convex functions of a single variable

Definitions

The twin notions of concavity and convexity are used widely in economic theory, and are also central to optimization theory. A function of a single variable is concave if every line segment joining two points on its graph does not lie above the graph at any point. Symmetrically, a function of a single variable is convex if every line segment joining two points on its graph does not lie below the graph at any point. These concepts are illustrated in the following figures.

A concave function:no line segment joiningtwo points on the graphlies above the graphat any point A convex function:no line segment joiningtwo points on the graphlies below the graphat any point A function that is neitherconcave nor convex:the line segment shown liesabove the graph at somepoints and below it at others

Here is a precise definition.

Definition

Let f be a function of a single variable defined on an interval. Then f is

- concave if every line segment joining two points on its graph is never above the graph
- convex if every line segment joining two points on its graph is never below the graph.

To make this definition more useful, we can translate it into an algebraic condition. Let f be a function defined on the interval $[x_1, x_2]$. This function is concave according to the definition if, for every pair of numbers a and b with x_1 d" a d" x_2 and x_1 d" b d" x_2, the line segment from (a, f(a)) to (b, f(b)) lies on or below the graph of the function, as illustrated in the following figure.

x1a(1 " ë)a + ëbbx2f(a)(1 " ë)f(a) + ëf(b)f((1 " ë)a + ëb)f(b)

Denote the height of the line segment from (a, f(a)) to (b, f(b)) at the point x by $h_{a,b}(x)$. Then according to the definition, the function f is concave if and only if for every pair of numbers a and b with x_1 d" a d" x_2 and x_1 d" b d" x_2 we have

$$f(x) \geq {}_b(x) \text{ for all } x \text{ with } a \leq x \leq b. \qquad (*)$$

Now, every point x with a d" x d" b may be written as x = (1 d ë)a + ëb, where ë is a real number from 0 to 1. (When ë = 0, we have x = a; when ë = 1 we have x = b.) The fact that $h_{a,b}$ is linear means that

$$h_{a,b}((1 - \lambda)a + \lambda b) = -\lambda)h_{a,b}(a) + \lambda h_{a,b}(b)$$

for any value of ë with 0 dd ë dd 1. Further, we have $h_{a,b}(a)$ = f(a) and $h_{a,b}(b)$ = f(b) (the line segment coincides with the function at its endpoints), so

$$h_{a,b}((1 - \lambda)a + \lambda b) = -\lambda)f(a) + \lambda f(b).$$

Thus the condition (*) is equivalent to

$f((1-\lambda)a + \lambda b)$		$-\lambda)f(a) + \lambda f(b)$ for all λ with $0 \leq \lambda \leq 1$.

We can make a symmetric argument for a convex function. Thus the definition of concave and convex functions may be rewritten as follows.

Definition

Let f be a function of a single variable defined on the interval I. Then f is

- concave if for all a " I, all b " I, and all ë " [0, 1] we have

f((1−λ)a + λb)		− λ)f(a) + λf(b)

- convex if for all a ″ I, all b ″ I, and all ë ″ [0, 1] we have

f((1−λ)a + λb)		− λ)f(a) + λf(b).

In an exercise you are asked to show that f is convex if and only if ″f is concave.

Note that a function may be both concave and convex. Let f be such a function. Then for all values of a and bwe have

f((1−λ)a + λb)		− λ)f(a) + λf(b) for all λ ∈ [0, 1]

That is, a function is both concave and convex if and only if it is linear (or, more properly, affine), taking the form f(x) = á + âx for all x, for some constants á and â.

Economists often assume that a firm's production function is increasing and concave. Examples of such a function for a firm that uses a single input are shown in the next two figures. The fact that such a production function is increasing means that more input generates more output. The fact that it is concave means that the increase in output generated by each one-unit increase in the input does not increase as more input is used. In economic jargon, there are "nonincreasing returns" to the input, or, given that the firm uses a single input, "nonincreasing returns to scale". In the example in the first of the following two figures, the increase in output generated by each one-unit increase in the input not only does not increase as more of the input is used, but in fact decreases, so that in economic jargon there are "diminishing returns", not merely "nonincreasing returns", to the input.

z ′!0f(z)Concave production function(z = input, f(z) = output) z ′!0f(z)Concave production function(z = input, f(z) = output)

The notions of concavity and convexity are important in optimization theory because, as we shall see, a simple condition is sufficient (as well as necessary) for a maximizer of a differentiable concave function and for a minimizer of a differentiable convex function. (Precisely, every point at which the derivative of a concave differentiable function is zero is a maximizer of the function, and every point at which the derivative of a convex differentiable function is zero is a minimizer of the function.)

The next result shows that a nondecreasing concave transformation of a concave function is concave.

Proposition

Let U be a concave function of a single variable and g a nondecreasing and concave function of a single variable. Define the function f by f(x) = g(U(x)) for all x. Then f is concave.

Proof

We need to show that f((1"ë)a + ëb) e″ (1"ë)f(a) + ëf(b) for all values of a and b with a d″ b.

By the definition of f we have

f((1"ë)a + ëb) = g(U((1"ë)a + ëb)).

Now, because U is concave we have

U((1"ë)a + ëb) e″ (1 d ë)U(a) + ëU(b).

Further, because g is nondecreasing, r e″ s implies g(r) e″ g(s). Hence g(U((1"ë)a + ëb)) e" g((1"ë)U(a) + ëU(b)).

But now by the concavity of g we have

g((1"ë)U(a) + ëU(b)) e" (1"ë)g(U(a)) + ëg(U(b)) = (1"ë)f(a) + ëf(b).

So f is concave.

Jensen's inequality: another characterization of concave and convex functions

If we let $ë_1$ = 1 " ë and $ë_2$ = ë in the earlier definition of a concave function and replace a by x_1 and b by x_2, the definition becomes: f is concave on the interval I if for all x_1 " I, all x_2 " I, and all $ë_1$ e" 0 and $ë_2$ e" 0 with $ë_1$ + $ë_2$ = 1 we have

f($ë_1x_1$ + $ë_2x_2$) e" $ë_1$f(x_1) + $ë_2$f(x_2).

The following result, due to Johan Jensen (1859–1925), shows that this characterization can be generalized. (The J in each of Jensen's names is, incidentally, pronounced the way an English speaker pronounces a Y.)

Proposition (Jensen's inequality) SOURCE

A function f of a single variable defined on the interval I is concave if and only if for all n ed 2

Concavity and the Second Derivative Test

You are learning that calculus is a valuable tool. One of the most important applications of differential calculus is to find extreme function values. The calculus methods for finding the maximum and minimum values of a function are the basic tools of optimization theory, a very active branch of mathematical research applied to nearly all fields of practical endeavor. Although modern optimization theory is considerably more advanced, its methods and fundamental ideas clearly show their historical relationship to the calculus. In this tutorial you will review how the second derivative of a function is related to the shape of its graph and how that information can be used to classify relative extreme values.

SOME FIRST DERIVATIVE FACTS

If you have not already done so, you should review the tutorial on the First Derivative. Click here to see a picture that summarizes the First Derivative Test.

Concavity

The Second Derivative Test provides a means of classifying relative extreme values by using the sign of the second derivative at the critical number. To appreciate this test, it is first necessary to understand the concept of concavity.

The graph of a function f is concave upward at the point (c,f(c)) if f,(c) exists and if for all x in some open interval containing c, the point (xf(x)) on the graph of f lies above the corresponding point on the graph of the tangent line to f at c. This is expressed by the inequality f(x)[f(c)+f(c)(x"c)] for all xin some open interval containing c. Imagine holding a ruler along the tangent line through the point(cf(c)): if the ruler supports the graph of f near (cf(c)), then the graph of the function is concave upward.

The graph of a function f is concave downward at the point (cf(c)) if f(c) exists and if for all x in some open interval containing c, the point (xf(x)) on the graph of f lies below the corresponding point on the graph of the tangent line to f at c. This is expressed by the inequality f(x)[f(c)+f(c)(x"c)] for all xin some open interval containing c. In this situation the graph of f supports the ruler. This is pictured below:

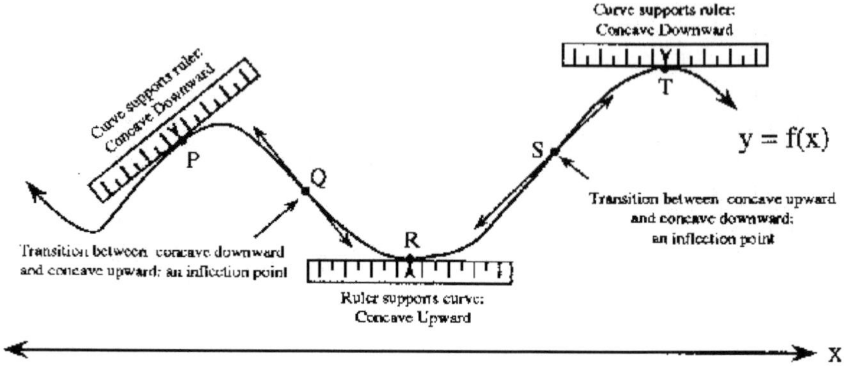

Concavity and the Second Derivative

The important result that relates the concavity of the graph of a function to its derivatives is the following one:

> **Concavity Theorem**: If the function f is twice differentiable at $x=c$, then the graph of f is concave upward at $(c, f(c))$ if $f''(c) > 0$ and concave downward if $f''(c) < 0$.

Example

Suppose f(x)=x3"3x2+x"2. Let's determine where the graph of f is concave up and where it is concave down. Since f is twice-differentiable for all x, we use the result given above and first determine that f(x)=6(x"1). Thus, f(x)0 if x1 and f(x)0 if x1. By the Concavity Theorem, the graph of f is concave up for x1 and concave down for x1.

Inflection Points

Notice in the example above, that the concavity of the graph of f changes sign at x=1. Points on the graph of f where the concavity changes from up-to-down or down-to-up are called inflection points of the graph. The following result connects the concept of inflection points to the derivatives properties of the function:

> **Inflection Point Theorem:** If $f'(c)$ exists and $f''(c)$ *changes sign at* $x=c$, then the point $(c, f(c))$ is an **inflection point** of the graph of f. If $f''(c)$exists at the inflection point, then $f''(c)=0$.

If we return to our example, where f(x)=x3"3x2+x"2, the Inflection Point Theorem verifies that the graph of f has an inflection point at x=1, since f(1)=0.

The Second Derivative Test

The Second Derivative Test relates the concepts of critical points, extreme values, and concavity to give a very useful tool for determining whether a critical point on the graph of a function is a relative minimum or maximum.

> **The Second Derivative Test:** Suppose that c is a critical point at which $f'(c)=0$, that $f'(x)$ exists in a neighborhood of c, and that $f''(c)$ exists. Then f has a relative maximum value at c if $f''(c) < 0$ and a relative minimum value at c if $f''(c) > 0$. If $f''(c)=0$, the test is not informative.

Example

Let's find and classify the extreme values for the function f with values f(x)=x3"3x2+x"2 that was introduced above. We find that f′(x)=3x2"6x+1, and so there are two critical numbers wheref(c)=0:

c1=3"6andc2=3+6

Notice that c11 and that f(c)0. Thus f has a relative maximum at x=3"6 . Since c21 and f(c2)0, the Second Derivative Test informs us that f has a relative minumum at x=3+6 .

Key Concepts

- Concavity Theorem:

If the function f is twice differentiable at x=c, then the graph of f is concave upward at (c,f(c)) if f′(c)0 and concave downward if f(c)0.

- The Second Derivative Test:

Suppose that c is a critical point at which f(c)=0, that f(x) exists in a neighborhood of c, and that f(c) exists. Then f has a relative maximum value at c if f(c)0 and a relative minimum value at c if f(c)0. If f(c)=0, the test is not informative.

Points of Inflexion

A point of inflexion is a point where the gradient of the curve stops falling and starts rising, or vice versa. If you imagine driving along the curve, it's a point where your steering wheel would be in the centre position.

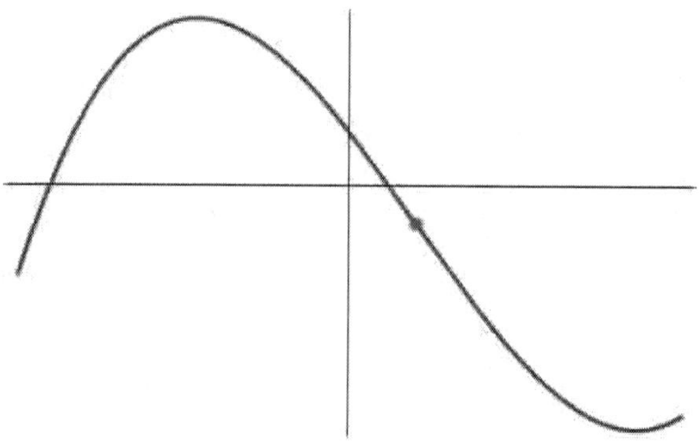

A point of inflexion

Another way of thinking about points of inflexion is that they are points where the gradient attains a maximum or a minimum. We therefore locate them by finding those places where the second derivative is zero.

In differential calculus, an inflection point, point of inflection, flex, or inflection (British English: inflexion) is a point on a continuously differentiableplane curve at which the curve crosses its tangent, that is, the curve changes from being concave (concave downward) to convex (concave upward), or vice versa.

If the curve is the graph of a function y = f(x), of differentiability class C^2, this means that the second derivative of f vanishes and changes sign at the point. A point where the second derivative vanishes but does not change sign is sometimes called a point of undulation or undulation point.

In algebraic geometry an inflection point is defined slightly more generally, as a regular point where the tangent meets the curve to order at least 3, and an undulation point or hyperflex is defined as a point where the tangent meets the curve to order at least 4.

Inflection points are the points of the curve where the curvature changes its sign while a tangent exists.

A differentiable function has an inflection point at (x, f(x)) if and only if its first derivative, f2 , has an isolated extremum at x. (This is not the same as saying that f has an extremum). That is, in some neighborhood, x is the one and only point at which f2 has a (local) minimum or maximum. If all extrema of f2 are isolated, then an inflection point is a point on the graph of f at which the tangentcrosses the curve.

A rising point of inflection is an inflection point where the derivative has a local minimum, and a falling point of inflection is a point where the derivative has a local maximum.

For an algebraic curve, a non singular point is an inflection point if and only if the multiplicity of the intersection of the tangent line and the curve (at the point of tangency) is odd and greater than 2.

For a curve given by parametric equations, a point is an inflection point if its signed curvature changes from plus to minus or from minus to plus, i.e., changes sign.

For a twice differentiable function, an inflection point is a point on the graph at which the second derivative has an isolated zero and changes sign.

A necessary but not sufficient condition

If x is an inflection point for f then the second derivative, f3 (x), is equal to zero if it exists, but this condition does not provide a sufficient definition of a point of inflection. One also needs the lowest-order (above the second) non-zero derivative to be of odd order (third, fifth, etc.). If the lowest-order non-zero derivative is of even order, the point is not a point of inflection, but an undulation point. However, in algebraic geometry, both inflection points and undulation points are usually called inflection points. An example of such an undulation point is x = 0 for the function f given by $f(x) = x^4$.

This definition assumes that f has some higher-order non-zero derivative at x, which is not necessarily the case, But if it has one, it follows from the definition that the sign of f2 (x) is the same on either side of x in a neighborhood of x. If this is positive, the point is a rising point of inflection; if it is negative, the point is a falling point of inflection.

Inflection points sufficient conditions:

(1) A sufficient existence condition for a point of inflection is:

If f(x) is k times continuously differentiable in a certain neighbourhood of a point x with k odd and k e" 3, while $f^{(n)}(x_0) = 0$ for n = 2,...,k " 1 and $f^{(k)}(x_0)$ "' 0 then f(x) has a point of inflection at x_0.

(2) Another sufficient existence condition requires f2 2 (x + å) and f2 2 (x " å) to have opposite signs in the neighborhood of x , if also a tangent exists here. (Bronshtein and Semendyayev2004, p. 231).

Categorization of points of inflection

Points of inflection can also be categorized according to whether f2 (x) is zero or not zero.

- if f2 (x) is zero, the point is a stationary point of inflection
- if f2 (x) is not zero, the point is a non-stationary point of inflection

A stationary point of inflection is not a local extremum. More generally, in the context of functions of several real variables, a stationary point that is not a local extremum is called a saddle point.

An example of a stationary point of inflection is the point (0,0) on the graph of $y = x^3$. The tangent is the x-axis, which cuts the graph at this point.

An example of a non-stationary point of inflection is the point (0,0) on the graph of $y = x^3 + ax$, for any nonzero a. The tangent at the origin is the line y = ax, which cuts the graph at this point.

FUNCTIONS WITH DISCONTINUITIES

Some functions change concavity without having points of inflection. Instead, they can change concavity around vertical asymptotes or discontinuities. For example, the function is concave for negative x and convex for positive x, but it has no points of inflection because 0 is not in the domain of the function.

Definition of an Inflection Point

Consider a function y=f(x), which is continuous at a point x0. The function f(x) can have a finite or infinite derivative f2 (x0) at this point. If, when passing through x0, the function changes the direction of convexity, i.e. there exists a number ´>0 such that the function is convex upward on one of the intervals (x0"´,x0) or (x0,x0+´), and is convex downward on the other, then x0 is called a point of inflection of the function y=f(x).

The geometric meaning of an inflection point is that the graph of the function f(x) passes from one side of the tangent line to the other at this point, i.e. the curve and the tangent line intersect.

Another interesting feature of an inflection point is that the graph of the function f(x) in the vicinity of the inflection point x0 is located within a pair of the vertical angles formed by the tangent and normal (Figure 2).

Necessary Condition for an Inflection Point (Second Derivative Test)

If x0 is a point of inflection of the function f(x), and this function has a second derivative in some neighborhood of x0, which is continuous at the point x0 itself, then

f2 2 (x0)=0.

Proof.

Suppose that the second derivative at the inflection point x0 is not zero: f2 2 (x0)′″0. Since it is continuous at x0, then there exists a ′-neighborhood of the point x0 where the second derivative preserves its sign, i.e.

f2 2 (x0)<0orf2 2 (x0)<0"x″(x0"′,x0+′).

In this case, the function is either strictly convex upward (when f2 2 (x)<0) or strictly convex downward (when f2 2 (x)>0). But then the point x0 is not an inflection point. Hence, the assumption is wrong and the second derivative of the inflection point must be equal to zero.

First Sufficient Condition for an Inflection Point (Second Derivative Test)

If the function f(x) is continuous and differentiable at a point x0, has a second derivative f2 2 (x0) in some deleted ′-neighborhood of the point x0 and if the second derivative changes when passing through the point x0, then x0 is a point of inflection of the function f(x).

Proof.

Suppose, for example, that the second derivative f2 2 (x) changes sign from plus to minus when passing through the point x0. Hence, in the left ′-neighborhood (x0"′,x0), the inequality f2 2 (x)>0, holds, and in the right ′-neighborhood (x0,x0+′), the inequality f2 2 (x)<0 is valid.

In this case, according to the sufficient conditions for convexity, the function f(x) is convex downward in the left ′-neighborhood of the point x0 and is convex upward in the right ′-neighborhood.

Consequently, the function changes the direction of convexity at the point x0, i.e. by definition, x0 is a point of inflection.

Second Sufficient Condition for an Inflection Point (Third Derivative Test)

Let f2 2 (x0)=0, f2 2 2 (x0)′″0. Then x0 is a point of inflection of the function f(x).

Proof.

As f2 2 2 (x0)′″0, the second derivative is either strictly increasing at x0 (if f2 2 2 (x0)>0) or strictly decreasing at this point (if f2 2 2 (x0)<0). Because f2 2 (x0)=0, then the second derivative for some ′>0 has different signs in the left and right ′-neighborhood of x0. Hence, on the basis of the previous theorem, it follows that x0 is a point of inflection of the function f(x).

Inflection points are where the function changes concavity. Since concave up corresponds to a positive second derivative and concave down corresponds to a negative second derivative, then when the function changes from concave up to concave down (or vise versa) the second derivative must equal zero at that point. So the second derivative must equal zero to be an inflection point. But don't get excited yet. You have to make sure that the concavity actually changes at that point.

Example 1 with $f(x) = x^3$.

Let's do an example to see what really happens. Given $f(x) = x^3$, find the inflection point(s). (Might as well find any local maximum and local minimums as well.)

Start with getting the first derivative:

$f'(x) = 3x^2$.

Then the second derivative is:

$f''(x) = 6x$.

Now set the second derivative equal to zero and solve for "x" to find possible inflection points.

6x = 0

x = 0.

We can see that if there is an inflection point it has to be at x = 0. But how do we know for sure if x = 0 is an inflection point? We have to make sure that the concavity actually changes. To do this pick a number on either side of x = 0 and check what the concavity is at those locations. Let's use x = -1 and x = 1 to check. At x = -1, the second derivative gives:

f "(-1) = -6

and the function is concave down at x = -1. If we check x = 1 we get:

f "(1) = 6

which means the function is concave up at x = 1.

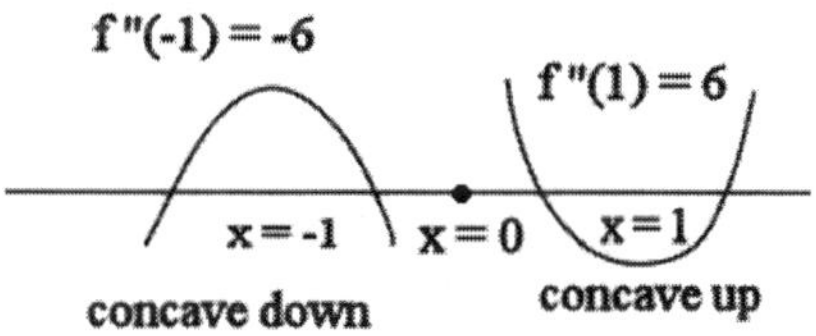

Thus we can see that the function has different concavities on either side of x =0 and the inflection point is at x=0. Note the inflection point is not necessarily where the function crosses the x-axis but is where the concavity actually changes.

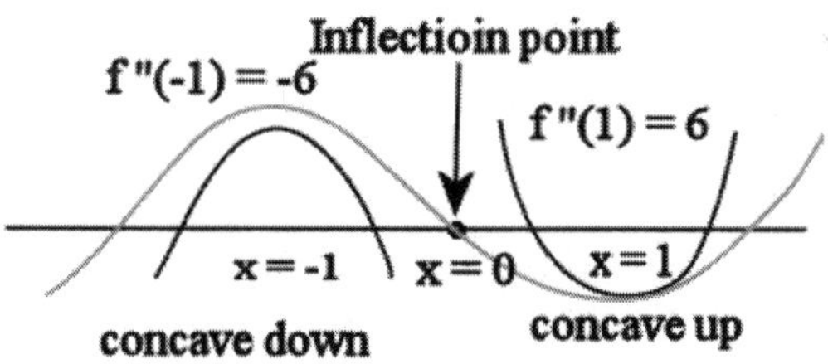

Let's now go back and find the local maximums and local minimums of this function. Start by finding the critical points.

f '(x) = $3x^2$

$3x^2$= 0

x = 0

We only have one critical point, x = 0. Is it a local max or a local min? Let's try using the second derivative test.

f "(x) = 6x

f "(0) = 6(0)

f "(0) = 0.

Well that's unfortunate because that means the function is neither concave up nor concave down. We still don't know if it is a local max or a local min. I guess we'll have to try another technique. Let's try the first derivative test.

Try using x=-1 and x=1 for numbers on either side of our critical point x=0. Plug them into the first derivative.

$f'(-1) = 3(-1)^2$

$f'(-1) = 3.$

$f'(1) = 3(1)^2$

$f'(1) = 3.$

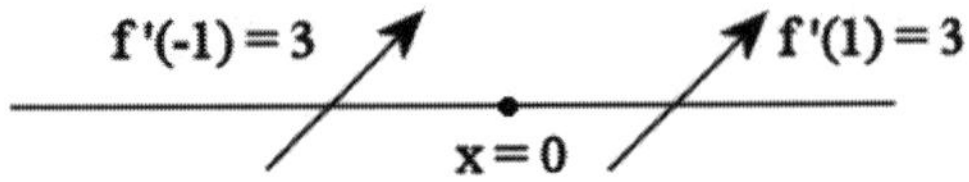

Since the derivative is positive in either side of the critical point, the function is increasing on both side of the critical point and there is no local maximum or local minimum.

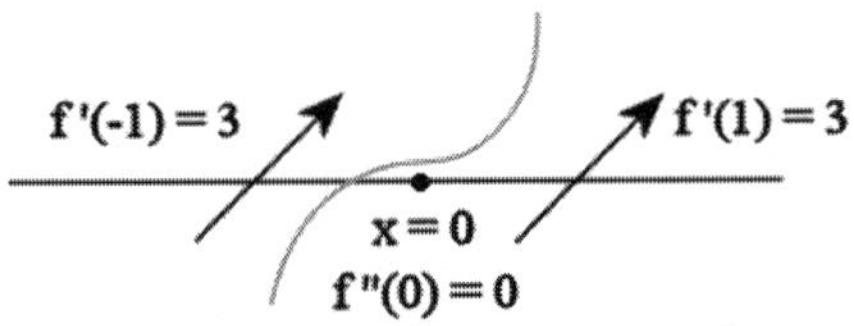

Inflection point only, no max or min.

Example 2 with $f(x) = x^4$.

Example 2:

Let's look at $f(x) = x^4$. Set the derivative equal to zero to find the critical point(s).

$f(x) = x^4$

$f'(x) = 4x^3 = 0$

$x^3 = 0$

$x = 0$

The only critical point is at x = 0. Let's try using the second derivative to test the concavity to see if it is a local maximum or a local minimum.

$F''(x) = 12x^2$

$f''(0) = 12(0)^2 = 0$

Since the second derivative is zero, the function is neither concave up nor concave down at x = 0. It could be still be a local maximum or a local minimum and it even could be an inflection point.

Let's test to see if it is an inflection point. We need to verify that the concavity is different on either side of x = 0. Let's test x = -1 and x = 1 in the second derivative.

$f''(-1) = 12(-1)^2 = 12$

$f''(1) = 12(1)^2 = 12$

No Inflectioin point

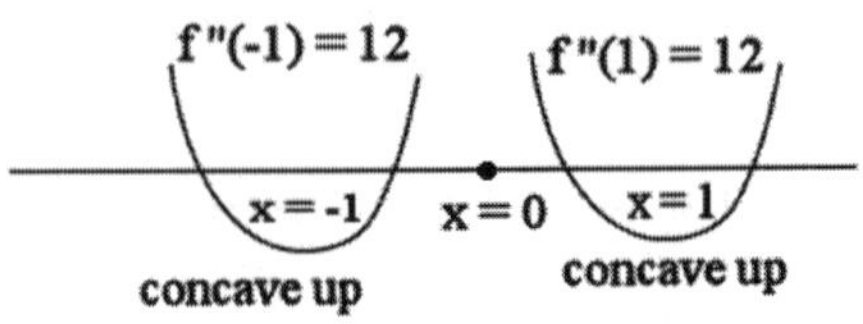

Since the second derivative is positive on either side of x = 0, then the concavity is up on both sides and x = 0 is not an inflection point (the concavity does not change). Well it could still be a local maximum or a local minimum so let's use the first derivative test to find out.

$f'(-1) = 4(-1)^3 = -4$

$f'(1) = 4(1)^3 = 4$

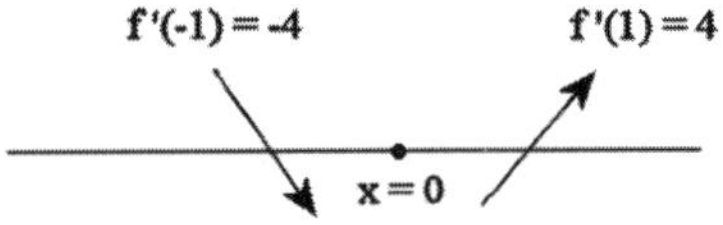

Since the function goes from decreasing to increasing on either side of x = 0, we can see that x = 0 is a local minimum.

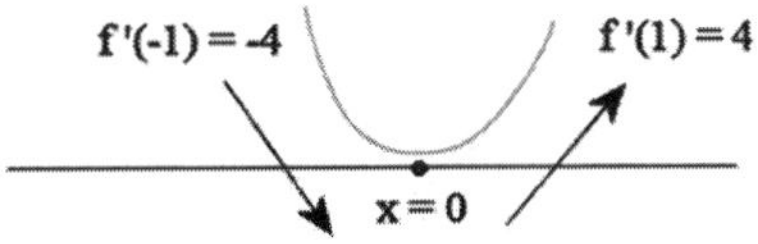

Local minimum with f "(0) = 0 momentarily at x = 0.

Even though $f(x) = x^4$ appears to be concave up every where, it is momentarily "flat" at x = 0 since the second derivative is zero at x = 0.

Distance Between Two Points; Circles

Given two points (x1,y1)(x1,y1) and (x2,y2)(x2,y2), recall that their horizontal distance from one another is ″x=x2"x1Äx=x2"x1 and their vertical distance from one another is ″y=y2"y1Äy=y2"y1. (Actually, the word "distance" normally denotes "positive distance". ″xÄx and ″yÄy are {\it signed\/} distances, but this is clear from context.) The actual (positive) distance from one point to the other is the length of the hypotenuse of a right triangle with legs |″x||Äx| and |″y||Äy|, as shown in figure 1.2.1. The Pythagorean theorem then says that the distance between the two points is the square root of the sum of the squares of the horizontal and vertical sides:

For example, the distance between points A(2,1)A(2,1) and B(3,3)B(3,3) is (3"2)2+(3"1)2"″″″″″″″″″″″″″″″″″″″″=5–″(3"2)2+(3"1)2=5.

(x1,y1)(x1,y1)

(x2,y2)(x2,y2)

″xÄx

″yÄy

Figure 1.2.1. Distance between two points, ″xÄx and ″yÄy positive.

As a special case of the distance formula, suppose we want to know the distance of a point (x,y)(x,y) to the origin. According to the distance formula, this is (x″0)2+(y″0)2"″″″″″″″″″″″″″″″″″=x2+y2"″″″″◀(x″0)2+(y″0)2=x2+y2.

A point (x,y)(x,y) is at a distance rr from the origin if and only if x2+y2"″″″″◀=rx2+y2=r, or, if we square both sides: x2+y2=r2x2+y2=r2. This is the equation of the circle of radius rr centered at the origin. The special case r=1r=1 is called the unit circle; its equation is x2+y2=1x2+y2=1.

Similarly, if C(h,k)C(h,k) is any fixed point, then a point (x,y)(x,y) is at a distance rr from the point CCif and only if (x″h)2+(y″k)2"""""""""""""""""""""=r(x″h)2+(y″k)2=r, i.e., if and only if

(x″h)2+(y″k)2=r2.(x″h)2+(y″k)2=r2.

FUNCTION OF A REAL VARIABLE

In mathematical analysis, and applications in geometry, applied mathematics, engineering, and natural sciences, a function of a real variable is a function whose domain is the real numbers ″!, or a subset of ″! that contains an interval of positive length. Most real functions that are considered and studied are differentiable in some interval. The most widely considered such functions are the real functions, which are the real-valued functions of a real variable, that is, the functions of a real variable whose codomain is the set of real numbers. Nevertheless, the codomain of a function of a real variable may be any set. However, it is often assumed to have a structure of ″!-vector space over the reals. That is, the codomain may be a Euclidean space, a coordinate vector, the set of matrices of real numbers of a given size, or an ″!-algebra, such as the complex numbers or the quaternions. The structure ″!-vector space of the codomain induces a structure of ″!-vector space on the functions. If the codomain has a structure of ″!-algebra, the same is true for the functions.

The image of a function of a real variable is a curve in the codomain. In this context, a function that defines curve is called a parametric equation of the curve. When the codomain of a function of a real variable is a finite-dimensional vector space, the function may be viewed as a sequence of real functions. This is often used in applications.

Real function

A real function is a function from a subset of denotes as usual the set of real numbers. That is, the domain of a real function is a subset, and its codomain is It is generally assumed that the domain contains an interval of positive length.

Basic examples

For many commonly used real functions, the domain is the whole set of real numbers, and the function is continuous and differentiable at every point of the domain. One says that these functions are defined, continuous and differentiable everywhere. This is the case of:

- All polynomial functions, including constant functions and linear functions
- Sine and cosine functions
- Exponential function

Some functions are defined everywhere, but not continuous at some points. For example

- The Heaviside step function is defined everywhere, but not continuous at zero.

Some functions are defined and continuous everywhere, but not everywhere differentiable. For example

- The absolute value is defined and continuous everywhere, and is differentiable everywhere, except for zero.

- The cubic root is defined and continuous everywhere, and is differentiable everywhere, except for zero.

Many common functions are not defined everywhere, but are continuous and differentiable everywhere where they are defined. For example:

- A rational function is a quotient of two polynomial functions, and is not defined at the zeros of the denominator.
- The tangent function is not defined for where k is any integer.
- The logarithm function is defined only for positive values of the variable.

Some functions are continuous in their whole domain, and not differentiable at some points. This is the case of:

- The square root is defined only for nonnegative values of the variable, and not differentiable at 0 (it is differentiable for all positive values of the variable).

GENERAL DEFINITION

A real-valued function of a real variable is a function that takes as input a real number, commonly represented by the variable x, for producing another real number, the value of the function, commonly denoted f(x). For simplicity, in this article a real-valued function of a real variable will be simply called a function. To avoid any ambiguity, the other types of functions that may occur will be explicitly specified.

Some functions are defined for all real values of the variables (one says that they are everywhere defined), but some other functions are defined only if the value of the variable is taken in a subset X of "!, the domain of the function, which is always supposed to contain an interval of positive length.

Domain

The domain of a function of several real variables is a subset of "! that is sometimes explicitly defined. In fact, if one restricts the domain X of a function f to a subset Y ," X, one gets formally a different function, the restriction of f to Y, which is denoted $f_{|Y}$. In practice, it is often not harmful to identify f and $f_{|Y}$, and to omit the subscript $_{|Y}$.

Conversely, it is sometimes possible to enlarge naturally the domain of a given function, for example by continuity or by analytic continuation. This means that it is not worthy to explicitly define the domain of a function of a real variable.

Real Functions

Real functions are the most important type of a mapping. After introducing this notion we look at inverse functions and functions defined by cases.

Definition.

By a real function of real variable we mean any mapping from some subset of the set of real numbers to the set of real numbers.

Note that people usually say just real functions and it is understood that we mean real functions of real variable. Other kinds of real functions (like real functions of complex variable) are always specified in full.

To a matematician, this definition of real functions is entirely satisfactory, since all properties known for mappings readily transfer to functions. However, for a practical user of mathematics this is less suitable, since one can happily and efficiently use real functions without even knowing about the existence of general mappings, and one can also do with less precision. We will therefore try to explain what a real function is without referring to mappings. This is best done by means of an example. We will use one inspired by physics, since indeed physics has been the inspiration for introducing functions in the first place. We will also use this example to review notions of domain, 1-1, onto, and inverse, so we definitely recommend that you look at it. We prefer to offer it as a separate page, since it is quite long and also since we will refer to it in other parts of Math Tutor.

Example.

Now we will briefly cover the basic notions.

A real function is a prescription which assignes values to arguments. The notation y = f (x) means that to the value x of the argument, the function f assigns the value y. Sometimes we also use the notationf : x ¦! y, in words, the function f sends x to y. The most usual way of specifying this assignment is by some formula, that is, the function value y can be obtained by substituting x to a specific formula that identifies the given function. For instance, the function f (x) = 2x + 3 would send the argument x = "1 to f ("1) = 2Å"("1) + 3 = 1. A very special case is the constant function, for instance the definitiong(x) = 7.3 sends every argument x to the value 7.3; this is sometimes confusing for beginners, so take this as a warning that one can "substitute" x into 7.3, and the outcome is 7.3 (like this: g(5) = 7.3).

There is an alternative way to express the idea hat we substitute some concrete number into a function. It uses a slightly lowered vertical bar and goes like this:

$$f(a) = f\big|_a = f\big|_{x=a}.$$

This notation is especially handy if the function is not ripe for substituting yet and we want to make some adjustments first. For example,

$$\left.\frac{x^2-1}{x-1}\right|_{x=3} = \left.\frac{(x-1)(x+1)}{x-1}\right|_{x=3} = (x+1)\big|_{x=3} = 4.$$

The best way to visualize functions is by means of the graph, where we mark in the two-dimensional plane (x, y) all couples (x, f (x)). We pointed out one such particular couple in the picture.

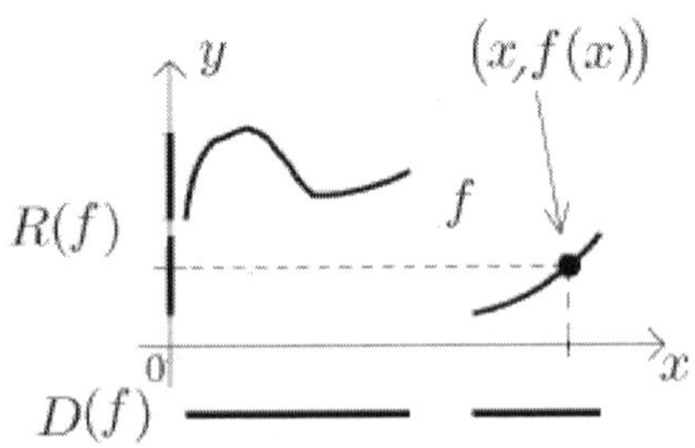

Note that if a function is given by a formula, we may be also interested in its algebraic make-up, which is another way to look at it, independent of the geometric point of view of the graph. For detail see thenote on order of evaluation.

The domain of the function is sometimes given when the function is defined. More often, the real function is just given as a prescription, a formula, in which case for domain we take the set of all numbers xthat can be substituted into the function, that is, for which the formula defining the function makes sense. It is usually denoted by D(f), D_f, or dom(f), here we will use the first notation. We can write

D(f) = {x″″!; f (x) makes sense}.

Typically, the domain would be a union of intervals.

The range of the function is the set of all values that can be obtained by substituting arguments from the domain. It is usually denoted by R(f), R_f, or ran(f), here we will use the first notation. We can write

R(f) = { f (x); x″D(f)}.

Sometimes the given function is considered only on a subset of the domain. We then say that the function was restricted to this subset, the new function is called the restriction. If M is a subset of the domain D(f) of a function f, then the restriction of f to the set M is denoted $f\,|_M$.

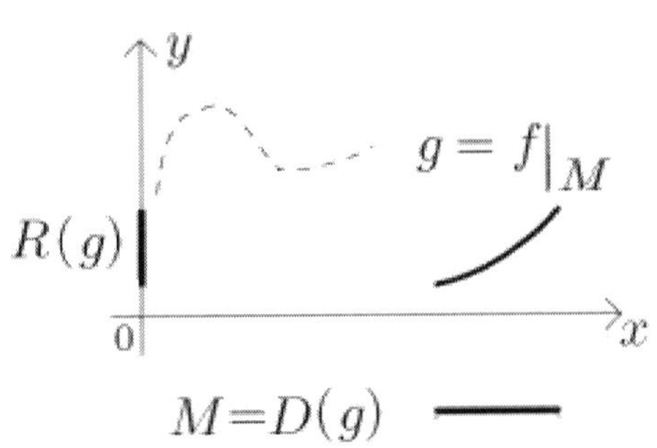

The restricted function has a different domain and might have a different range (although this is not necessary, sometimes a function can cover all its values on that subset to which we restrict it and only repeats them in the parts that were ignored).

The properties of functions depend a lot on their domains, so one in fact always has to think of a function and its domain as a pair. This is also reflected in the following

Definition.

We say that two functions, f and g, are equal, denoted f = g, if they have the same domain D and for all x from D we have f (x) = g(x).

Consider the following example. This equality is obviously true:

$$\ln\left(\tfrac{x}{x-1}\right) = \ln(x) - \ln(x-1).$$

Now we will define two functions.

$$f(x) = \ln\left(\tfrac{x}{x-1}\right), \qquad g(x) = \ln(x) - \ln(x-1).$$

Although algebraically these two formulas are the same, the two functions are not, we cannot write f = g, because their domains are not the same:

$$D(f) = (-\infty, 0) \cup (1, \infty), \qquad D(g) = (1, \infty).$$

Sometimes we want to disregard values of f outside a specific set M, but we still want to be able to work on its whole domain. Then the notion of restriction does not help, instead we can "kill" the values of foutside M like this:

$$g(x) = \begin{cases} f(x), & x \in M; \\ 0, & x \in D(f) - M. \end{cases}$$

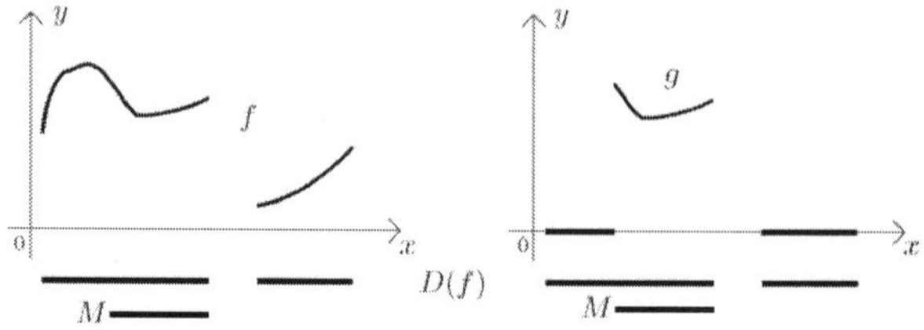

This approach has several advantages, one is that the function g can be expressed algebraically, see the section on characteristic functions in Theory - Elementary functions.

There is an alternative way to visualize functions. It carries less actual information about the given function, but it is sometimes helpful in emphasizing the function as a procedure for sending points from one set to points in another set. A real function actually sends points from one copy of reals to another (different) copy of reals, the difference is emphasized by different letters for elements of the two sets.

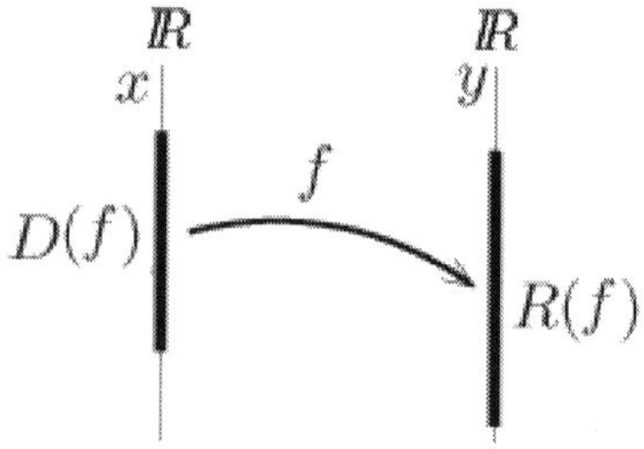

Sometimes we may even indicate how some points are sent, for instance the function $f(x) = 2x + 3$ can be shown as follows:

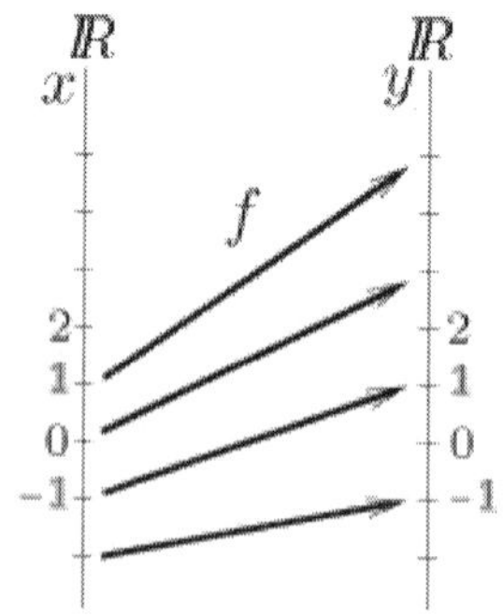

1-1 functions.

A function is said to be 1-1 if it does not happen that two different arguments would lead to the same value. That is, for a 1-1 function f it cannot happen that for distinct x_1, x_2 from the domain one would have $f(x_1) = f(x_2)$. In a graph, a function is 1-1 if it does not happen that some horizontal line intersects the graph of the function at more than one place. As you can see in the picture, the original function f is not 1-1, but its restriction g is 1-1.

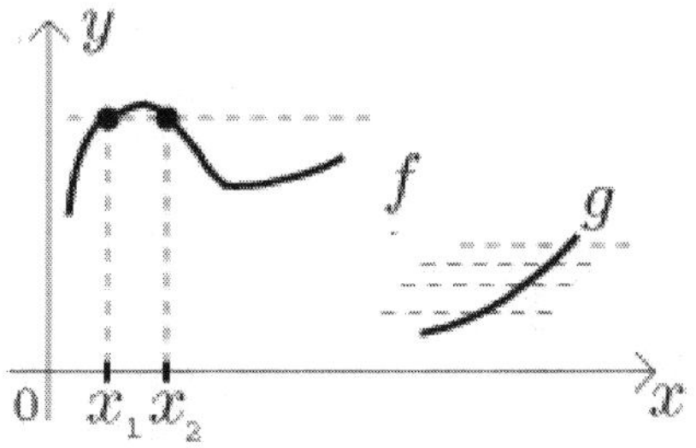

In practice it is easier to work with a different condition, formally the counterpositive of the above conditon: If after substituting two points we get the same value, then it must have been the same point.

Definition.

We say that a function f is 1-1 (or injective, or an injection) if for all x_1,x_2 from its domain the following implication is true: If f (x_1) = f (x_2), then $x_1 = x_2$.

If we were also given a target set B with the function f, we can ask whether f is onto (or surjective, or a surjection), that is, whether the target set is equal to the range, B = R(f). However, this almost never happens with real functions. Usually we are not given the target set, in which case we take the range for the target set and the function is automatically onto, we do not investigate this property.

A function is bijective (a bijection) if it is 1-1 and onto. For real functions this means that we only worry about being 1-1.

Inverse function

By an inverse function of a function f (or its inverse for short) we mean another function g which satisfies the condition g(f (x)) = x for all x from D(f) and f (g(y)) = y for all y from R(f)—if you are uneasy about what these formulas mean, check out composition in the next section on operations. The first equality means that when the function f sends the argument x to some point f (x), then applying the inverse function g to this number f (x) ″undoes″ this “sending”, returns you back to where you started. The second equality says that also f undoes the effect of g. One can prove that if the first equality is true (for all x from the domain of f), then the second equality automatically follows.

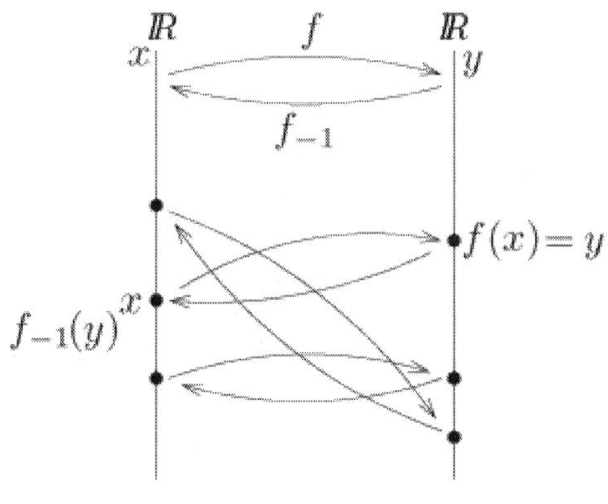

If such an inverse function exists, we denote it $f_{''1}$. Actually, most textbooks prefer the highly unsuitable notation $f^{''1}$, see the note in the above Example.

There is a well-known theorem which states that an inverse exists exactly when the given function is a bijection. For real functions we however have a special form:

Theorem.

Assume that f is a real function with domain D(f) and range R(f). It has an inverse if an only if it is 1-1. Then the inverse is unique and satisfies

$D(f_{''1}) = R(f), \quad R(f_{''1}) = D(f).$

The graph of an inverse function can be obtained by flipping the graph of f around the main diagonal. This, by the way, shows the importance of being 1-1 for the existence of an inverse. In the following picture we first show an example of a function that has an inverse. The second picture shows a function that is not 1-1 (we get to the indicated value y from two different values of argument). You can see that when we attempt to flip around its graph, we obtain something that is definitely not a graph of a function, since a function cannot have two values for one argument.

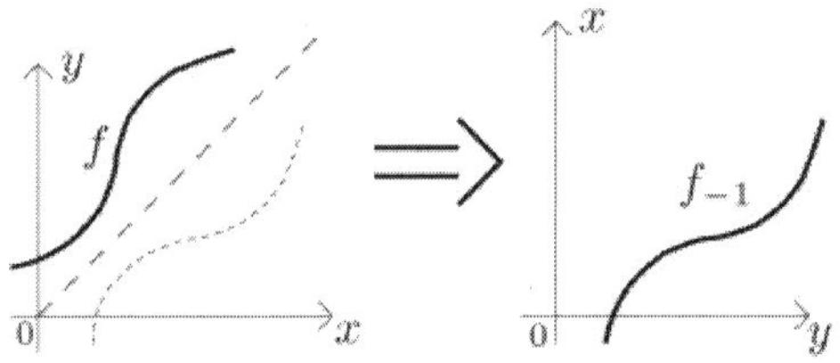

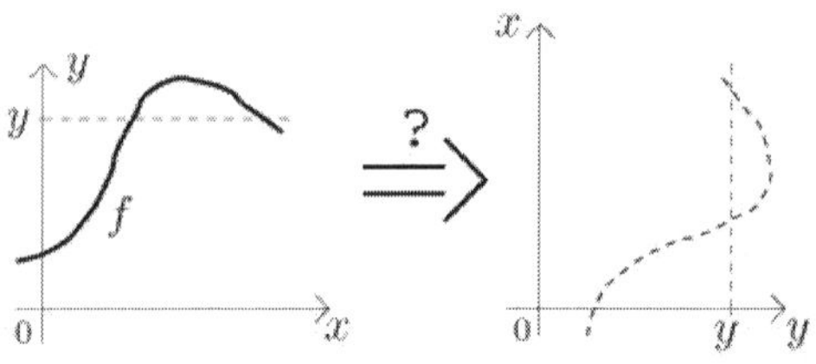

We will return to the inverse in the next section, Operations with real functions, namely when talking about composition.

Example. Consider the function

$f(x) = 4 '' (1 '' x)^2$.

First we need to look at the domain. What numbers x can be substituted into this formula? We know that any real number can be squared, so the answer is: All real numbers. Therefore D(f) = ″!.

Range is usually much less important, it is also often difficult to determine, so in most cases we do not bother. However, here it is quite simple, so we try it. What values can be obtained using the formula forf ? We know from experience that squares are always positive or at least zero, so the formula $4 '' (1 '' x)^2$ cannot yield a number larger than 4. This shows that the range should be a subset of the interval(″,4]. Could it actually be equal to this interval? In other words, can we get an arbitrary number from this interval using the given formula? This translates to the following question: Can we get any non-negative number using $(1 '' x)^2$? Since the answer is positive, we conclude that the range is R(f) = (″,4].

What about being 1-1 and having an inverse? We know that these two questions are equivalent. To solve the first one, we try to prove the implication for the definition. We assume that f (x_1) = f (x_2) and see where it takes us:

4 ″ (1 ″ x_1)^2 = 4 ″ (1 ″ x_2)^2,

(1 ″ x_1)^2 = (1 ″ x_2)^2,

We know that this equation has two possible solutions. One is 1 ″ x_1 = 1 ″ x_2, that is, x_1 = x_2, which is the trivial solution we always get (it says that to get to the same place, one can start twice from the same place, which is obviously always true). But there is also another solution, (1 ″ x_1) = ″(1 ″ x_2), that is, x_2 = 2 ″ x_1. This is more interesting, because it shows that there might be two different numbers going to the same place. We will try it with some concrete numbers. For instance, choosing x_1 = 0 we get x_2 = 2. Check:

f (x_1) = f (0) = 4 ″ (1 ″ 0)^2 = 3, f (x_2) = f (2) = 4 ″ (1 ″ 2)^2 = 3.

We confirmed that two different arguments, 0 and 2, are assigned the same value, namely 3. Therefore the function f is not 1-1 and does not have an inverse.

We can also try to find the inverse function and see if we succeed. For the inverse to exist, we should be able to reverse the assignment x ¦! y, that is, given y from the range of f, we should be able to solve the equation f (x) = y uniquely for x. We try it:

f (x) = y, 4 ″ (1 ″ x)^2 = y, (1 ″ x)^2 = 4 ″ y.

We know that this equation has two distinct solutions, namely we can have x equal to 1 minus root of 4 ″ y or 1 plus root of 4 ″ y, which again shows that there is no inverse.

Drawing a graph properly is covered later in Graphing functions in Derivative - Theory, so to get at least some idea of the graph of our f, we start by getting some points on it, that is, by calculating the values of f at many values of x. Of course, the more the better, but for simplicity we will try just some here:

x:	−2	−1	-.5	0	0.25	0.5	0.75	1	1.25	1.5	1.75	2	2.5	3	4
f (*x*):	−5	0	1.75	3	3.4375	3.75	3.9375	4	3.9375	3.75	3.4375	3	1.75	0	−5

We put these values in the graph:

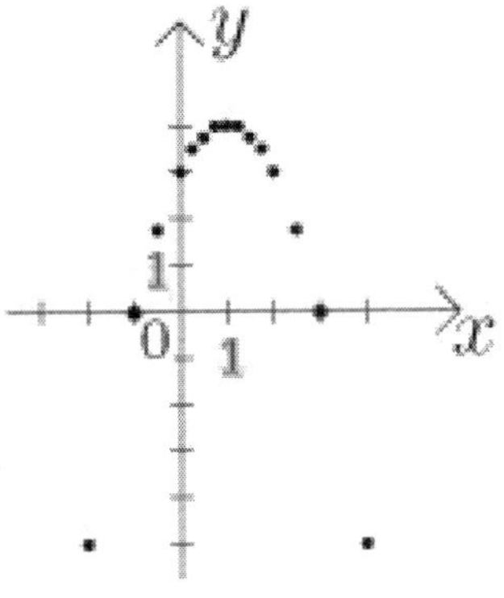

They suggest the general shape. In fact, the exact shape is a parabola turned down, like this:

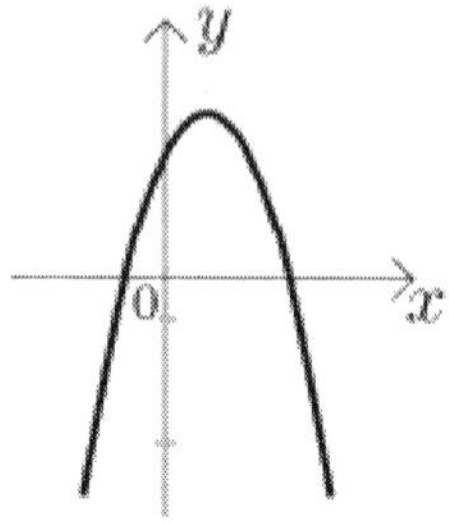

We see right away that this function is definitely not 1-1, since apart from the vertex (1,4), all levels on it can be reached from two different values of argument, in other words, it fails the necessary condition about as badly as possible.

We can also see from the picture that if we restrict this function to just one branch of the parabola (or even just a part of it), it becomes 1-1. For instance, we can consider the function g obtained by restricting fto the open interval (d,0).

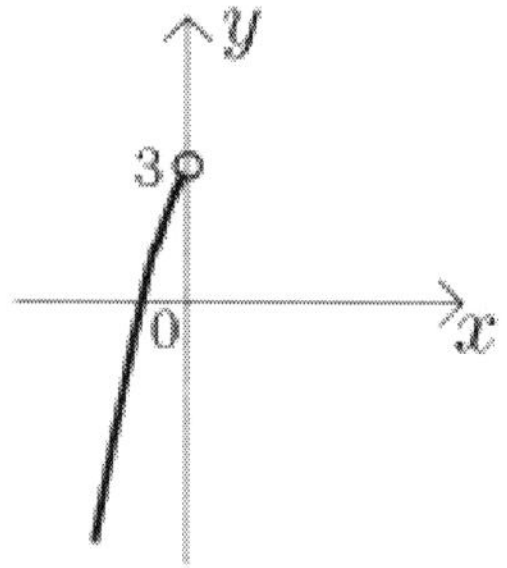

This function is now 1-1 and has an inverse, we confirm this by succesfully finding it. We need to solve the equation g(x) = y, which, as we have seen, leads to the equation $(1 ″ x)^2 = 4 ″ y$. But now the only possible x's are from the interval (d,0), which means that (1 ″ x) should be at least 1, in particular positive. Therefore, when taking roots in the equation, we have to put (1 ″ x) equal to the positive root of4 ″ y. Consequently we obtain

$$x = 1 - \sqrt{4-y} \qquad \Longrightarrow \qquad g_{-1}(y) = 1 - \sqrt{4-y}.$$

Thus we found the inverse function, incidentally proving that it exists and that the function g is 1-1. The inverse function has domain $D(g_{″1})$ = (d,3), range $R(g_{″1})$ = (d,0) and graph

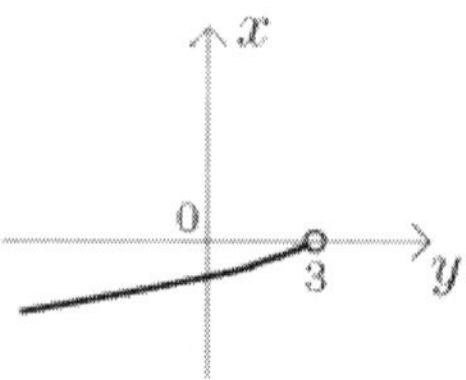

Note how it came naturally that the inverse function has y as its variable. Since in most cases we prefer x as variable, people often, when finding the inverse function, switch variables and eventually write$g_{″1}$(x) = 1 - root of (4 ″ x). While formally this is

not entirely wrong—we may use any letter for the free variable when working with an individual function, it is not quite correct either. The reason is that $g_{''1}$ does not exist by itself, but in the context of g. That function is given, it sends numbers from the copy of reals denoted by x to another copy of reals denoted using the letter y, while by definition $g_{''1}(y)$ sends numbers the other way around. Therefore $g_{''1}(y)$ has to use y as its variable, otherwise we are mixing different worlds, formally $g_{''1}$ cannot accept numbers from the world of x as its domain is somewhere else. This distinction becomes all the more important in applications when the variables have some physical meaning, switching them would lead to nonsense.

Note that if we chose another restriction on the left half of the parabola, for instance to (″,1), (″,1], or perhaps to (″,″13], we would get different inverse functions. They would be given by the same formula, but their domains and ranges would differ. If we restricted f to some part of the right branch, for instance to [1,″), we would now even get a different formula for the inverse. And if we restricted f to, say, the interval (0,2), then the restriction would not be 1-1 and it would have no inverse.

Remark: When talking about the inverse $g_{''1}$, we said that we can change the letter denoting the variable. What does such a change mean? If we change all the appearances of one letter into another letter, from mathematical point of view nothing changes. In fact we can use any letter we wish, as long as we do it consistently and do not use the same letter for some other thing in the same situation. In particular, changing a letter in the description of a function does not change it; the object is still the same. For instance, the function f (w) = 4 ″ (1 ″ w)^2 is the same function as the one in the Example above. We would also change the label at the horizontal axis in the graph and then everything that we did in the example with x could be done with w, yielding the same answers. In other words, this new function is exactly the same, and for instance f (k) = 4 ″ (1 ″ k)^2 is still the same function, what is important is not the letter we use for the variable, but the expression defining the function. While it is good to stick to letters that are traditional for the argument (that is, for the variable), it is sometimes to our advantage to use a different letter, for instance because it may better fit with some more complicated setup.

One has to be a bit careful if there are more letters in play, say a function may have its definition dependent on some parameter p. For instance, the formula x^p defines, for a fixed value of p, a function - namely a power, with x as the variable. When more letters are in place, one has to be careful not to mix variables and parameters, it is also necessary to be a bit more careful when deciding on changing the name of the variable so that there is no mix-up.

Functions defined by cases

Some functions are not given by just one formula, but by several, some arguments are being substituted into one formula, others into a different one. We say that such a function is defined "by cases", or that it is "piecewise-defined", or that it is a "split function" (which we prefer here), and typical example would be this:

$$f(x) = \begin{cases} x^2, & x \in [1,2]; \\ 4 - x, & x \in (2.5, 4). \end{cases}$$

How does this work? Arguments for which we know f can be taken from the closed interval [1,2] or from the open interval (2.5,4), so the domain of this function is

D(f) = [1,2] *″ (2.5,4).

When we want to substitute some x from the domain, we check into which of the alternatives it falls and then use the appropriate formula. For instance, x = 1.5 belongs to the first named interval, so we should use the first formula: f (1.5) = 1.5^2 = 2.25, while 3 belongs to the second interval and therefore f (3) = 4 ″ 3 = 1. When we try to visualize the graph of f by calculating many points, we should get the correct feeling that the graph of f looks like the piece of the graph of x^2 corresponding to the interval [1,2] and the piece of the graph of 4 ″ x corresponding to the interval (2.5,4) put together:

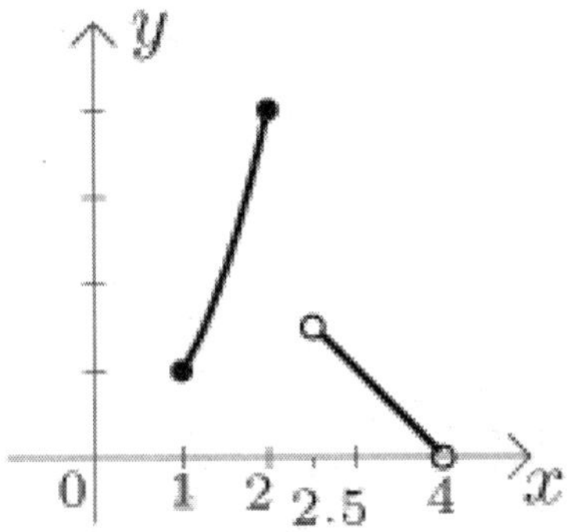

And that is exactly how functions defined by cases are investigated. Each case is checked separately and the graphs and answers are then put together. One can also have more cases with more interesting conditions, for instance the function

$$g(x) = \begin{cases} x - 4, & x \geq 5; \\ 3, & x = 4; \\ 1, & x \in (0,3); \\ x^6, & x = -1. \end{cases}$$

It has domain

D(g) = {″1} *″ (0,3) *″ {4} *″ [5,″)

and the graph

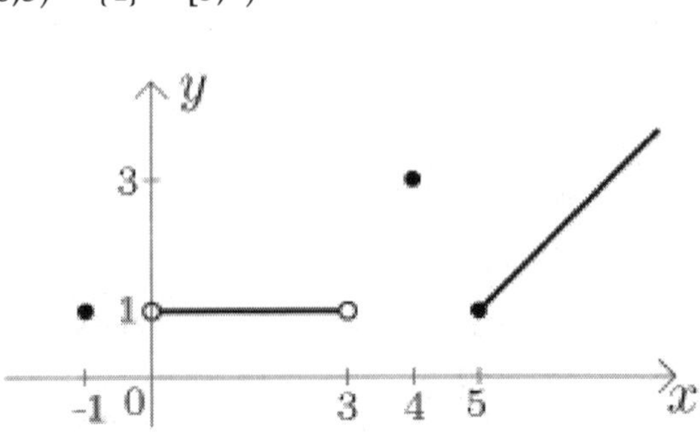

Some values: g(″1) = (″1)^6 = 1 (the last alternative), g(1) = 1 (the third alternative), g(2.3) = 1 (the third alternative), g(4) = 3 (the second alternative), g(6) = 6 ″ 4 = 2 (the first alternative), g(7) = 7 ″ 4 = 3(the first alternative).

This idea of putting together pieces of functions is used not only with functions defined by cases, but also with functions that are defined by one formula, but their domains split into several parts. For instance, the function h(x) = 1/x has domain D(h) = (″,0) *″ (0,″) since one can put all numbers but zero into this formula. Thus one can investigate separately the parts of the graph on the right and on the left and then put the information together.

Sometimes a function is given by one formula, but definition by cases is hidden in it, which is above all the case of the absolute value.

OPERATIONS WITH REAL FUNCTIONS

In this section we will look at different ways of combining functions together. We start with algebraic operations, then we look at composition and at the end we briefly survey transformations of functions. We conclude this section with a handy notation for comparison of functions.

Consider two functions, f and g. In order to use algebraic operations with them (addition, multiplication,...) we first need to check that they have the same domain. What happens if they do not? Then there are two possibilities. If their domains are disjoint, then we cannot put them together algebraically, since the two functions "live" at entirely different places. Otherwise we just look at the set where their domains intersect and the result of the operation will have this intersection as its domain.

Definition (algebraic operations).

As a result of these operations we always get a new function, defined on the intersection as stated above (for division sometimes on a smaller set). The properties of such a new function are usually some sort of a combination of properties of the two original functions, but the way they combine depends on the particular property and on operation, so we prefer to leave this discussion to parts where the various properties are discussed. One thing should be noted here, since this property was already covered: the property of being 1-1 does not get preserved by these operations. If you start with two functions that are 1-1 and perform one of these operations, you may end up with a function that is not 1-1.

For example, imagine that we have two 1-1 function f and g whose values are, among others, f (1) = 1, f (2) = ″1, g(1) = ″1, and g(2) = 1. How much is f + g at 1 and 2? We check:

(f + g)(1) = f (1) + g(1) = 1 ″ 1 = 0,

(f + g)(2) = f (1) + g(2) = ″1 + 1 = 0.

This shows that the function f + g is not 1-1. Similar calculations show that also the functions fÅ″g and f / g are not 1-1, a slight modification of this example would also show that a difference of two 1-1 functions need not be 1-1.

Since the 1-1 property si not preserved, one also cannot say anything nice about inverse to outcomes of these operations. There are no nice rules like $(f + g)_{″1} = f_{″1} + g_{″1}$, when it comes to the inverse, an outcome of these operations must be investigated as a new problem; knowledge of f and g does not help.

Once we have multiplication of functions, we can also define integer powers in the usual way, for instance f^3 = fÅ″fÅ″f and $f^{″2}$ = 1/(fÅ″f). This definition can then be extended to rational exponent using roots, and limit procedure allows us to define f^a for any real a. Fro details see Powers in Theory - Elementary functions.

The last step is to define the general power f^g by
$(f^g)(x) = e^{g(x)Å''\ln(f(x))}$ for x"D(f))" D(g) for which f (x) > 0.
For details, see General powers in Theory - Elementary functions.
Composition
For the next operation it is useful to recall another visualization of a a function. The fact that f sends points from one set (a subset of the real line) to points in another set can be expressed like this:

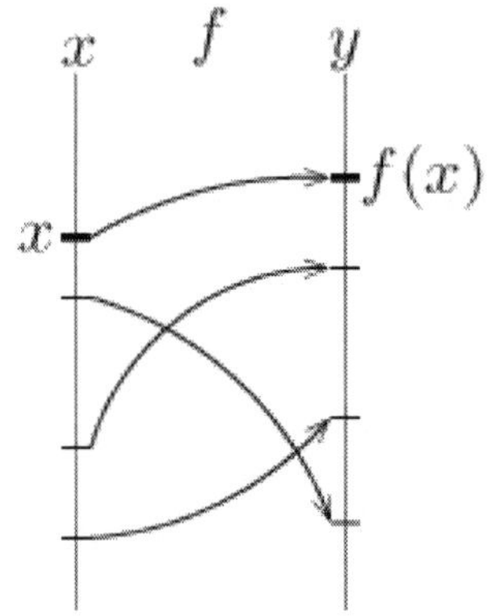

Now the values of the function f are numbers again, so they can be used as arguments for another function g like this:

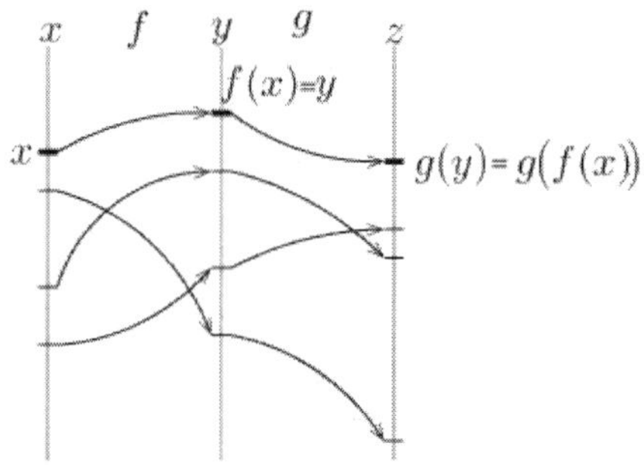

As a whole we get a procedure that sends numbers from one copy of the set of real numbers (denoted using x) into another copy denoted using z, we can forget that there was a middle step and consider the whole thing to be just another function.

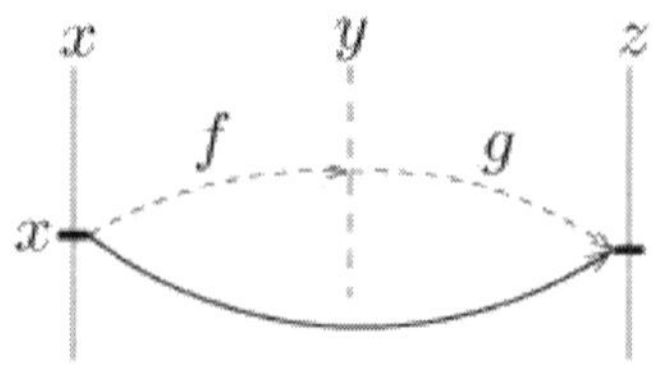

This new function is called the composition of the given two functions and for this idea to work, one condition is needed: That the values of f lie in the domain of g (all or at least some). What is the notation? One is natural. We start with x, then apply f and obtain the number y = f (x). To this number we then apply g, thus obtaining z = g(y) = g(f (x)). That is, g(f (x)) is the notation for the composition applied to number x; note that when facing such a composition, we evaluate from the inside out, that is, we start with x and work our way out by performing necessary operations. After all, this is not surprising, when somebody asks you to calculate (3 + 2)^2, you start by adding 2 and 3, not by squaring (since you do not know what to square unless you add first). Naturally, f is called the inner function or inside function, and g is called the outer function or outside function. Since x ¦! g(f (x)) gives a prescription for sending numbers x somewhere, it is in fact another function and deserves a name. One possible notation is g(f). Note that when evaluating functions denoted like this, the inner function is applied first, so for instance if 2 is a number that can be substituted, then g(f)(2) = g(f (2)). To put it another way,

g(f): x ¦! g(f (x)).

This is one notation for composition, but mathematicians often prefer a notation of a different kind, since they like to express operations by some symbol between the two functions involved. Therefore we also use this notation: g Ë% f. Note that this is read right to left! That is, the function on the right is applied first and then the function on the left. Why? So that it fits with the previous notation. One has(g Ë% f)(x) = g(f (x)), note that the order is the same. Fine, ten points for making sense, but still it is pretty confusing for many beginners. Beware and be careful.

Example. Consider the functions f (x) = x + 2 and g(x) = x^2. How much is (g Ë% f)(1)?

(g Ë% f)(1) = g(f (1)) = g(1 + 2) = g(3) = 3^2 = 9.

How does it work? The number x = 1 is substituted into f and gives y = f (1) = 3. Since g acts on the values of f denoted by y, it may be instructive to rewrite g as g(y) = y^2. Substituting for y we now getz = g(3) = 9.

It is easy to check that (g Ë% f)(x) = (x + 2)^2 for any x. Note that we used functions whose domains are the whole real line to avid any troubles with allowable arguments.

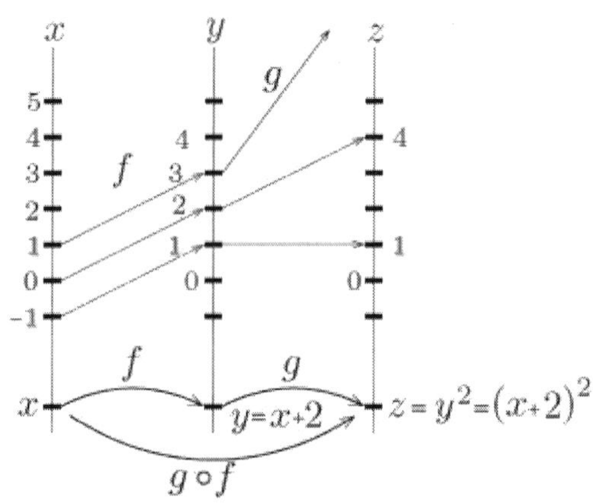

Since there is no trouble with domains, we may also try to change the order in which we compose these two functions. Since variable names are just for our convenience, we should not get troubled by the fact that f "expects" x, but the outcomes of g were called z. We simply have a new situation, so we can use different letters for variables to reflect this. This is in fact typical for mathematics. However, there are situations where variable names have a specific meaning, then it is better to stick to the given ones (for instance when working with inverse functions, in applications etc.)

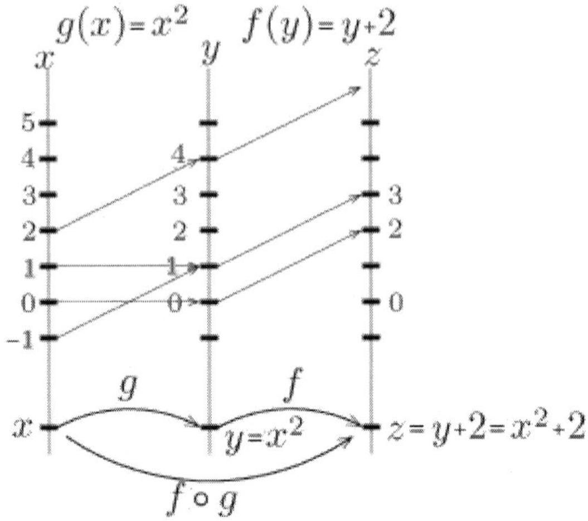

As you can see, (f Ë% g)(x) = f (x²) = x² + 2. This shows that composition of functions is definitely not commutative, that is, f Ë% g need not be the same as g Ë% f (and typically it is not).

Now we know how it works, so for the sake of completeness we include a formal definition.

Definition.

Let f and g be functions such that R(f) is not disjoint with D(g). We define the composition of f and g as a function defined for x from the set

D(g Ë% f) = {x″D(f); f (x)″D(g)}

by

(g Ë% f)(x) = g(f (x)).

Of course, having more functions, one can compose them like a chain assuming that their ranges and domains fit:

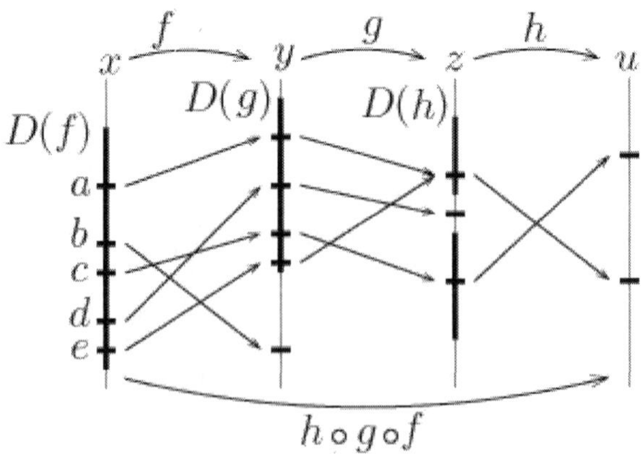

We marked the domains of respective functions. Of the five indicated values of x, only a, c, and e belong to the domain of the composition h Ë% g Ë% f. Indeed, the number b gets sent by f to some number f (b)which is not in the domain of g, so this function cannot be applied to it. The number d was initially fine, since f (d) belongs to D(g); thus d would belong to the domain of the composition g Ë% f. However, when g is applied to f (d), the value thus obtained is not in the domain of h and thus the last step in the composition fails.

To show why this domain fitting is important, consider this example:

Example.

Consider f (x) = "1 " x^2 and g(x) = $x^{1/2}$ (that is, g is the square root). We see that the domain of f is the whole real line (no problem with squaring), its range is the interval (","1]. The domain of g is [0,"),since we cannot take the square root of a negative number. Since the range of f does not intersect the domain of g, we cannot compose them in the order "f first, then g", that is, g Ë% f. Indeed, no matter what number x we try, substituting into (g Ë% f)(x) = ("1 " $x^2)^{1/2}$ we get into trouble. For instance, (g Ë% f)(2) should be the square root of "5, which does not exist (in the realm of real numbers).

We can, however, compose without problem in the other way, since the range of g (non-negative numbers) is a subset of the domain of f. We can calculate (f Ë% g)(x) = f ($x^{1/2}$) = "1 " ($x^{1/2})^2$. This new function has domain [0,") because of the root. Algebraically, this formula can be simplified as "1 " x, but one still has to keep in mind that as a result of composition, this is only considered on the interval [0,"). Note that if we were given this formula as an independent function, not influenced by any background or past calculations, we would determine the domain of "1 " x as the whole real line.

Composition behaves nicely with respect to the 1-1 property. If f and g are both 1-1, then their composition (in any order that makes sense) is also 1-1 (we will make a formal statement below). If one or both functions are not 1-1, then nothing can be said about their composition.

Recall that the 1-1 property is closely related to the existence of an inverse function. Now that we have the notion of composition, we can rewrite the definition of inverse function. Given a 1-1 function f, its inverse $f_{"1}$ is characterized by these equalities:

- $f_{"1}$(f (x)) = x for all x from D(f),
- f ($f_{"1}$(y)) = y for all y from R(f).

In the language of composition, an inverse function is a function that satisfies $f_{"1}$ Ë% f = $Id_{D(f)}$ and f Ë% $f_{"1}$ = $Id_{R(f)}$, where Id_M is the identity function on a set M, for further info see Identity function in Theory - Elementary functions.

Theorem (composition and inverse).

Let f and g be functions that are 1-1 and such that the composition g Ë% f makes sense. Then the function g Ë% f is also 1-1 and its inverse satisfies

(g Ë% f $)_{"1}$ = $f_{"1}$ Ë% $g_{"1}$.

Note that inverse reverses the order in composition, this is important as we know that composition is definitely not commutative. The part about "composition making sense" means that the range of fintersects the domain of g so that their composition is defined for at least some values of x. Since the domain and range of an inverse function are the range and domain of the original function, also the composition of the inverses will make sense, that is, the range of $g_{"1}$ will intersect the domain of $f_{"1}$.

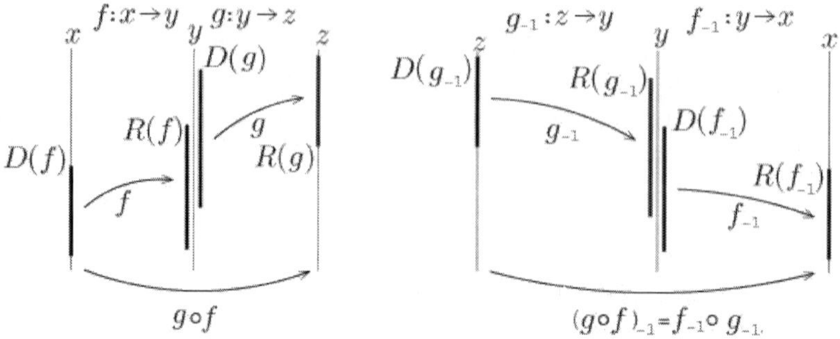

If we tried to compose the inverses in the other way, this is no longer true. In fact, this is an example where keeping the names of variables is rather helpful. The original composed function should send x's to z's. Thus the inverse should send z's to x's. It does work when we did it the right way above, but fails when we try it the other way; now the variables do not even match, do not form a chain, which should suggest that something is wrong here.

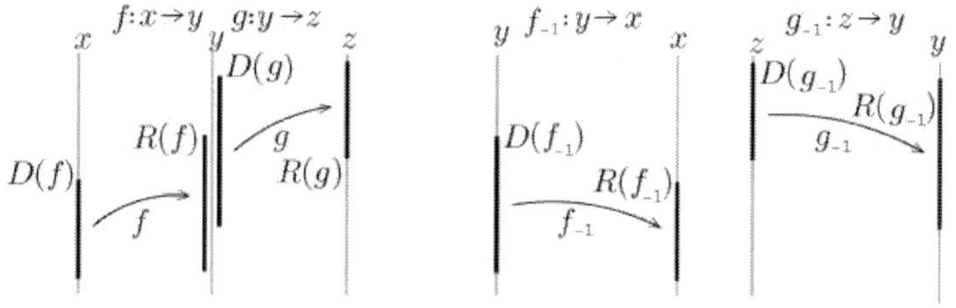

Transformations

Operations with functions have a very interesting special case: When one of the functions is a constant function. The other function then gets transformed. Therefore consider a function f (x) and a real number c.

Case 1: Define g(x) = f (x) + c, or g = f + c for short. The function g has all its values shifted by the same amount, so its shape stays the same. The graph of g can be obtained by shifting the graph of f up by cif c is positive or down by |c| if c is negative. In the following picture, g = f + 3, therefore the graph of g is obtained by shifting the graph of f up by 3.

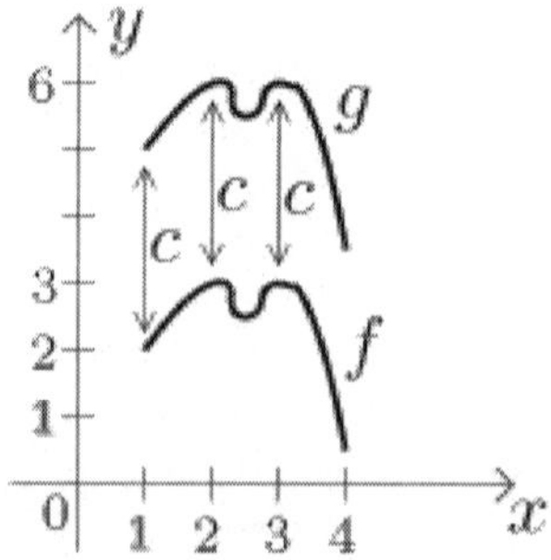

Case 2: Define g(x) = f (x + c). The function g has argument increased always by the same amount. If we think of x as time and imagine that c is positive, then g at time x is obtained by looking at f at a time somewhat to the future, which means to the right on

the x-axis. In other words, we obtain the value of g by looking at the graph of f a bit to the right; thus the graph of g will be just like the graph of f, but shifted to the left. Similarly, for negative c we will be looking at f at earlier times. Thus in general for case 2, the shape stays the same, the graph of g can be obtained by shifting the graph of f left by c if c is positive or right by |c| if c is negative. In the following picture we will show the graph of g(x) = f (x ″ 5), so it will be just the graph of f shifted right by 5.

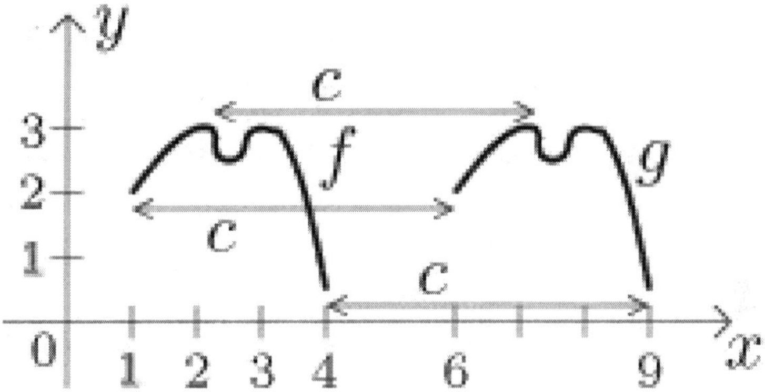

Case 3: Define g(x) = cÅ″f (x), or g = cÅ″f for short. The function g has all its values multiplied by the same amount. If c > 1, then at places where g is large, it gets much larger, while at places where g is small, it gets larger, but only by a little bit. For instance, if c = 2, then twice a large number is a big change, while twice almost zero is still almost zero. In the following picture one has g(4) = 2Å″f (4) = 2Å″0 = 0, which means no change at all, whereas g(2) = 2Å″f (2) = 2Å″3 = 6, a large change.

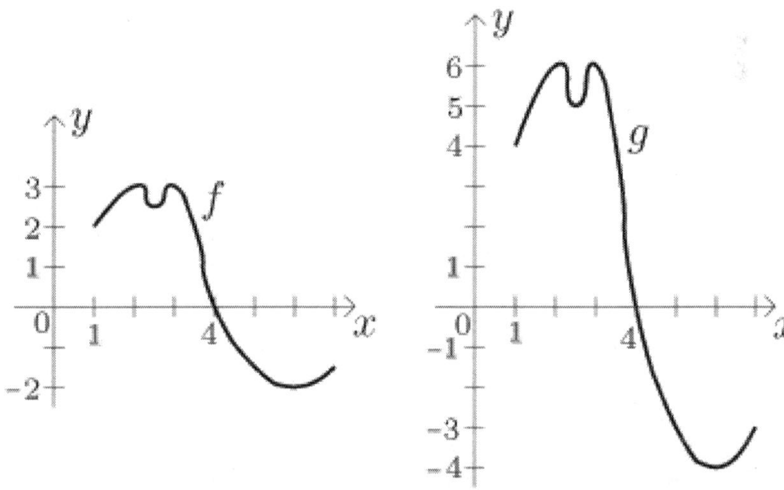

In words, the bigger are the values of f - that is, the further is the graph of f from the x-axis - the bigger is the change when going to g = 2Å″f. From this one would guess that the graph of g can be obtained by stretching proportionally with respect to the x-axis (that is, by changing the amplitude). If c > 1, values get larger and the graph expands, if 0 < c < 1, values get smaller and the graph shrinks. If c is negative, then there is again the expanding/shrinking alternative depending how |c|

compares to 1, but this time the graph also flips around the x-axis. In the following example we first recall the function f as above, then show the graph of g = 2Å"f and then the graph of g = Š1/2Å"f.

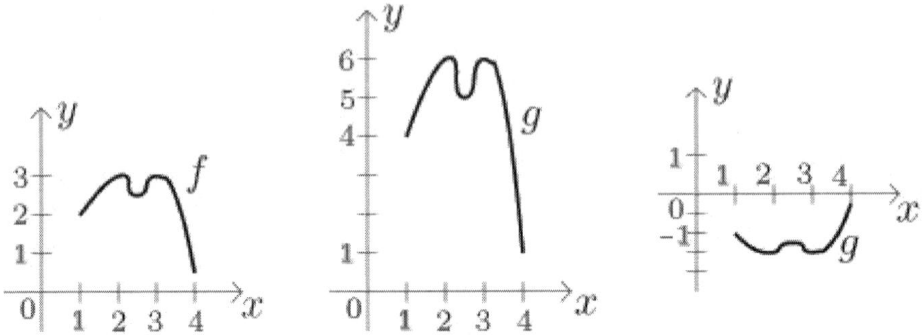

Special case: g = df has graph equal to the graph of f flipped symmetrically about the x-axis.

Case 4: Define g(x) = f (cÅ"x). To see the effect of this transformation, imagine that you drive on a road and f is the height of the hills along the road. What does it mean when we consider g(x) = f (3Å"x)? To seeg at time x, we look at f at time three times bigger. That is, this transformation corresponds to making the time go faster, and when we go faster, the hills seem to come more quickly, as if they were squeezed closer together. Thus the graph of this g will be the same as the graph of f but squeezed three times in the horizontal direction with respect to the origin. Just like the previous case, the further one gets from the y-axis, the larger the change. If 0 < c < 1, the time slows down and the graph stretches to go slower. In the following picture we show g(x) = f (x/2).

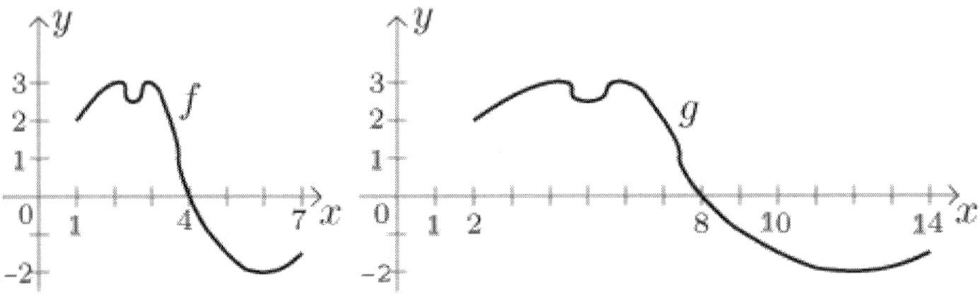

For instance, g(10) = f (10/2) = f (5) = "3/2.

It sort of makes sense, before it took us 4 hours to get past the 2-hump hill, with g it takes twice as long.

Therefore, in general, the graph of the function g can be obtained by stretching the graph of f c times in horizontal direction, with respect to the y-axis, in case c > 1, and shrinking the graph of f 1/c times in horizontal direction, with respect to the y-axis, in case 0 < c < 1. If c is negative, the graph is again transformed horizontally depending on whether |c| is larger or smaller then 1, but now the graph is also flipped around the y-axis. In the following picture we recall f as above, then show the graphs of g = f (2Å"x) and g = f (("1/2)Å"x) (compare to the last example).

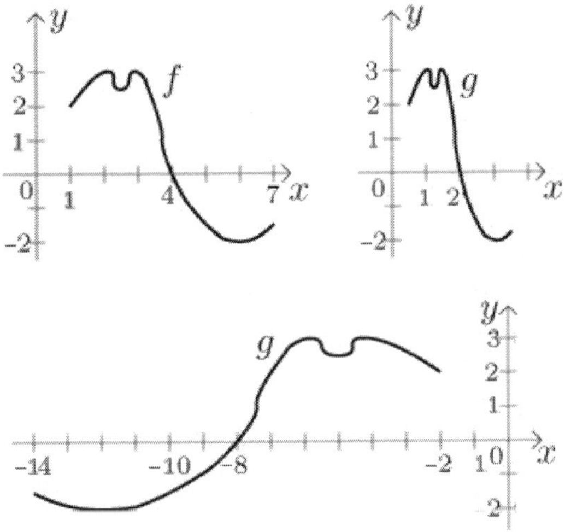

Special case: g(x) = f (“x) has graph equal to the graph of f flipped symmetrically about the y-axis.

One can guess graphs of quite complicated functions by knowing the graphs of elementary functions and applying transformations to them. See Transformations and graph guessing in Methods Survey.

In general, it is almost impossible to guess the shape of the graph of a function created from two functions using algebraic operations. Sometimes we get lucky, the easiest case is f + g. At every relevant x, the value of g is added to (put on the top of) the value of f. One can imagine the values of g as bars of appropriate height that are moved up and down so that their bases fit with the shape of the graph of f.

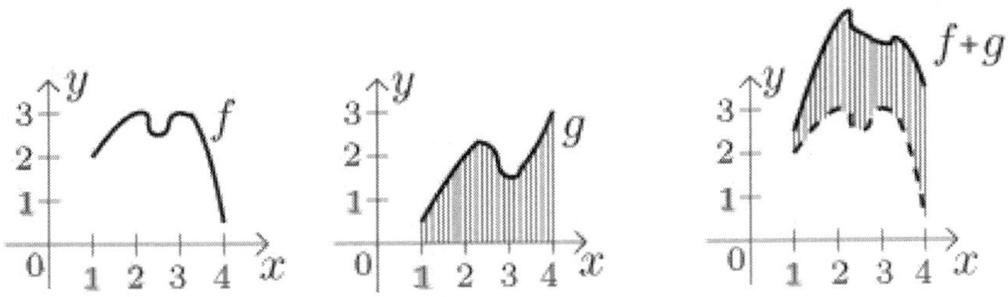

However, this requires quite a bit of experience and imagination and so it is rarely used.

SHORT NOTATION, COMPARISON

We already saw an example of a shortened notation, g = f + c meant g(x) = f (x) + c for all x. In general, when we write an equality or inequality featuring function(s), but without specifying their argument(s), it is automatically assumed that this equality/inequality is considered for all possible values of the argument. Thus h = f + g means h(x) = f (x) + g(x) for all x from the intersection of the domains of the functions f and g.

Similarly, f d” g means f (x) d” g(x) for all x from the intersection of the domains of f and g. This is how we compare functions. There are five possibilities. One can have f d" g as explained above, or f < gmeaning f (x) < g(x) for all x from the intersection of domains; in the other direction there are inequalities f e" g or f > g; finally, it can happen that no such comparison is possible, in other words, for some x the function f is larger, and for other x the function g is larger.

Uniform Continuity

When we defined continuity, it was a local notion. Even when defining continuity on a set, we did it by checking continuity at every point of the set individually. There is also a purely global notion that strengthens continuity.

Definition.

Let f be a function, let M be a subset of its domain D(f). We say that f is uniformly continuous on M if for every å > 0 there exists ä > 0 such that for all a,x from M satisfying | x " a | < ä we have | f (x) " f (a) | < å.

This is almost the same as the definition of continuity we saw, there is one crucial difference though. If we wanted to show ordinary continuity on M, we would start by fixing a point a from M and then we would show that we can win every epsilon-delta game. That is, we play the game independently for every point, which means that for the same given epsilon we have the freedom of choosing different deltas for different points. For uniform continuity we play the game simultaneously for all points from M. Given one particular tolerance, we want to find delta so that it wins this epsilon-game at all points a from M.

Example: Consider the function f (x) = 1/x. We know that it is continuous on (0,"). However, it is not uniformly continuous on this interval. Indeed, in the picture we tried to show that given a certain epsilon, we are forced to make delta smaller and smaller to win games for this epsilon as we move closer to 0 with a and it seems that this process does not stop, no delta will work universally.

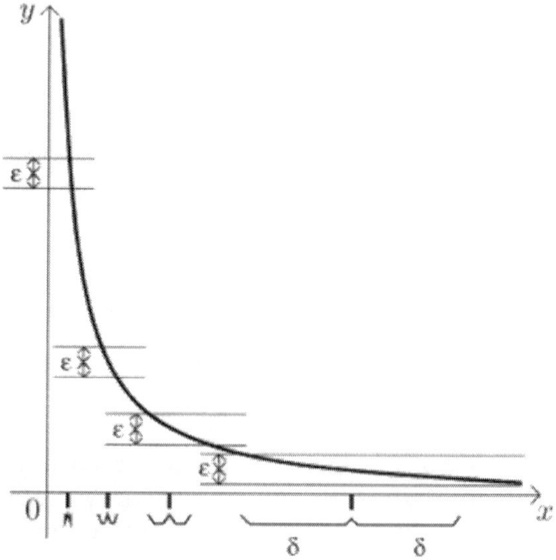

On the other hand, if we restrict ourselves just to the interval [1,"), this function is already uniformly continuous there. The delta we find "at the left end" will work everywhere.

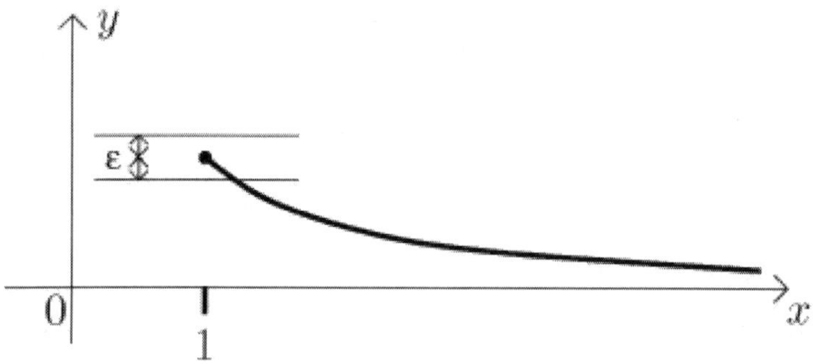

We can look at it from another point of view. If we decide to make steps of a certain length, then in the second situation, no matter where we make such a step, the function always grows or falls by an amount that does not exceed certain universal quantity. In other words, for a uniformly continuous function, given the size of a step, we can make a global upper estimate of how much the function changes, no matter where we make that step in the given set. On the other hand, in the first picture we see that if we start making even very small steps closer and closer to 0, the function will jump by more and more, without any bound for such jump.

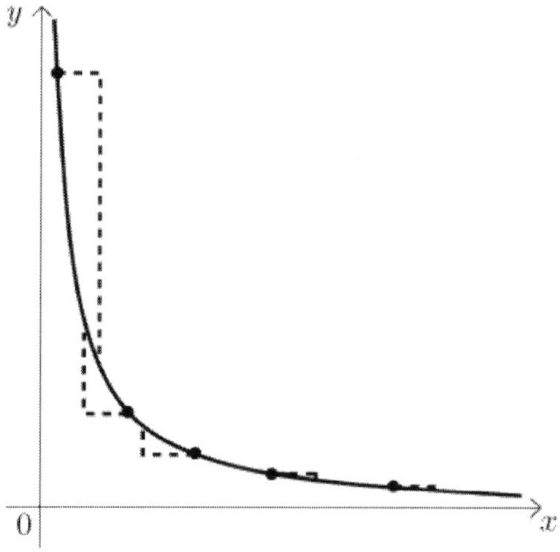

In fact, the point a stops being special, we deal with pairs of points. Thus we can restate the definition in this way. A function is uniformly continuous on a given set if, given a bound epsilon, we can find a delta so that any two points from the set that are near enough (closer than delta) when substituted into the function will not cause a jump larger than the given epsilon.

As the name and the above paragraphs suggest, uniform continuity is more than continuity.

Theorem.

A uniformly continuous function on a set is necessarily continuous on that set.

On the other hand, the above exampe shows that not every continuous function is uniformly continuous. Even boundedness would not help, the function sin(1/x) is bounded and continuous on the bounded interval (0,1), yet it is not uniformly continuous, since it has infinitely many waves of height 2 squeezed next to the origin, which means that with even very (arbitrarily) tiny steps one can go from the top of one wave to the bottom of another in Theory - Elementary function).

There are several ways to recognize uniformly continuous functions without hunting for deltas.

Theorem.

A function that is continuous on a bounded closed set is necessarily uniformly continuous on that set.

Note that both conditions on the set are needed. The function 1/x is not uniformly contiuous even on the bounded set (0,1). For an example on an unbounded closed set we may consider x^2 on [0,"), uniform continuity fails in a similar way as in the above example, just there we had problems around 0, now with x^2 the speed of growth keeps increasing as we go to infinity.

Theorem.

If a function is continuous on an interval, differentiable in its interior and the derivative is bounded there, then the function is uniformly continuous on that interval.

Note that the condition on boundedness of derivative is not necessary, because there are uniformly continuous functions whose derivative in not bounded on the set of uniform continuity. One such example is the root of x on [0,1].

Uniformly continuous functions are nicer to work with. However, this is somewhat beyond the scope of Math Tutor, so we leave it here.

Monotonicity at a point, local extrema

Given a function f and a point a from its domain, one can look at f around a and compare values with f (a). There are essentially four useful cases.

Definition (local monotonicity).

Let a belong to the domain of a function f.

We say that f is increasing at a if there is a neighborhood U of a such that for all x from U)" D(f) the following is true:

- If x > a, then f (x) > f (a),
- if x < a, then f (x) < f (a).

We say that f is non-decreasing at a if there is a neighborhood U of a such that for all x from U)" D(f) the following is true:

- If x > a, then f (x) e" f (a),
- if x < a, then f (x) d" f (a).

We say that f is decreasing at a if there is a neighborhood U of a such that for all x from U)" D(f) the following is true:

- If x > a, then f (x) < f (a),
- if x < a, then f (x) > f (a).

We say that f is non-increasing at a if there is a neighborhood U of a such that for all x from U)" D(f) the following is true:

- If x > a, then f (x) d" f (a),
- if x < a, then f (x) e" f (a).

What is the meaning of these definitions? We look at the first one. First of all, the definition is local. It starts by saying that there is a neighborhood of a on which some condition holds, no information outside this neighborhood is relevant. There is also no guarantee concerning the size of this neighborhood, it might be extremely small, which may sometimes cause problems. So now we will look at the neighborhood and see what is happening there. We start at a, where the function has value f (a).

When we look at f to the right from a, the function gets larger. When we look to the left, the function gets smaller. It should then seem natural that f is called increasing at a.

The second case is similar, we just also allow the function to be equal to f (a) on U. The next two cases are exactly the opposite tendency for f. Here are typical examples of these notions:

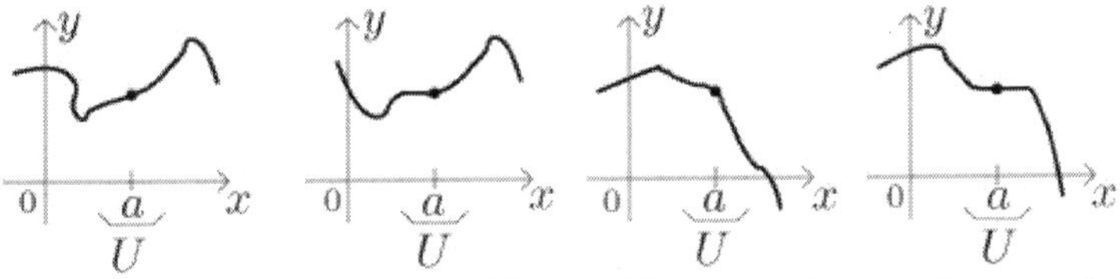

Note that increasing automatically implies non-decreasing, and decreasing automatically implies non-increasing. Also, increasing and decreasing cannot be true at the same time. On the other hand, it is possible to have a function that is simultaneously non-decreasing and non-increasing at some point. A function that is constant on a neighborhood of some point satisfies both these properties there.

Note that we cannot expect any tendency or nice properties at other points than a, not even in the neightborhood U. Here is a relatively tame example of a function non-decreasing at a,

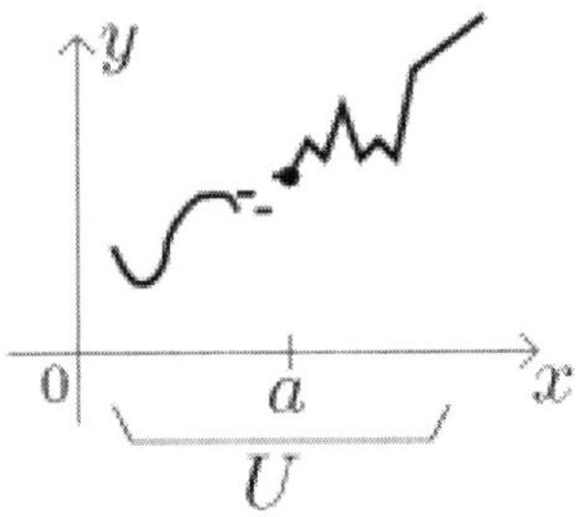

but things can get much worse. In "Saw-like" functions in Theory - Elementary functions we show an example of a function that is increasing at a certain point but there is no neighborhood of this point on which this function would be monotone or continuous.

So the best one can say based on local monotonicity is that on some neighborhood of a, the graph of the function must stay in appropriate "quadrants centered at (a, f (a))" as suggested in the picture, on the left is the idea of increasing and non-decreasing, on the right the idea of decreasing and non-increasing.

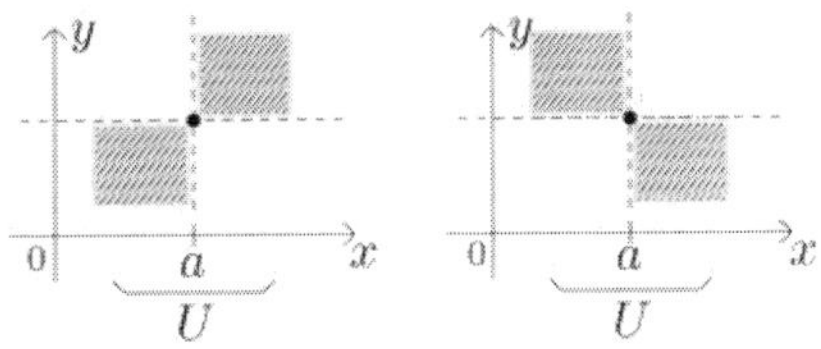

Important note.

Unfortunately, this terminology is not generally accepted. Some people would not use the names increasing, non-decreasing, decreasing, non-increasing, but instead they would say (in the same order) strictly increasing, increasing, strictly decreasing, decreasing. Both terminologies have about the same number of followers and equally persuasive arguments, so it is hard to take sides. Generally, people pass on what they learned from their teachers, which is what I am doing here, I also think that single-word names are a bit better and the names we defined here are more fitting :-). Usually this is not a problem, but notice that people who use this other notation would have to also make changes in some theorems. That's where it gets confusing a bit, but fortunately it does not happen too often. The usual advice: Check what terminology your prof is using and stick to it.

Another note.

It is also possible to ask about one-sided local monotony. For instance, to be increasing at a from the right, one would have to find a right reduced neighborhood U of a such that f (x) > f (a) for all x from that right reduced neighborhood; recall that a right reduced neighborhood is an interval of the form (a,b) for some b > a. Similarly one defines other notions from the right and from the left.

Local extrema

Recall the picture with quadrants above. It can also happen that when we restrict the function to a suitable neighborhood, its graph stays in certain quadrants, but they are not of the form (when going from left to right) "below-above" or "above-below" as we saw above. What are the other alternatives? There are two, "below-below" and "above-above".

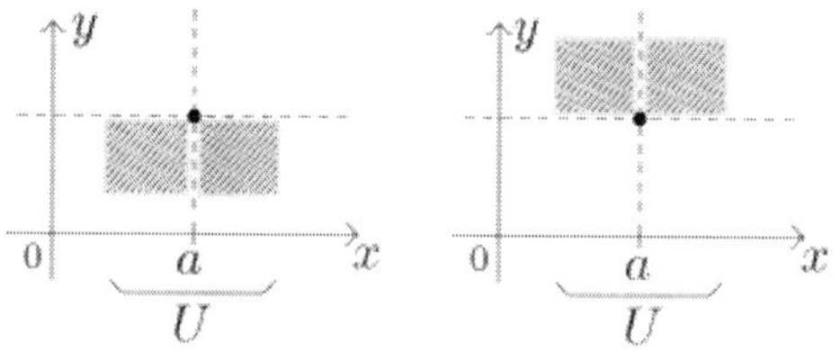

Typical examples (when the function is very nice) are

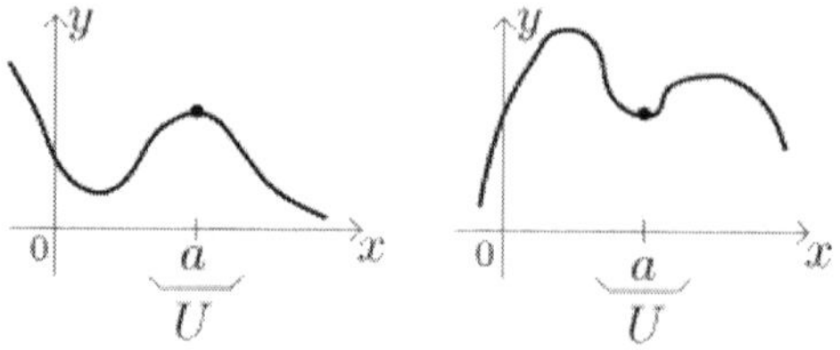

Since the "hump" and the "dip" show that the function is a bit extreme at the point a, the name "extreme" seems in place, and since we only have information about a small neighborhood, it is local.

Definition (local extrema).

Let f be a function defined on some neighborhood of a point a.

We say that f has a local maximum at a, or that f (a) is a local maximum of f, if there is a neighborhood U of a such that for all x from U)" D(f) the following is true: f (x) d" f (a).

We say that f has a local minimum at a, or that f (a) is a local minimum of f, if there is a neighborhood U of a such that for all x from U)" D(f) the following is true: f (x) e" f (a).

The name local extreme is a general handle for a local maximum or a local minimum. Some people also distinguish sharp local extrema, which basically means that they exclude the constant situation. Definition of a sharp local maximum is just like local maximum above, only the inequality is taken sharp; similarly for sharp local minimum.

Again, this is a local notion, so having a local extreme, we have no guarantee that the function behaves reasonably around that point, not even in the neighborhood U.

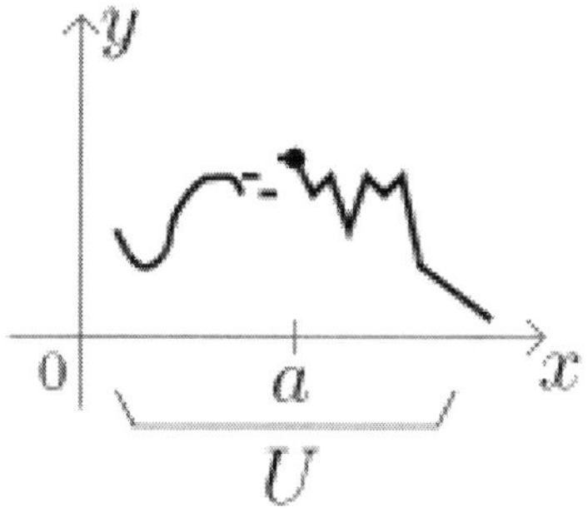

Can a point be at the same time a local minimum and a local maximum? Sounds strange but yes, if the function is constant on a neighborhood of this point.

Global monotonicity

Here we will look at the function as a whole, on some subset M of its domain. Typically, M would be an interval (including an infinite one) or a union of intervals.

Definition (global monotonicity).

Let M be a subset of the domain of a function f.

We say that f is increasing on M if for all x,y from M satisfying x < y we have f (x) < f (y).

We say that f is non-decreasing on M if for all x,y from M satisfying x < y we have f (x) d″ f (y).

We say that f is decreasing on M if for all x,y from M satisfying x < y we have f (x) > f (y).

We say that f is non-increasing on M if for all x,y from M satisfying x < y we have f (x) e″ f (y).

We say that f is monotone on M if it satisfies one of the above four conditions.

We say that f is strictly monotone on M if it is increasing on M or decreasing on M.

These conditions are actually quite simple. For increasing functions, when we move right, the function must go up. For decreasing functions it must go down. The "non" versions are similar, but we also allow the function to stay the same for a while.

This notion does not really make too much sense unless the set M is reasonable, typically a non-degenerate interval or a union of such intervals. Here are three typical examples of monotone functions on an interval; the first function is increasing, the second is non-decreasing, and the third is decreasing. The fourth function is not monotone on the indicated interval.

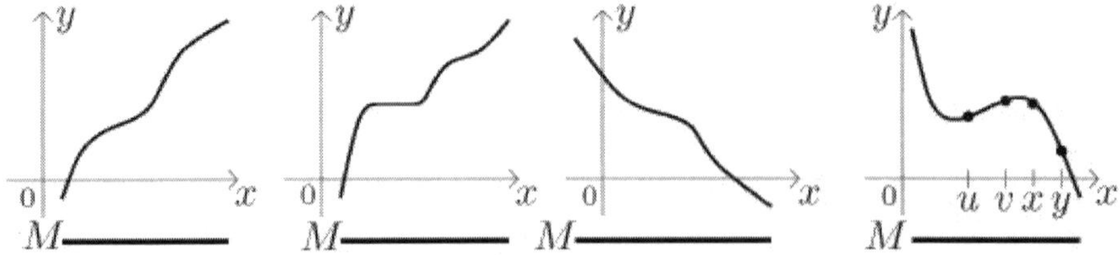

Recall that the conditions in the above definition are of the form "for every couple x, y..." and in order to violate such a condition, it is enough to find one counter-example, that is, a pair of points that do not satisfy the relevant condition. In the fourth example we indicated a pair x, y which is a counter-example to that function being increasing or non-decreasing; indeed, when going from x to y, the function does decrease and thus violates the two relevant inequalities. We also marked a pair u,v which can serve as a counter-example to the function being decreasing or non-increasing, because when going from u to v, the function does increase.

Note again that increasing automatically implies non-decreasing, and decreasing automatically implies non-increasing. Increasing and decreasing are mutually exclusive notions. However, there are functions that are simultaneously non-increasing and non-decreasing on M - namely all constant functions on M.

Important note.

Just like in the local case, this terminology is not generally accepted. Some people would not use the names increasing, non-decreasing, decreasing, non-increasing, but instead they would say (in the same order) strictly increasing, increasing, strictly decreasing, decreasing. The good news is that these people use monotone and strictly monotone just like we do, so if these notions are used in a theorem, it works for both terminologies.

Connecting intervals of monotonicity

Monotonicity on an interval is easiest to investigate and we have good tools for it. Is it possible to put information about individual intervals together? Consider the following situation. We have a function fand two non-degenerate intervals, I and J. For simplicity, assume that I is to the left of J, that is, max(I) < min(J). Assume also that f is monotone on each of these intervals.

Can we say something about monotonicity of f on their union?

Depends on what monotonicity we have. Clearly, if f is increasing on one interval and decreasing on another interval, then it cannot be monotone on the set formed by their union (draw a picture). The only interesting case is when we have the same kind of monotonicity on both intervals. We will now look at the following situation: Imagine that we know that f is increasing on I and increasing on J. Is there any monotonicity on the set obtained by their union? This is a very good question, in particular because students find it very boring to write two monotonicities as we wrote them and they tend to write it like this: "f is increasing on I *" J". Unfortunately, very often this is wrong and students lose points for this. Here's one such example, in the picture on the right we show two points that are a counter-example to f being increasing on the union as one set.

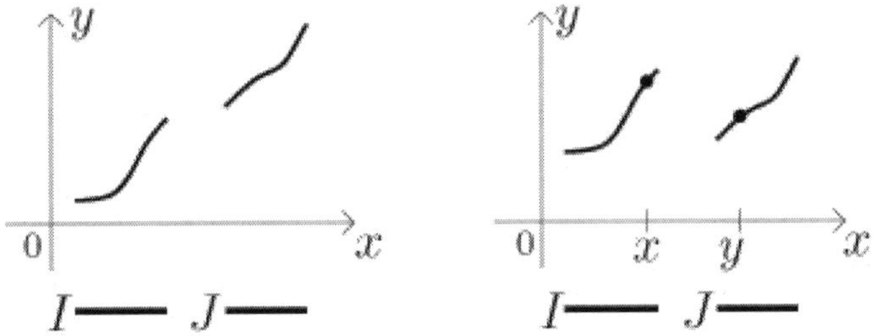

We are getting to the first important point of this part: If you write it as we did before, "f is increasing on I and increasing on J", it will be always correct. If it seems too long, try "f is increasing on I and on J". Unless you are forced to look closer, leave it this way.

But what if we really want to know more, for instance if we are asked for "maximal intervals of monotonicity"? First, there is one simple rule: If the two intervals intersect, then we do get monotonicity on their union, since the intersection serves as a nice bridge from one part of the graph to another. What if they are disjoint as on the picture above? The picture suggests what should be done. We should somehow compare values on the adjacent ends of intervals and make sure that they do not jump the wrong way. We get the following rules:

Let an interval I be to the left from an interval J, in particular they are disjoint.

1. If f is non-decreasing on I and non-decreasing on J, and $\sup_I$(f) d″ $\inf_J$(f), then f is non-decreasing on I *″ J.

2. If f is increasing on I and increasing on J, we have $\sup_I$(f) d″ $\inf_J$(f) but $\max_I$(f)=$\min_J$(f) is not true, then f is increasing on I *″ J.

3. If f is non-increasing on I and non-increasing on J, and $\inf_I$(f) e″ $\sup_J$(f), then f is non-increasing on I *″ J.

4. If f is decreasing on I and decreasing on J, we have $\inf_I$(f) e″ $\sup_J$(f) but $\min_I$(f)=$\max_J$(f) is not true, then f is decreasing on I *″ J

Draw a few pictures to convince yourself that these statements make sense. They are a bit complicated, since we have to take into account both open and closed situations, but in fact it is all just a common sense, in practice it is much easier than the precise (and a bit undigestible) mathematical statements here.

Local and global: relationship.

If the set M is just some general subset of the domain of the given function f, there is little hope of relating local and global monotonicity. However, once we pass to one interval, things become nice.

Theorem.

If f is a function monotone on an interval, then it is monotone (in the same way) at all interior points of this interval.

If this interval includes some of its endpoints, then at such points we have one-sided monotonicity, but this notion is rarely used, so we did not put it in the theorem. Now we try it the other way.

Theorem.

If f is a function monotone in the same way at every point of an open interval, then it is monotone (in the same way) on this interval.

Why "open"? Because the opposite of the previous theorem (where we allowed just any interval) would not work. Indeed, the function in the following picture is increasing at all interior points of the indicated interval, but it is not increasing on that interval (try going from a to any other point of that interval and the function will drop).

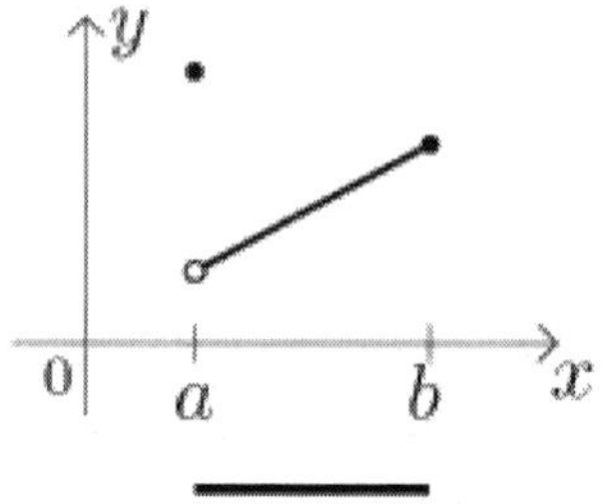

We can fix this problem by requiring one-sided monotonicity at endpoints, but this notion is rather obscure and most people never heard of it. Therefore the preference goes to a slightly weaker but way more practical statement that uses continuity:

Theorem.

If f is a function that is monotone in the same way at every interior point of an interval, and continuous on the whole interval, then it is monotone (in the same way) on this interval.

Note that continuity on an interval including endpoints actually does include a one-sided notion, but continuity from the right/left is a standard notion, so this is no problem.

Now we will look at how monotonicity helps with other properties.

Theorem.

If a function is monotone on an interval, then all its one-sided limits at interior points of that interval converge. Also appropriate one-sided limits at endpoints exist, regardless of whether these endpoints belong to the interval.

So the limits at the endpoints exist, but they might be infinite.

Theorem.

If a function is strictly monotone on a set, then it must be necessarily 1-1 there and the corresponding inverse is also monotone (in the same way).

If we also have continuity and work on an interval, then this implication becomes an equivalence.

Theorem.

Let f be a function continuous on an interval I. Then f is 1-1 on I if and only if it is strictly monotone on I. If it is so, then the corresponding inverse function is also continuous and monotone on its domain.

All these notions - local and global monotonicity, local extrema - can be investigated very conveniently using derivatives, which is explained in the section Monotonicity in Derivative - Theory - Graphing functions.

Global extrema

We know that when we consider a function f on a set M, we can find its supremum and infimum over this set, and if we are lucky, we can also find its maximum and minimum over this set. Maximum a minimum are global extrema over that set. While we had a close relationship between local and global monotonicity, with extrema it is a bit more complicated. For instance, in this example we have a local minimum and a local maximum, but there is no global maximum and the global minimum is not local minimum.

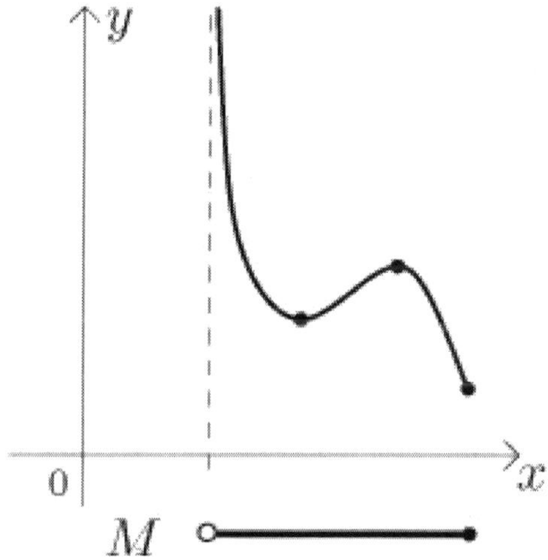

In fact, there is some interplay between local and global extremes.

Theorem.

Let f be a defined on a bounded closed interval I. If f attains its global extreme at a point c, then either f has a local extreme at c or c is an endpoint of I.

However, this still does not imply that there actually are some global extrema. Is there any way to guaarantee it? We do have one useful theorem.

Theorem (Extreme Value Theorem, EVT).

Every function continuous on a bounded closed set attains its maximum and minimum over this set.

We now look closer at this theorem. First of all, the theorem is an implication, we may have global extrema even in other cases. In the following picture we have a function that is not continuous, but it does have global extrema over a set that is not bounded nor closed. Incidentally, the picture also shows that we can have more global extrema, they need not be unique.

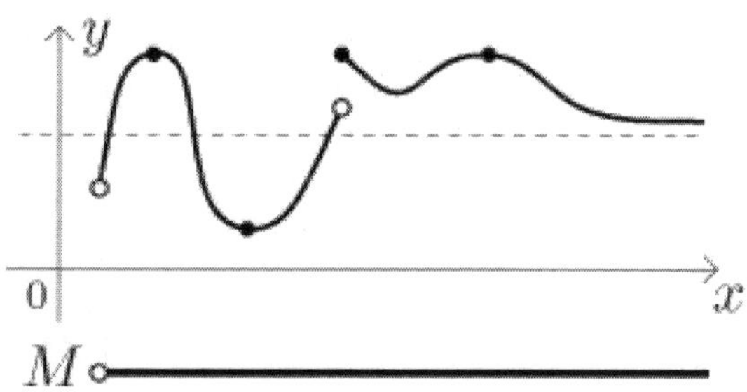

The theorem lists the necessary conditions without which global extrema are no longer guaranteed and having them is a matter if luck as in the previous picture. Now look at the following picture:

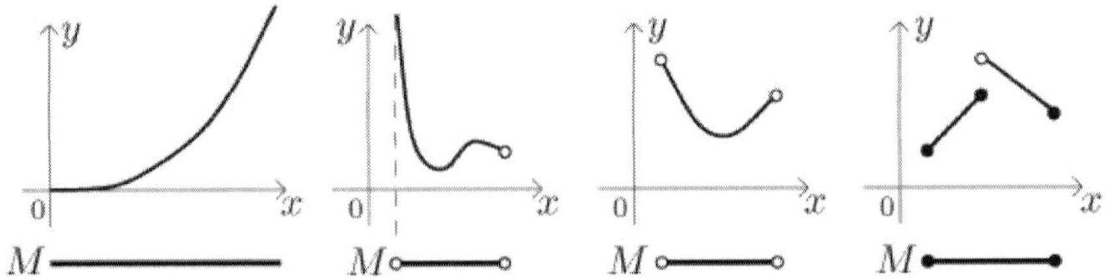

The picture on the left shows that a function that is continuous on a closed interval need not have a maximum, the boundedness is missing sorely. The next three pictures have all bounded intervals, but they have other problems. The two pictures in the middle have open intervals but no maxima, although the the middle right function is actually bounded. This shows that without closedness we also cannot guarantee anything. The picture on the right shows that even with bounded closed intervals we cannot win if we lack continuity. So we need all three properties, just two of them won't be enough.

17

Lagrange's Theorem

LAGRANGE'S THEOREM IS THE FIRST GREAT RESULT OF GROUP THEORY. IT STATES

THEOREM: The order of a subgroup H of group G divides the order of G.

First we need to define the order of a group or subgroup

Definition: If G is a finite group (or subgroup) then the order of G is the number of elements of G.

Lagrange's Theorem simply states that the number of elements in any subgroup of a finite group must divide evenly into the number of elements in the group. Note that the {A, B} subgroup of the Atayun-HOOT! group has 2 elements while the Atayun-HOOT! group has 4 members. Also we can recall that the subgroups of S3, the permutation group on 3 objects, that we found cosets of in the previous chapter had either 2 or 3 elements — 2 and 3 divide evenly into 6.

A consequence of Lagrange's Theorem would be, for example, that a group with 45 elements couldn't have a subgroup of 8 elements since 8 does not divide 45. It could have subgroups with 3, 5, 9, or 15 elements since these numbers are all divisors of 45.

Now that we know what Lagrange's Theorem says let's prove it. We'll prove it by extablishing that the cosets of a subgroup are

- disjoint — different cosets have no member in common, and
- each have the same number of members as the subgroup.

This leads to the conclusion that a subset with n elements has k cosets each with n elements. These cosets do not overlap and together they contain every element in the group. This means that the group must have kn elements — a multiple of n. We'll accomplish our proof with two lemmas.

Lemma: If H is a finite subgroup of a group G and H contains n elements then any right coset of H contains n elements.

Proof: For any element x of G, Hx = {h • x | h is in H} defines a right coset of H. By the cancellation law each h in H will give a different product when multiplied on the left onto x. Thus each element of Hwill create a corresponding unique element of Hx. Thus Hx will have the same number of elements as H.

Lemma: Two right cosets of a subgroup H of a group G are either identical or disjoint.

Proof: Suppose Hx and Hy have an element in common. Then for some elements h_1 and h_2 of H

$h_1 \bullet x = h_2 \bullet y$

This implies that $x = h_1^{-1} \bullet h_2 \bullet y$. Since H is closed this means there is some element h_3 (which equals $h_1^{-1} \bullet h_2$) of H such that $x = h_3 \bullet y$. This means that every element of Hx can be written as an element of Hy by the correspondence

$h \bullet x = (h \bullet h_3) \bullet y$

for every h in H. We have shown that if Hx and Hy have a single element in common then every element of Hx is in Hy. By a symmetrical argument it follows that every element of Hy is in Hx and therefore the "two" cosets must be the same coset.

Since every element g of G is in some coset (namely it's in Hg since e, the identity element is in H) the elements of G can be distributed among H and its right cosets without duplication. If k is the number of right cosets and n is the number of elements in each coset then $|G| = kn$.

Alternate Proof: In the last chapter we showed that $a \bullet b^{-1}$ being an element of H was equivalent to a and b being in the same right coset of H. We can use this Idea establish Lagrange's Theorem.

Define a relation on G with $a \sim b$ if and only if $a \bullet b^{-1}$ is in H. Lemma: The relation $a \sim b$ is an equivalence relation.

Proof: We need to establish the three properties of an equivalence relation — reflexive, symmetrical and transitive.

(1) Reflexive: Since $a \bullet a^{-1} = e$ and e is in H it follows that for any a in G

$a \sim a$

(2) Symmetrical: If $a \sim b$ then $a \bullet b^{-1}$ is in H. Then the inverse of $a \bullet b^{-1}$ is in H. But the inverse of $a \bullet b^{-1}$ is $b \bullet a^{-1}$ so

$b \sim a$

(3) Transitive: If $a \sim b$ and $b \sim c$ then both $a \bullet b^{-1}$ and $b \bullet c^{-1}$ are in H. Therefore their product $(a \bullet b^{-1}) \bullet (b \bullet c^{-1})$ is in H. But the product is simply $a \bullet c^{-1}$. Thus

$a \sim c$

And we have shown that the relation is an equivalence relation.

It remains to show that the (disjoint) equivalence classes each have as many elements as H.

Lemma: The number of elements in each equivalence class is the same as the number of elements in H.

Proof: For any a in G the elements of the equivalence class containing a are exactly the solutions of the equation

$a \bullet x^{-1} = h$

Where h is any element of H. By the cancellation law each member h of H will give a different solution. Thus the equivalence classes have the same number of elements as H.

One of the imediate results of Lagrange's Theorem is that a group with a prime number of members has no nontrivial subgroups. (why?)

Definition: if H is a subgroup of G then the number of left cosets of H is called the index of H in G and is symbolized by (G:H). From our development of Lagrange's theorem we know that

$|G| = |H|\ (G:H)$

Converse of Lagrange's Theorem One of the most interesting questions in group theory deals with considering the converse of Lagrange's theorem. That is if a number n divides the order of group G does that mean that G must have a subgroup of order n? The answer is no in general but the special cases where it does work out are many and interesting. They are dealt with in detail in the Sylow Theoremswhich we will treat later. As a tidbit we look at the following

Theorem: If the order of a group G is divisible by 2 then G has a subgroup of two elements.

Proof: The proof is left as an exercise for the student. [Hint: If an element other than the identity of a group is its own inverse then that element together with the identity forms a subgroup of two elements (Prove!). The identity is its own inverse. If we remove the identity from a group of even order must at least one of the remaining elements be its own inverse?]

Cyclic Groups and Subgroups

Let's start with the number 1. We'll allow ourselves to add or subtract the number 1 to get to new numbers.

Question: what integers will we be able to reach by this process?

Answer: all of them.

To get to 17 simply add 1 16 times. To get to -42 simply subtract 1 43 times. The fact that the integers can be "built" by adding and subtracting 1 means that the additive group of integers is a cyclic group.

Now let's look at the Attayun-HOOT! group. The entire Attayun-HOOT! group can be "built" from repeated applications of R:

- $R = (R^1) = R$
- $R \bullet R = (R^2) = B$
- $R \bullet R \bullet R = (R^3) = L$
- $R \bullet R \bullet R \bullet R = (R^4) = A$

Groups that can be generated in their entirety from one member are called cyclic groups. For infinite groups we have to clarify what we mean by "generated from." For example in the additive group of Integers starting with 1 and adding it over and over to itself will never get a negative number nor the identity zero. Thus for a cyclic group we have the definition that all the elements may be generated from a single element together with its inverse. For finite cyclic groups the addition of "together with its inverse" is not needed. That statement probably should be stated as a theorem.

Theorem: If a finite group G can be generated from one of its elements, a together with its inverse a^{-1} then it can be generated from a alone.

Proof:

The strategy of the proof is to show that we can generate a^{-1} from repeated applications of a.

If G is finite then the various powers of a cannot all be distinct [note that this would not necessarily follow if G is infinite]. Therefore for some two different positive integers, j and k a • a • a • . . . • a • a(j times) = a • a • a • . . . • a • a(k times)

Assume without loss of generality that j is greater than k (Why can we do this?). Now if we multiply both sides of the equation by a^{-1} • a^{-1} • a^{-1} • . . . • a^{-1} • a^{-1}(k times) then we end up with

a • a • . . . • a(j-k times) = e

That means that a^{-1} = a • a • . . . • a(j-k-1 times).

Let's get away from all the " . . . " and "(j-k times)" by using exponential notation. Just as we do for multiplication of regular old numbers we can express repeated application of the group operation by a given element with exponential notation. So we'll agree that for an element a of a group G with operation •

- a^n means a • a • a • . . . • a (n times)
- a^{-n} means a^{-1} • a^{-1} • a^{-1} • . . . • a^{-1} (n times)
- a^0 means the identityb element e

We find that with this notation the basic law of exponents all work as in regular arithmetic namely

- $a^m \bullet a^n = a^{m+n}$
- The inverse of a^n is a^{-n}
- $(a^m)^n = a^{mn}$

With this notation we mean that a power of a is a^n where n is any integer: positive, negative or zero.

Cyclic Subgroups

If we pick some element a from a group G then we can consider the subset of all elements of G that are powers of a. This subset forms a subgroup of G and is called the cyclic subgroup generated by a. If forms a subgroup since it is

- Closed. If you multiply powers of a you end up with powers of a
- Has the identity. a • $a^{-1} = a^0$ = e
- Has inverses. The inverse of any product of a's is a similar product of a^{-1}'s.

But this is the long way of proving subgrouphood. Let's use our theorem that says if x and y are in the subset implies that x • y^{-1} is in the subset then the subset is a group. This is simple here. If y is a power of a then so is y^{-1} and so, therefore, is x • y^{-1}.

A few facts about cyclic groups and cyclic subgroups:

1. Cyclic groups are Abelian.
2. All groups of prime order are cyclic.
3. The subgroup of a group G generated by a is the intersection of all subgroups of G containing a
4. All infinite cyclic groups look like the additive group of integers.

The second fact is a result of Lagrange's theorem. If G is a group of prime order p then take any non-identity element a and look at the subgroup generated by it. The order of the subgroup must be a divisor of the order of G (Lagrange's Theorem) but since G is of prime p order the only divisors of p are 1 and p itself. Since a isn't the identity the subgroup it generates must have more than one element. Therefore its order must be pand it must be G itself — G is cyclic. Furthermore since the choice of a was only limited by its not being the identity it follows that G can be generated by any of its elements other than e.

The third fact is rather simple. Any subgroup containing a contains a^{-1} and therefore contains any products of a's and a^{-1}. Thus any subgroup containing a contains the cyclic subgroup generated by a. Since the cyclic subgroup generated by a is a subroup the fact follows.

The fourth fact is poorly stated. "looks like" is not a well understood piece of mathematical terminology. We'll tighten things up as we go along.

Suppose I is an infinite cyclic group generated by the element a. Then any element of the group is of the form a^n where n is an integer. Then the multiplication rule for this group ends up being

$[a^m] \bullet [a^n] = a^{m+n}$

We can tighten up the notation of this group by letting the integer n stand for the element a^n. Then the "multiplication" rule for the group becomes

$m \bullet n = (m+n)$

But this is just the behavior of the additive group of integers.

We have a one to one correspondence [a^n corresponds to n] which is preserved by the group operation [You can either perform the operation in one group and then find the corresponding element in the other or you can first find the corresponding elements in the other group and then perform the operation on them. The result is the same. And it works both ways. This is a group isomorphism, a concept we have looked at but haven't rigorously defined yet.

Definition: Two groups G and H are isomorphic if there exists a one-to-one relation $f : G \to H$ between them such that for any x and y in G, $f(x) \bullet f(y) = f(x * y)$. The one-to-one relation f is called an isomorphism. Thus two groups are said to be isomorphic if an isomorphism exists between them.

[Note on my notation I've used two different symbols for the group operations in that last passage. • represents the group operation in H and * represents the group operation in G This was to emphasize that we are dealing with two different groups each with its own group operation.]

[Exercise: show that the inverse relation works too. That is, show that if $g : H \to G$ is the inverse of f then for any u and v in H it is true that $g(u) * g(v) = g(u \bullet v)$.]

One of the most common examples of a group isomorphism is found in the theory of logarithms. Let $f : G \to H$ be the logarithm function and let G be the group of positive real numbers under multiplication and let H be the real numbers under addition. It turns out that

$\text{Log}(x) + \text{Log}(y) = \text{Log}(xy)$

This together with the fact that Log is a one-to-one function (it has an inverse — the exponential or "antilog" function) makes this a group isomorphism. This used to have great practical importance back in the days before pocket calculators. Long numerical calculations involving multiplication (and taking of roots and powers) could be simplified by doing addition of the Logs of the numbers. In those days no scientist or engineer was far from his table of Logarithms.

The following are basic facts about group isomorphism:

If $f: G \to H$ is a group isomorphism then:

- If e is the identity in G then f(e) is the identity in H>.
- The inverse of f(a) in H is $F(a^{-1})$.

Now we can restate our poorly stated fourth fact:

4. All infinite cyclic groups are isomorphic to the additive group of integers.

The isomorphism is simply $a^n \to n$ as we have already seen.

The main point about cyclic groups and isomorphism is that all finite cyclic groups of the same order are isomorphic. Thus there is only one cyclic group (up to isomorphism) of order, say, 6. If we let a be a generator then the elements of the cyclic group of order 6 are a, a^2, a^3, a^4, a^5, and $a^6 = e$.

The canonical example of a cyclic group of order n is the additive group of integers mod n: Z/nZ. The cayley table for Z/6Z is

+	0	1	2	3	4	5
0	0	1	2	3	4	5
1	1	2	3	4	5	0
2	2	3	4	5	0	1
3	3	4	5	0	1	2
4	4	5	0	1	2	3
5	5	0	1	2	3	4

Note how each row (or column) is simply the previous row (or column) cycled forward one space.

The set of integers mod n is not a group under multiplication because 0 has no inverse – there is nothing that multiplies 0 to give 1, the identity element for multiplication. If n is a prime number and we throw out zero the remaining elements of Z/nZ does form a group – a cyclic group. Here's the Cayley table for the non-zero elements of Z (mod 7)

×	1	2	3	4	5	6
1	1	2	3	4	5	6
2	2	4	6	1	3	5
3	3	6	2	5	1	4
4	4	1	5	2	6	3
5	5	3	1	6	4	2
6	6	5	4	3	2	1

The cyclic nature of the group isn't apparent from a glance at the table. However if we consider the powers of 5 then we find that 5 generates the entire group. Similarly 3 also generates the group. By rearranging the elements of the Cayley table the cyclic nature is readily seen.

×	1	5	4	6	2	3
1	1	5	4	6	2	3
5	5	4	6	2	3	1
4	4	6	2	3	1	5
6	6	2	3	1	5	4
2	2	3	1	5	4	6
3	3	1	5	4	6	2

[Exercise: The above table was rearranged by listing the 0th power of 5 first, then the first power of 5 then the 2nd power of 5 etc. Rearrange the table in order of powers of 3 such that its cyclic structure is evident.]

We can form a cyclic subgroup of any group by grabbing an element a of the group an looking at the set of all its powers. This gives the subgroup generated by a notated as $< a >$. Now by Lagrange's theorem the number of elements in $< a >$ (the order of the subgroup) must be a divisor of the order of the group. Also if k is the order of $< a >$ then k is the least positive integer with $a^k = e$. For any element of a group the least positive integer that gives the identity when the element is raised to that power is called the order of the element. Thus the order of a cyclic subgroup and the order of its generator are the same number.

We might remark at this point that any subgroup of a cyclic group must itself be cyclic. If G is generated by a and S is a subgroup of G Then S must be cyclic. To see this consider all the elements of S. They can all be expressed as powers of a. Let k be the least positive power of a that is in S. If every element is not a power of a^k then there is some element a^m where m is positive and not a mulitple of k. Let p be the greatest multiple of k that is less than m. Then m - p is less than k. But this means that $a^m \bullet a^{-p} = a^{m-p}$ is in S. This contradicts the choice of k as the smallest positive power of a in the subgroup. Thus all elements of S are generated by a^k and S is cyclic.

Now if for any element its order divides the order of the group then it follows that if n is the order of the group then $a^n = e$ for any element a of any finite group. [Why must this be so?]

Now we have a constrained converse of Lagrange's theorem. It is true that for cyclic groups if n is a divisor of the order of the group then the group has a subgroup of order n. To see why this must be so recall that a cyclic group is generated by the powers of one of its elements. Let's see why this must be for the cyclic group of order 12.

$C_{12} = \{e, a, a^2, a^3, a^4, a^5, a^6, a^7, a^8, a^9, a^{10}, a^{11}\}$

Since 3 divides 12 (i.e. $3 \times 4 = 12$) then $(a^3)^4 = a^{12} = e$. Thus a^3 is an element with order 4. (If it were less then a power of a less than 12 would equal e). Thus a^3 generates a subgroup of order 4. Similarly a^4generates a subgroup of order 3 and a^2 generates a subgroup of order 6 while a^6 generates a subgroup of order 2. There are always the trivial subgroups of orders 1 and 12 (The subgroup containing only the identity and the whole group considered as a subgroup of itself).

There is nothing special about the number 12. Let a cyclic group have order n. If r is a divisor of n then there is some number s such that rs = n. Let a be an element of the group with order n. (it must have one if it is cyclic) Then a^r is of order s and therefore generates a cyclic subgroup of order s while a^s generates a subgroup of order r.

We can go a bit further in looking at subgroups of cyclic groups. Not only does a cyclic group have a subgroup of order r for every r that divides the order of the group but it has exactly one subgroup for each distinct divisor of n. To see this suppose that S is a subgroup of a cyclic group of order n and a is a generator. Suppose S has order r where rs = n. Then S is generated by one of its elements, a^k for some k. This means that a^k has order r which in turn means that k times r equals some multiple of n. Now since rs = n, s must divide k. Therefore a^k is in the group generated by a^s which is of order r. However a^kgenerates a group of order r. Thus r elements of the subgroup generated by a^k are in the subgroup generated by a^s which has only r elements. The subgroups must be identical.

Finally let's use group theory to deduce a famous result of number theory. Consider the non-zero members of integers mod p where p is prime. If we let our binary operation be multiplication mod p then we get a group with p-1 elements (we've thrown out zero). [Is it really a group?

- It's closed. Since p is prime any two non-zero elements (means not divisible by p) multiply to give a non-zero element.

- It's associative. Multiplication is associative in regular integers and therefore in integers mod p.
- 1 is the identity
- Since p is prime the cancellation law holds. [If ax = ay then ax - ay = a(x-y) is divisible by p. Since a is not divisible by p (or else it would be 0 mod p and that's been thrown out) x - y must be a multiple of p. This means that x = y mod p.] So the same element cannot appear twice in any row or column of the Cayley table of this system which means that every element appears once. More specifically 1 appears exactly once in every row and in every column and thus every element has an inverse.]]

If a is in this group then $a^{p-1} = 1$. [Why?] That means that if n is an integer that is not divisible by the prime p then

$n^{p-1} = 1 \bmod p$

Permutations

The "old shell game" is an example of a permutation. Three shells, one containing a pea, are in a row in front of the sucker — er, I mean, client. The operator then changes the order of the shells and challenges the client to tell which one contains the pea. The changing of the order constitutes a permutation. In the shell game there are 6 possible permutations. The operator may do nothing. This would be the identity permutation where shell 1 stays in its position, shell 2 stays in its position and shell 3 also stays in its position. We can notate this permutation as

| 1 2 3 |
| 1 2 3 |

The top row of numbers indicates the starting position of a shell. The bottom row indicates the final position of the shell. Thus to find where an object went under a permutation look up its number in the top row and its final place will be under that number. Thus if the shell-game operator switches shells 1 and 2 we can notate this permutation by

| 1 2 3 |
| 2 1 3 |

Shell 1 went to the place formerly occupied by shell 2 (the 2 in the bottom row is below the 1) and shell 2 went to the place formerly occupied by shell 1 (the 1 in the bottom row is below the 2). Since there are six different orders that the numerals "1, 2" and "3" can be ordered in the bottom row [the 1 2 3 in the top row is fixed] there are 6 permutations of 3 objects.

The main reasons for interest in permutations are

- They form a group when we use the operation "followed by."
- Understanding them helps to solve Rubik's Cube.
- There is a theorem due to Arthur Cayley that says that every group is isomorphic to a permutation group.
- Permutations are everywhere whether we know it or not.

For a given set the permutations on it form a group with the operation followed by. This is because (a) permutations are closed under the operation followed by, (b) since the operation is "followed by" it is associative, (c) the "leave everything where it is" permutation is the identity and (c) every permutation can be undone by "running the film backwards" thus all permutations have inverses.

In a Cayley table for a group each element appears exactly once in each row and exactly once in each column (Latin square property). This is a "pictorial" representation of the cancellation property. If I take a group element a and multiply it on the right to every element of a group I'll get all the elements of the group back again. For any element x of G x • a will be in G (This is just the closure property).

Furthermore any element of G will be one of the x • a's since for any b in G the equation x • a = b has a solution. Thus multiplication on the right by a permutes the elements of G. What if we were to first permute the elements of G by multiplying on the right by a followed by multiplying on the right by b? That would give us a new permutation which would (by the associative property) be the same as multiplying on the right by a • b.

Thus each element of G corresponds to a unique permutation of the elements of g. And the permutation corresponding to the product of two elements of G is the composition under followed by of the permutations corresponding to the two elements singly. This fits the definition of an isomorphism. Thus every group is isomorphic to a permutation group. This result is known as Cayley's Theorem.

Rubik's Cube and similar puzzles are examples of permutation puzzles. The challenge is to form a certain permutation from the composition of a limited set of permutations. The permutation you are challenged to find is the one that permutes everything into the solved state. Exactly which permutation that is depends on the initial state. We will have much more to say concerning Rubik's Cube in a later chapter.

Even and Odd Permutations

An old bar bet involves three glasses on the bar. They are in a row in front of the sucker — er, I mean client, but the first and third glasses are mouth-down on the bar. Let "D" stand for a mouth-down glass and "U" stand for a mouth-up glass. Then the client sees the following row of glasses:

DUD

"The object of the game," you tell him, "is to get all the glasses mouth-up in exactly three moves. A move consists of turning two glasses over together, one in each hand." Then you demonstrate by turning over the first two glasses to reach

UDD

Then you turn over the 1st and 3rd glasses

DDU

Finally you turn over the first two glasses again

UUU

All very clean and simple. You now turn over the middle glass and bet your "friend" that he can't do what he just saw you do. He, having had a few drinks, pulls out his wallet and bets you. He begins facing

UDU

and no matter what he tries he can't get them all mouth-up in exactly three moves.

The secret of the trick is parity. The number of mouth-up glasses is either even or odd. Turning over one glass will change the number of mouth-up glasses from odd to even or else from even to odd since you will be either adding a mouth-up glass or subtracting a mouth-up glass. In the game you must turn over two glasses simultaneously and therefore the parity, either odd or even of the number of mouth-up glasses does not change. Now the desired final position has 3 (an odd number) of mouth up glasses. The trick is that you start with 1 (an odd number) of mouth-up glasses but you have your "friend" begin with 2 (an even number) of mouth up glasses. He can't win.

Permutations also have a parity which is changed by a transposition. A transposition is a permutation which exchanges the place of two objects whilst leaving all the other objects unmoved. Thus

| 1 2 3 4 5 |
| 1 5 3 4 2 |

is a transposition as it swaps elements 2 and 5 and does not change the positions of 1, 3 and 4. An important fact is that any permutation on n objects can be done with a series of transpositions. It is a further fact that if it can be done in an even number of permutations then it can't be done in an odd number and vice versa.

Theorem: Every permutation is equivalent to a product of transpositions.

Proof: A simple way to see this is to think of how you might reorder a dozen eggs in an egg carton. Suppose the eggs were numbered 1 to 12. If you were given a permutation of the eggs and told to rearrange the eggs into that permutation you might begin with the first place in the egg carton, find the numbered egg that went into that place and traded the egg in the first place with the egg that belonged there. Repeat this process with the second place then the third place etc. After a minimum of 11 transpositions you would be done. (Why a minimum of ll and not 12?)

The same could be done with any permutation on n objects. First note the object that 1 goes to. If necessary exchange it for object 1. Next note the object that 2 goes to. If necessary exchange it for object 2 etc. After, at most, n-1 such transpositions you will have the desired permutation.

[Exercise: Use the above procedure to express the permutation

| 1 2 3 4 5 6 7 8 |
| 4 5 6 3 8 7 1 2 |

as the product of transpositions.]

The second fact is that if a permutation can result from an odd number of transpositions then it can't result from an even number of permutations and vice versa. To see this we will define the parity of a permutation: For each element count up the number of elements that were before it in the order before the permutation which are after it after the permutation has been applied. For instance in the permutation

| 1 2 3 4 5 6 7 |
| 1 4 6 3 5 2 7 |

there are no elements before 1 before the permutation so the total for 1 is zero. There are two elements after 4 that were before 4 to begin with, namely 2 and 3 so the total for 4 is 2. Continuing in this manner we find that the total for 6 is 3, the total for 3 is 1, the total for 5 is 1, the total for 2 is zero, and the total for 7 is zero. Adding up these totals we get a grand total:

$0 + 2 + 3 + 1 + 1 + 0 + 0 = 7$

If the grand total is an odd number then the permutation is of odd parity. If the grand total is an even number then the permutation is of even parity.

7 is an odd number so the example is an odd permutation. If you apply any transposition to this permutation you will end up with an even permutation. (Try it!).

Why is this so? Why does each transposition change the parity of a permutation? Let's look at what a transposition does to the parity. Let the following represent an ordering of the n numbers 1 to n.

$a_1\ a_2 \ldots x \ldots y \ldots a_{n-1}\ a_1$

Each number in the string of numbers can affect the grand total in two ways. First when it is the number we're using to count how many smaller numbers are beyond it and second when it is one of the numbers "beyond" whatever number we're using to count how many smaller numbers are beyond.

Now if we exchange elements labeled x and y how will the parity of the grand total change? Let's break up the numbers into three groups.

1. The numbers either before both x and y or after both x and y
2. The numbers between x and y.
3. x and y.

The same numbers will be before and after the numbers in group 1 so their contribution to the grand total will not change. The numbers in group 2 will each contribute to the grand total twice once for the change in x's position (either they were less than x or x was less than them.) and again for the change in y's position. Thus their total contribution will be to change the grand total by an even number. This will not affect the parity of the permutation. Finally, x and y switching places will add or subtract 1 from the grand total depending on whether x or y is greater. Thus the parity of the grand total (odd or even) will change.

Now we are in a position to appreciate Samuel Loyd's famous challenge with his 14-15 puzzle back in the 1800's. The puzzle consisted of 15 sliding tiles in a square frame with an empty space which a tile could slide into. The tiles were numbered from 1 to 15. The task was to get the tiles into the normal order:

1	2	3	4
5	6	7	8
9	10	11	12

Rubik's Cube

In the early 1980's a puzzle invented by Ernö Rubik of Hungary captivated the world's imagination. Millions of the puzzles were sold, television programs appeared devoted to it and the medical world added "Rubik's thumb," —a repetitive stress, overuse injury from constant twiddling of the cube—to it's jargon.

The cube appears to be a 3 x 3 x 3 "stack" of smaller cubes (which we will call cubelets) with an ingenious mechanical design that permits any of the 6 faces of the cube to be rotated.

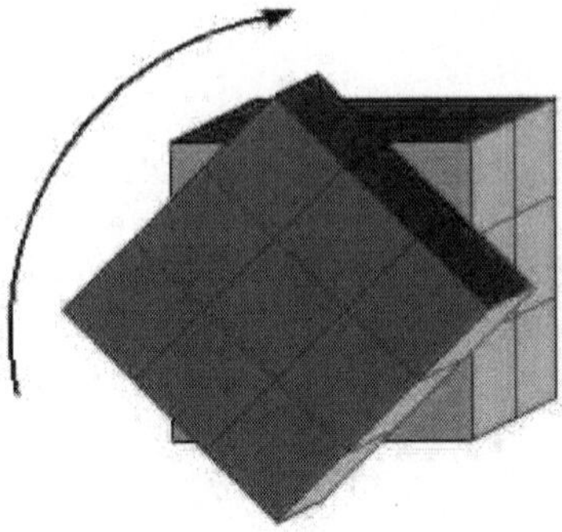

Turning the front face

After a succession of such face rotations the cube can become quite mixed up. The basic problem of the puzzle is to restore the cube to its original condition with all the faces of the cube the same color as shown at the top of this page. In analyzing the cube we find that the "little cubes" that it is mad of, cubelets, are of three varieties. The cubelets that are in the center of a face, center cubelets, are not moved by a rotation of that, or any, face.

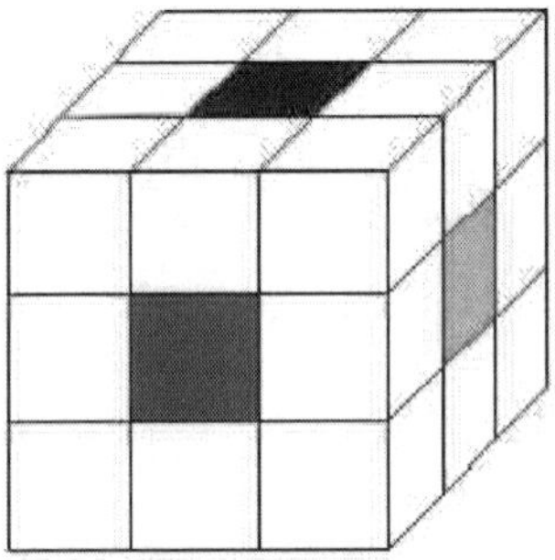

Center Cubelets

The remaining cubelets are either edge cubelets which have two colors showing:

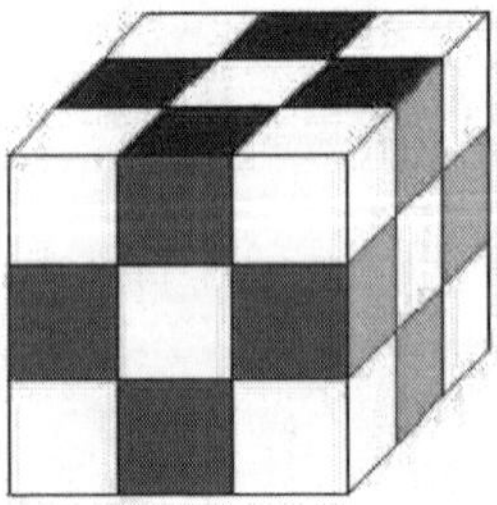

Edge cubelets

or corner cubelets which have three colors showing:

Corner cubelets

The first step in analyzing the cube is to devise a notation. The standard notation consists of naming the faces of the cube according to its orientation in space (with respect to you). The obvious starting point is to name the faces of the cube by the colors of that face (or at least by the color of the corner cublet which never chianges its position relative to the rest of the cube). This scheme has the fatal problem that there is no standard way which the faces of the cube are colored. Therefore we will be using the standard notation based on the cubes orientation.

With the cube on the table in front of you name the face of the cube that is facing you the "Front," call the face opposite the front the "Back;" the face on the right side of the cube is called the "Right;" The face on the left side is the "Left;" the fact pointing upwards is called the "Up" face and the face on the bottom (the one in contact with the table) is the "Down" face.

This naming has the advantage that each face's name begins with a different letter. Thus we can abbreviate each face with one letter—F, B, R, L, U, D— without confusion. Furthermore we will use the abbreviation for each face to symbolize the process of rotating that face. Thus in the illustraton above the Front face is being rotated clockwise (as viewed when looking at the face) by 90°. This move is symbolized by "F." A "kinesthetic" description of the moves F, B, R, L, U and D would be to grasp the cube in the left hand, grasp the face in question with the right hand and turn the named face in the direction pointed to by the right thumb. This will result in a clockwise rotation viewed as looking at the face.

The opposite or inverse of F for instance would be to turn the front face counter-clockwise 90° — the reverse of F. This will be symbolized as F^{-1} which is the same notation used for group inverses. Similarly B^{-1}, R^{-1}, L^{-1}, U^{-1} and D^{-1} are the inverses of B R L U and D.

With this bit of notation out of the way we can begin to discuss some of the mathematics of the cube. For instance we notice that any move permutes four corner cubelets and four edge cubelets. In fact it the four corner cubelets perform a four cycle as do the four edge cubelets. The Up-Front-Left (ufl)corner goes to the place formerly occupied by the Up-Front-Right (ufr) cubelet; the Up-Front-Right cubelet goes to the place formerly occupied by the Down-Front-Right (dfr) cubelet; the Down-Front-Right cubelet goes to the place formerly occupied by the Down-Front-Left (dfl) cubelet and the Down-Front-Left cubelet completes the four-cycle by moving to the place formerly occupied by the Up-Front-Left cubelet. We will symbolize this cyclic permutation of the four corner cubelets by

(ufl, rfu, dfr, lfd)

Notice also the order of letters in each three-letter symbol. The fact that ufl is followed by rfu (rather than by, say, ufr) denotes the fact that

- The upper facelet of the ufr cubelet (u) goes to the right facelet (r) of the rfu cubelet.
- The front facelet of the ufr cubelet (f) goes to the front facelet (f) of the rfu cubelet.
- The left facelet of the ufr cubelet (l) goes to the upper facelet (u) of the rfu cubelet.

[notice the use of lower-case letters for the facelets and cubelets and upper-case letters for the moves and faces of the cube.]

The move F then symbolizes not only a four-cycle of four corner cubelets and a four-cycle of four edge cubelets, but also three four-cycles of the facelets of the four corner cubelets and two four-cycles of the facelets of the edge cubelets. This leads to some interesting conclusions about what is and isn't possible with the cube.

If we consider only the cubelets (and ignore, for the moment, the facelets of the cubelets) then every move consists of two disjoint four-cycles: one of corner cubelets and one of edge cubelets. Each of these four-cycles is an odd permutation as we have seen in the last chapter. Now let's consider the system of corner cubelets and the system of edge cubelets as two separate systems. Any sequence of moves that has the effect of switching two corner cubelets would be equivalent (as far as the corner cubelets are concerned) with a transposition of two corner cubelets—an odd permutation. Thus any such process would have to contain an odd number of moves. This has to follow since every move is an odd permutation and an odd permutation followed by an odd permutation is an even permutation. Now what happens to the edge cubelets when this hypothetical sequence of moves that transposes two corners is being performed. More specifically we want to know if such a process can leave the corner cubelets unmolested from their original places. This would amount to the identity permutation (an even permutation) operating on the system of edge cubelets. But any process with an odd numbers of moves cannot represent an even permutation of the edge cubelets since every move is a four-cycle (an odd permutation) of edge cubelets. Thus we have arrived at our first cube theorem:

Theorem: There is no process whose only effect is to switch two corner cubelets.

If we run through the reasoning that led to this result but reverse the roles of corner cubelets and edge cubelets we have a dual of the above theorem:

Theorem: There is no process whose only effect is to switch two edge cubelets.

Thus if one were to come up with a process whose only effect on the corner cubelets is to switch two of them then it would have to perform an odd permutation on the edge cubelets (and vice versa for the dual result).

Both of these theorems are corollaries of a more basic theorem. Since every move is equivalent to the product of two disjoint four-cycles it is the product of two odd permutations and is therefore an even permutation. Thus:

Theorem: The only possible processes on the cube are those whose effect is an even permutation of the cubelets.

This limits the possible reconstructions of the cube if it is mechanically disassembled. If a disassembled cube is reassembled with an odd permutation of the cubelets from the solved state then it will be impossible to solve the cube without resorting to cheating.

If we look more deeply at effect of a move on the cube and consider the facelets of each cubelet then things become a bit more involved. With this viewpoint a move consists of five disjoint four-cycles. There are three four-cycles of corner facelets (three on each corner cubelet) and two four-cycles of edge facelets (two on each edge cubelet). This means that every move is an odd permutation of corner facelets but an even permutation of edge facelets.

From this we can conclude that (1) there is no process that can create an odd permutation of the edge facelets and that (2) an odd permutation of the corner facelets requires an odd number of moves and an even permutation of the corner facelets requires an even number of moves.

From this it follows that

Theorem: There is no process whose only effect on the edge cubelets is to flip one edge cubelet.

Where "flip" refers to having the facelets exchange places. Note that this theorem says nothing about what happens to the corner cubelets. It says that if you consider only the effect on the edge cubelets there is no way to flip exactly one of them without disturbing some of the other edge cubelets. In fact we can extend this theorem to say that there is no process that will flip exactly an odd number of edge cubelets without disturbing any of the other edge cubelets.

With corner cubelets it is not a matter of flipping them as it is twisting them. A corner cubelet can have three different orientations in any given corner. We'll study the parity of the corner cubelets using integers mod 3 rather than mod 2 (even / odd) that we used on the edge cubelets.

Grasp a corner cubelet with the thumb and first two fingers of the right hand and imagine turning it as one turns a screwdriver when driving a right-handed screw. This will be the positive direction. The opposite direction will be called the negative direction. To understand the parity of corner cubelets we'll need to look at the cube in a slightly different light. Imagine a cube that is white on all its faces except the Up face and the Down face which are colored in. Below are two illustrations of this cube with (1) a corner cubelet twisted in the negative direction and (2) a corner cubelet twisted in the positive direction. Since any corner cubelet must be part of either the Up face or the Down face we can look at any corner cubelet at any time and decide whether it has +1 twist, -1 twist or 0 twist.

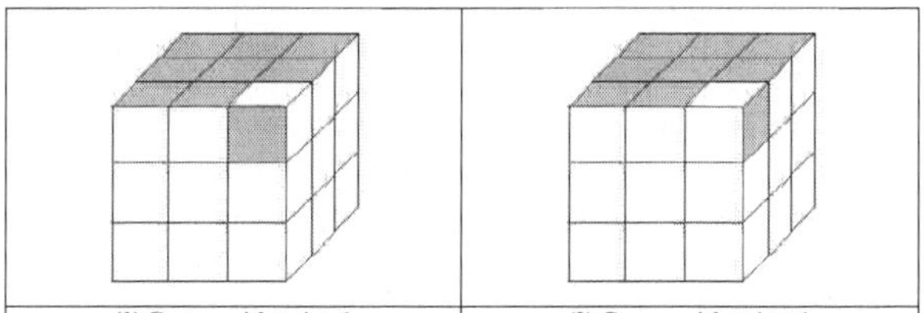

(1) Corner with twist -1 (2) Corner with twist +1

Now a -1 twist followed by a +1 twist is the same as 0 twist which is also the same as +3 twist (the identity twist?). Consider how a move such as F affects the twist parity of the corner cubelets. If F is applied to a cube the ufl cubelet goes to the rfu position. In doing so it now has been given a +1 twist. It appears as the corner cubelet in illustration (2) with respect to the UP face. Similarly the dfl cubelet will become the lfu cubelet and will now have been given a -1 twist. Thus the twists on the same side (L in this case) will cancel each other out so the total twist parity due to these two cubelets is 0. The same thing will happen with the corner cubelets on the right front. One will be given a +1 twist and the other will be given a -1 twist.

So a single move cannot change the total twist parity of the corners it affects. Also for this simplified cube (with only the top and bottom colored) any turn of the Up or Down faces doesn't change the parity of any cubelets. We can conclude that

Theorem: The total twist parity of the cube cannot change.

So what kinds of twists of the corners can be acheived with the cube? Well, it's possible to twist two corners in opposite directions wiht 0ne getting a +1 twist and the other a -1 twist for a total twist parity change of 0. It is also possible to twist three corner cubelets in the same direction either each +1 or each -1 which still gives a total twist parity change of 0 (1 + 1 + 1 = 0 mod 3).

Putting all these results together we find that if we were to disassemble a cube we would have some significant choices on how to reassemble it. (1) we could either reassemble it with an edge cubelet flipped or not and (2) we could reassemble it with a corner cubelet given a +1 twist, a -1 twist or a zero twist and finally (3) we could reassemble it with a pair of corner cubelets exchanged [or instead a pair of edge cubelets exchanged]. There are two independent ways do (1). Either we flip an edge or not. There are three independent ways to do (2): +1, -1 or 0. [It doesn't matter which corner we give the +1 twist to, we can transfer that +1 twist to another corner by cube processes later so all reassemblings with a +1 corner twist given to some corner are equivalent] and finally there are two ways to do (3): either switch two cubelets or not. [again which two cubelets we switch doesn't matter since we can use cube processes to go from any two switched edge cubelets to any two other switched edge cubelets and the same for switched corner cubelets].

Putting it all together we find that there are 12 (= 2 x 3 x 2) independent ways to reassemble the cube. Thus if we were to take a cube apart and reassemble it in the dark we would have a 1 out of 12 chance of having a solvable cube.

Rubik's Cube Groups

If we consider all sequences of moves on a Rubik's Cube we notice the following:

- One sequence followed by another sequence is a sequence: Closure.
- Followed by is associative.
- The do nothing sequence does nothing: Identity.
- Every sequence of moves can be done backwards and therefore undone: Inverses.

This makes the set of all sequences of moves on a Rubik's Cube a group. However since a sequence of moves can be of any length (you can do RLU over and over until the cows come home) let's cut it down to a more manageable and useful set. Two sequences will be considered equivalent if they have the same final effect on the cube. We'll call the final effect of a sequence of moves a process. Thus a process is a possible permutation of the cubelets together with any twists of the corners or flips of the edges.

One of the first questions we ask is, How many processes are there in the group of the cube? In other words, what is the order of the group of the cube. We can solve this question by counting the permissible permutations of edges and the permissible permutations of corners.

There are 8 corner cubelets and 12 edge cubelets. Each corner cubelet can have three possible orientations ("twists") and each edge cubelet can have two possible orientations ("flips"). Thus the possible states of the cube can be no more than

Upper Bound on Order of Cube Group = $(8!)(12!)(3^8)(2^{12})$

However, as we have seen in the previous chapter not all of these configurations are reachable from the initial (solved) state. There we learned that only 1/12 of all conceivable states are actually reachable so to get the order of the cube we must divide the upper bound given above by 12

Order of the Cube Group = $(8!)(12!)(3^8)(2^{12}) \div 12$

This is a rather large number: 43,252,003,274,489,856,000 (or a little over 43 quintillion). At the rate of 1,000,000 configurations per second it would take well over a million years to run through them all. From this we can safely conclude that coupling random twiddling with patience is a poor strategy for solving the cube.

One of the first things we notice is that the cube group is not Abelian. F followed by R is not the same as R followed by F. (To simplify the notation we will leave out the term "followed by" and simply juxtapose the moves in a sequence.) To have some simpler fun we'll look at some of the easier subgroups of the cube group.

The F subgroup

This subgroup consists of all possible processes that are the result of turning only the front face F. This has the disadvantage of being a very boring subgroup whose only elements are E (the identity), F, $F^2$and F^{-1}. This is just the Attayun—HOOT group that we've seen before in various disguises. Of course there is nothing special about F. There are other copies of the Attayu—HOOT group using other faces or even central slices.

The SLICE SQUARED subgroup

This is about the simplest non-trivial subgroup of the cube. It consists of all those processes that are the result of limiting ourselves to the following moves: X = (R^2L^2), Y= (U^2D^2) and Z = (F^2B^2). This subgroup gets its name from the fact that each move is equivalent to moving a central "slice" by 180°. Since a move is 90° a move of 180° is the same as a normal move followed by itself or "squared."

One of the first things we notice about this group is that it's Abelian; XY has the same effect on the cube as YX etc. Also we notice that XYZ applied to a solved cube results in a cute "checkerboard" pattern:

The "Checkerboard"

Since X, Y and Z are not only commute but are their own inverses we can write out all the elements of the slice squared group: E (the identity), X, Y, Z, XY, XZ, YZ, and XYZ. This group has the Cayley Table given here:

	E	X	Y	Z	XY	XZ	YZ	XYZ
E	E	X	Y	Z	XY	XZ	YZ	XYZ
X	X	E	XY	XZ	Y	Z	XYZ	YZ
Y	Y	XY	E	YZ	X	XYZ	Z	XZ
Z	Z	XZ	YZ	E	XYZ	X	Y	XY
XY	XY	Y	X	XYZ	E	YZ	XZ	Z
XZ	XZ	Z	XYZ	X	YZ	E	XY	Y
YZ	YZ	XYZ	Z	Y	XZ	XY	E	X
XYZ	XYZ	YZ	XZ	XY	Z	Y	X	E

CAYLEY TABLE OF THE SLICE SQUARED SUBGROUP

The SLICE subgroup

This subgroup is large enough to be challenging to the beginning cubologist but not too big to be overwhelming. It consists of all processes generated by the moves X = (RL^{-1}), Y= (UD^{-1}) and Z = (FB^{-1}). This group gets its name from the fact that each of the allowed moves is equivalent to a move of a middle "slice." In the slice group the patterns on each face exhibit a certain symmetry. The corner facelets of each face are always the same color. Also, opposite edge facelets of each face are the same color. This gives a pattern similar to this:

Pattern in the SLICE group

One cute pattern in the slice group is formed by the move sequence $LR^{-1}BF^{-1}UD^{-1}LR^{-1}$

This pattern goes by the name of "dots"

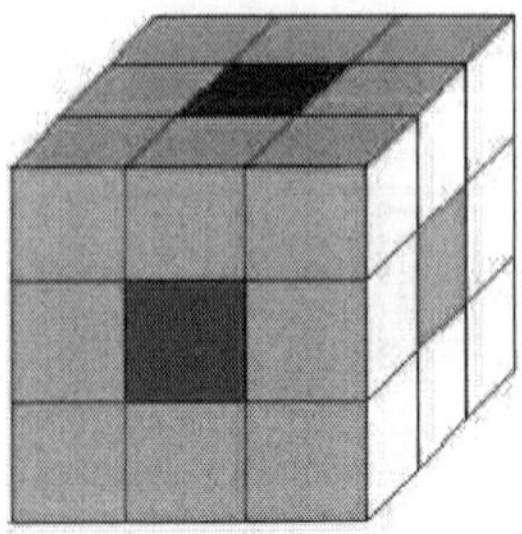

Dots

Sanity Check: A would-be cubologist tells you that the SLICE group has exactly 108 elements. How do you know, imediately, that this is incorrect?

Solution to Sanity Check: The SLICE group contains the SLICE SQUARED group as a subgroup. Thus the order of the SLICE SQUARED group must be a divisor of the order of the SLICE group according to Lagrange's Theorem. But 8 doesn't divide evenly into 108.

The (F^2R^2) subgroup

This group consists of all processes which consist of the moves (F^2R^2) repeated over and over again. This gives a cyclic subgroup of the cube group. A little experimenting shows that performing (F^2R^2)six times in succession brings the cube back to its original position. Thus the cube element (F^2R^2) has cyclic order of 6 and the (F^2R^2) subgroup is an abelian group of order 6. If you perform (F^2R^2) three times in succession you find that you have swapped the uf edge cubelet with the df edge cubelet and also swapped the ur edge cubelet with the dr edge cubelet. This is a useful process to remember when trying to restore the cube.

[Exercise: Find processes which have the effect of

- swapping the ul cubelet with the dl cubelet and the ub cubelet with the db cubelet.
- swapping the ur cubelet with the dr cubelet and the ub cubelet with the db cubelet.
- swapping the uf cubelet with the df cubelet and the ul cubelet with the dl cubelet.
- swapping the fl cubelet with the bl cubelet and the uf cubelet with the bf cubelet.
- swapping the fr cubelet with the br cubelet and the df cubelet with the db cubelet.
- swapping the fr cubelet with the fl cubelet and the dr cubelet with the dl cubelet.]

The (F^2R^2) subgroup is a legitimate subgroup of the group of the cube. Let's use this fact to create some more processes that could be useful in restoring the cube. Let's create some cosets of the (F^2R^2)subgroup. If we perform a move before applying a process from the (F^2R^2) group we no longer have an element from the (F^2R^2) subgroup but we have instead an element from one of its left cosets. If we perform the move after performing a process from the (F^2R^2) subgroup then we will be left with an element of one of its right cosets. Now we can extend this idea to come up with some interesting and useful processes.

Multiplying a subgroup on the left by an element of the group creates a left coset. Multiplying a subgroup on the right by an element of the group creates a right coset. Now we'll look at multiplying something (not necessarily a subgroup) on both the left and the right by an element and its inverse. In other symbols we'll look at elements of the form $a \bullet h \bullet a^{-1}$ Group elements of this form are called conjugates. $a \bullet h \bullet a^{-1}$ is called the conjugate of h by a. If we consider all the conjugates by a of all the elements of a group G we note the following:

These properties of conjugates by a show that they are closed under the group operation, that the identity e is a conjugate by a and that the inverse of a conjugate by a is also a conjugate by a. These properties together with associativity show that conjugates by a form a subgroup.

Let's apply our idea of conjugation to our (R^2F^2)-repeated-three-times [or $(F^2R^2)^3$] process. We found that the $(F^2R^2)^3$ process swapped the uf edge cubelet with the df edge cubelet and also swapped the uredge cubelet with the dr edge cubelet. We also found in the exercise that by using faces other than the F and R faces we could various other pairs of opposite edge cubelets. Now let's perfom a D movement before performing the $(F^2R^2)^3$ movement then performing D^{-1} afterwards. This gives the conjugate by D of the $(F^2R^2)^3$ element. This will swap two pairs of edge cubelets but no longer is two pair of opposite cubelets that are swapped. This idea of using conjugates to create new useful processes from old useful processes is standard in cubology and we'll find it useful in creating a plan for solving the cube in the next chapter.

The (FRBL) ["furball"] subgroup

This is the cyclic subgroup of all processes that are repetitions of (FRBL). A question arises: what is the order of this group? In other words, how many times in a row must you perform the (FRBL) move in order to return to the original position? A little experimenting together with a little arithmetic gives the answer.

Pick up a cube and begin repeating the process (FRBL) over and over. One of the first things you notice will be that after 5 repetitions (20 total moves) the top four edge cubelets come back to their original positions. Also after 3 repetitions (12 total moves) the top four corner cubelets come back to their original positions. This means that after 5, 10, 15 repetitions etc. (20, 40, 60 moves etc.) the top edges will come back into place again. Also after 3, 6, 9, 12 repetitions etc. (12, 24, 36, 48, moves etc.) the top corners will come back again. Thus after exactly 15 repetitions (60 moves) of the (FRBL) process the top layer of the cube will be restored. This is because 15 is the smallest multiple of both 5 and 3 — the least common multiple.

Back to the beginning and let's consider the middle layer of cubelets.. After 5 repetitions of (FRBL) (20 moves) the lf cubelet comes back to its home position and after 7 repetitions (28 moves) the other middle-layer cubelets, rf, lb and rb return together to their home positions. Thus the middle layer will be restored after 35 repetitions of (FRBL) or after a total of 140 moves.

Now let's consider the bottom face. It is found that after 7 repetitions (28 moves) the bottom edge cubelets are restored and after 9 repetitions (36 moves) the bottom corner cubelets are restored. Now we are in a position to find the order of the (FRBL) group. The top will be restored every 15 repetitions, the middle layer will be restored every 35 repetitions and the bottom will be restored every 63 repetitions (least common multiple of 7 and 9). Thus the whole cube will be restored after any number of repetitions which is a common multiple of 7, 9, 15 and 35. The least common multiple of these numbers is 315. Thus it will take 315 repetitions (1260 moves) to restore the cube and therefore the (FRBL) group is a subgroup of order 315.

Bibliography

- Allenby, R. B. J. T. (1991), Rings, Fields and Groups, Butterworth-Heinemann, • Artin, Michael (1991), Algebra, Prentice Hall, ISBN 978-0-89871-510-1
- Burris, Stanley N.; Sankappanavar, H. P. (1999) [1981], A Course in Universal Algebra
- Gilbert, Jimmie; Gilbert, Linda (2005), Elements of Modern Algebra, Thomson Brooks/Cole,
- Lang, Serge (2002), Algebra, Graduate Texts in Mathematics, 211 (Revised third ed.), New York: Springer-Verlag, Sethuraman, B. A. (1996), Rings, Fields, Vector Spaces, and Group Theory: An Introduction to Abstract Algebra via Geometric Constructibility, Berlin, New York: Springer-Verlag,
- Whitehead, C. (2002), Guide to Abstract Algebra (2nd ed.), Houndmills: Palgrave,
- W. Keith Nicholson (2012) Introduction to Abstract Algebra, 4th edition, John Wiley & Sons.
- John R. Durbin (1992) Modern Algebra : an introduction, John Wiley & Sons
- Zeidler, A. Bernhard, Abstract Algebra (PDF), Web-Book published under Open Public License, 646 pages, Work in Progess.
- Cantor, Georg (1874), "Ueber eine Eigenschaft des Inbegriffes aller reellen algebraischen Zahlen", Journal für die reine und angewandte Mathematik (in German), 77: 258–262, doi:10.1515/crll.1874.77.258
- Johnson, Philip (1972), A History of Set Theory, Prindle, Weber & Schmidt.

4 Bolzano, Bernard (1975), Berg, Jan (ed.), Einleitung zur Größenlehre und erste Begriffe der allgemeinen Größenlehre, Bernard-Bolzano-Gesamtausgabe, edited by Eduard Winter et al., Vol. II, A, 7, Stuttgart, Bad Cannstatt: Friedrich Frommann Verlag, p. 152.

- Dauben, Joseph (1979), Georg Cantor: His Mathematics and Philosophy of the Infinite, Harvard University Press, pp. 30–54.
- Young, William; Young, Grace Chisholm (1906), Theory of Sets of Points, Cambridge University Press
- Kolmogorov, A.N.; Fomin, S.V. (1970), Introductory Real Analysis (Rev. English ed.), New York: Dover Publications, pp. 2–3.

- Jech, Thomas (2003), Set Theory, Springer Monographs in Mathematics (Third Millennium ed.), Berlin, New York: Springer-Verlag, p. 642.
- Bishop, Errett (1967), Foundations of Constructive Analysis, New York: Academic Press.
- Feferman, Solomon (1998), In the Light of Logic, New York: Oxford University Press, pp. 280–283, 293–294, .
- Wittgenstein, Ludwig (1975), Philosophical Remarks, §129, §174, Oxford: Basil Blackwell.
- Ferro, Alfredo; Omodeo, Eugenio G.; Schwartz, Jacob T. (September 1980), "Decision Procedures for Elementary Sublanguages of Set Theory. I. Multi-Level Syllogistic and Some Extensions", Communications on Pure and Applied Mathematics, 33 (5): 599–608, doi:10.1002/cpa.3160330503
- Cantone, Domenico; Ferro, Alfredo; Omodeo, Eugenio G. (1989), Computable Set Theory, International Series of Monographs on Computer Science, Oxford Science Publications, Oxford, UK: Clarendon Press, pp. xii, 347.
  for Advanced Study.

Index

A

Adequality, 137, 225

Algebraically, 35-36, 38, 40, 45, 76, 94, 107-110, 205, 207-210, 233, 273-274, 281

Annihilators, 61, 67, 93

Antiderivative, 255

Apollonius, 52, 123, 137, 184

Arithmetic, 2, 5-8, 26, 33, 35, 43, 69, 74, 90-91, 93, 127, 304, 320

Ascending, 61, 68, 89, 92

Automorphic, 104, 114, 118-119

Axioms, 1, 3-4, 7, 10-11, 13, 15-20, 25-26, 30, 32, 79-82, 85-87

Azimuth, 143, 145

B

Beetroot, 224

Bijection, 19, 21, 39, 110-111, 275-276

Bimodule, 64, 87

Binomial, 72, 180

Birational, 38-39, 53, 208

Bivariate, 44, 205-206

C

Canonical, 24, 42, 48, 87, 97, 103, 107, 153, 306

Cardinality, 22, 87

Complexity, 5, 40, 43, 50-51

Concave, 232, 244-245, 258-269

Congruent, 16

Conjugacy, 3, 22, 78, 111, 113

Contravariant, 66, 83, 90

Converse, 6, 30, 61-62, 127, 303, 307

Cosmology, 155

Cotangent, 129

Covariance, 81, 223

Covariant, 64, 66, 82-83, 90, 103, 169

Crystallography, 27, 108

Cubelet, 313-317, 319-320

Curvature, 153, 155-157, 159, 161, 163, 165, 167, 169, 171, 173, 175, 177, 179, 181, 183

Cyclotomic, 76, 78

D

Degenerate, 5, 48, 239, 295-296

Denominator, 38, 157, 185, 189-196, 198, 210-211, 271

Descartes, 53, 123, 133, 137-138, 148

Deviate, 82

Diagonal, 17-18, 110, 185, 276

Dihedral, 17, 24, 28, 30, 113

Diophantine, 35, 53

Dirichlet, 76

Discriminant, 70, 77-78, 160, 170

Disjoint, 22, 281, 284, 297, 301-302, 314-315

Diverse, 34, 103-104, 115, 118, 171

Division, 1, 7, 9, 11, 19, 25-26, 30, 33, 42-43, 55-61, 64, 67, 69, 107, 193, 281

Divisors, 11, 41, 301, 304

Domenico, 322

Driving, 263, 315

Dthenh, 82

Dualized, 83

Dually, 7, 89

Dyadic, 8

E

Ellipses, 34, 135, 138, 170, 182

Elliptic, 34, 54, 78, 208, 210, 251

Emanating, 171

Encapsulated, 161

Encountered, 13, 185, 238

Endomorphism, 57, 67, 82, 87, 90, 111

Endomorphisms, 55, 57, 82, 90, 105, 107

Endowed, 23, 31, 79, 96, 110

Epimorphism, 82, 97-98, 100-101

Epsilon, 290-291

Erdmann, 113

Erential, 255

Erlangen, 14, 104, 118

Ernst, 3, 14, 92

Euclidean, 29, 31, 34-35, 42, 53, 116, 123, 153, 155, 163, 170, 172, 204-205, 270

Exponent, 170, 304

Extablishing, 301

Extensionl, 73

Extremum, 225, 236, 244-245, 247, 264-265

Extremumis, 236

Extrinsic, 153, 156, 159, 161, 167-168, 170

F

Facelet, 314-315, 318

Farright, 229

Fermat, 2, 12, 14, 33, 35, 53-54, 123, 137, 210, 225-226, 230-231, 236, 244

Finitely, 64, 66, 89

Folium, 133

Fontana, 53, 70, 144

Fourier, 60, 104, 114-116

François, 70, 137

Functor, 11, 60, 63-66, 81-84, 90, 96, 103

G

Gluing, 34, 51, 94

Graphat, 259

Greek, 53, 115, 123, 143, 149, 184

Groupoid, 5, 32, 81

Guerino, 85

H

Harmonic, 31, 104, 108, 114-116, 119

Harvard, 144, 321

Henrik, 16, 71, 74, 102

Hermite, 177-178, 180

Hilbert, 3, 8, 33-37, 40, 43, 45-46, 49, 53, 62, 67, 76-78, 92, 94-95, 104, 108, 113, 115, 119

Hmodules, 120

Holomorphic, 127, 209

Holonomy, 118, 156

Homogeneous, 39-40, 53, 67, 114, 155, 172, 206-207, 209

Homomorphism, 11, 20, 23, 29, 37, 57, 60, 80, 87, 89-90, 97, 99-101, 106, 120-121

Hyperbolic, 14, 24, 28, 127-130, 155, 210

Hypotenuse, 149, 183, 269

I

Icosahedral, 28

Idempotent, 6, 10, 64-65, 111-112

Implication, 32, 275, 277, 298-299

Implicit, 47, 49, 132, 173, 183, 187, 196, 200, 205, 207

Inaugurating, 53

Inclination, 142, 187, 200

Indecomposable, 56, 89, 107, 111-113

Inevitable, 102, 172

Inextricably, 118

Inseparable, 74

Interpolates, 180, 182

Invariance, 31

Invariant, 28, 31, 65, 67-68, 76, 86, 88, 92, 97, 104, 108, 113-114, 116, 118, 120, 182, 209

Irreducible, 25, 37, 40, 42, 48-49, 56, 75, 77-78, 89, 106-108, 110-114, 116, 207-210

Isomorphic, 3, 20-21, 27, 38, 57, 59, 61, 63-64, 66, 72-73, 75-76, 78, 83, 88-89, 97, 100, 106, 111-112, 208, 210, 305-306, 308-309

J

Jargon, 260, 311

K

Kinematic, 1, 155

Kinship, 13

Klein, 14, 53, 67, 73, 78, 100, 118, 210

L

Laboriously, 158, 170

Lacroix, 144

Lagrange, 2, 14, 21, 23, 30, 53, 70, 241, 301-307, 309, 311, 313, 315, 317, 319

Laguerre, 53

Laplacian, 116

Latter, 32, 78, 92, 108, 110, 113, 187, 200, 243

Lattice, 6, 28, 78, 88, 102, 210

Lexicographic, 43-45

Lexicographical, 41-42, 45, 47

Locus, 36, 168, 170, 208

Logarithm, 131, 255, 271, 305

Lorentz, 32, 114

Louis, 2, 14, 70, 123, 153

Lurking, 132

M

Magnitude, 148, 150, 153-154, 158, 164, 172, 180

Manifold, 8, 12, 30-31, 34, 54, 81, 88, 117, 127, 153, 155, 171-172, 208-209, 222

Mathieu, 28, 75-76

Methodology, 3

Metric, 8, 53, 156, 161, 163-164, 166-167, 169-171, 225

Minima, 225-227, 229, 231, 233, 235, 237, 239-243, 245-247, 249, 251

Modular, 2, 6, 14, 26, 30, 43, 78, 89-91, 108-111, 118-119

Modulo, 2, 22-23, 26, 75, 87, 210

Monads, 52, 80

Monoid, 6, 32, 55, 87

Monomial, 40-43, 45-47, 49

Monomorphism, 82, 97-98, 100-101

Monotone, 126, 205-206, 293, 295-298

Morita, 55, 60-61, 63-67

Morphisms, 11, 30, 37, 79-84, 98-99, 102, 209

Mumford, 52, 94, 118

N

Ndimensional, 153

Negligible, 139

Neumann, 8, 57, 65

Neuroscience, 115

Nilpotent, 51, 67

Noetherian, 33, 57, 59, 61, 65, 68, 89, 92-94

Noncommutative, 33, 50, 52, 55, 58, 61, 63, 67-68, 91-92, 95

Nonhomogeneous, 51

Nonsingular, 38, 65, 112, 208, 210

Nontrivial, 4, 30, 51, 58, 89, 107, 302

Nurbs, 182

O

Oating, 255

Obeying, 6

Oblique, 184-187, 193-194, 196-200, 202-204, 211

Obscure, 71, 298

Octahedral, 28

Orientation, 5, 18, 144, 154, 185, 313

Orthogonal, 28, 108, 112, 140, 156-157, 159, 161, 164, 170, 172

Oscillatory, 115-116

P

Parabola, 38-39, 44, 52, 137, 233, 236, 248-249, 278-279

Pascal, 53, 123, 144

Perga, 123, 184

Permutation, 2-3, 14, 59, 67, 69, 71-77, 301, 308-311, 314-316

Phenomena, 13, 30, 122, 150, 172

Phenomenon, 24, 107, 115, 150, 156

Photographed, 115

Piecewise, 226, 279

Pigmentation, 224

Polynomial, 1, 13-14, 27, 29, 33-51, 53, 62, 67, 69-78, 87, 91-94, 118, 159, 182, 187, 200, 202-208, 210, 256, 270-271

Posthumously, 14

Prolongation, 39

Prominent, 51, 53, 61, 94, 104, 116

Prototypical, 5, 54, 127

Puiseux, 207-208

Pythagorean, 169, 269

Q

Quantum, 27, 32-33, 113, 115, 119

Quaternion, 113

Quintillion, 317

Quotient, 11, 13, 19-21, 23-24, 39, 42, 56, 58, 76, 86, 91, 97, 107, 118-119, 125, 139, 194, 216-217, 271

R

Rcosa, 135, 148-149

Reciprocal, 153, 173, 185, 206, 211

Reducible, 42, 56, 89, 107-109, 207

Reinterpreted, 125

Rigorous, 54, 123, 139

Ringed, 51, 90, 94

Rmodule, 57, 59, 88

Ruffini, 2, 14, 69, 71, 74

Ruvuv, 167-168

S

Sagemath, 50

Saintvincent, 144

Sankappanavar, 321

Schwartz, 322

Seidenberg, 50

Semantic, 85

Semiprime, 61, 65

Semiprimitive, 57-58, 65

Sheaf, 34-35, 52, 54, 85, 90, 103

Sinusoid, 138, 206

Slant, 185-186, 194, 198, 202, 211

Slope, 38-39, 122, 125-126, 136, 139, 142, 173, 184, 210, 214-216, 220, 232-234, 243, 247

Solvability, 2, 14, 29, 71, 74

Solvable, 69, 71, 74-76, 78, 117-118, 316

Sporadic, 30, 75-76

Springer, 321-322

Summand, 56, 110, 112

Surjective, 24, 98-99, 101, 275

Syllogistic, 322

Systematic, 60, 74, 80, 109

Syzygies, 95

T

Tanch, 129

Tangent, 39, 117, 122-123, 125-127, 129, 136-143, 148, 152-162, 164, 168, 170, 172-173, 175, 177, 180, 182-184, 202, 205-206, 209, 232, 234, 236, 239, 244-245, 253, 261, 264-265, 271

Tedious, 133, 237-238

Tensor, 60, 66, 88-89, 119-121, 153, 155, 161, 166-171

Terminology, 14, 58, 293-294, 296, 305

Thumb, 311, 313, 315

Topological, 8, 11, 16, 24, 29-32, 37, 39, 81, 93, 96-97, 103, 108, 114, 116, 208-210, 225

Transcendental, 76, 208

Tremendous, 3, 102

Trigonometric, 127-128, 142

Trivial, 21, 30, 34, 67, 72, 78, 88, 97, 99, 107, 111-113, 120, 211, 253, 277, 307, 317

Tschirnhausen, 205

U

Ubiquity, 29, 67

Ueber, 321

Unambiguity, 18

Unary, 5-6, 16

Unicursal, 210

Unitary, 59, 108, 113-114, 116-117

Univariate, 42, 47, 87, 207, 214-215, 217, 219

Uredge, 319

V

Vague, 81

Verlag, 321-322

W

Wheng, 21

Wiener, 115

Wolfgang, 41, 92

Y

Yoneda, 51, 84, 94

Z

Zahlen, 321

Zeidler, 321